Integrated Ecosystem Assessments for Fisheries Management in the Yellow Sea, the East China Sea, and the East/Japan Sea

Integrated Ecosystem Assessments for Fisheries Management in the Yellow Sea, the East China Sea, and the East/Japan Sea

Editors

Sang Heon Lee
Seok-Hyun Youn

MDPI • Basel • Beijing • Wuhan • Barcelona • Belgrade • Manchester • Tokyo • Cluj • Tianjin

Editors

Sang Heon Lee
Department of Oceanography
Pusan National University
Busan
Korea, South

Seok-Hyun Youn
Oceanic Climate & Ecology
Research Division
National Institute of
Fisheries Science
Busan
Korea, South

Editorial Office
MDPI
St. Alban-Anlage 66
4052 Basel, Switzerland

This is a reprint of articles from the Special Issue published online in the open access journal *Journal of Marine Science and Engineering* (ISSN 2077-1312) (available at: www.mdpi.com/journal/jmse/special_issues/bz_fisheries_management).

For citation purposes, cite each article independently as indicated on the article page online and as indicated below:

LastName, A.A.; LastName, B.B.; LastName, C.C. Article Title. *Journal Name* **Year**, *Volume Number*, Page Range.

ISBN 978-3-0365-7535-3 (Hbk)
ISBN 978-3-0365-7534-6 (PDF)

Contents

Journal of
*Marine Science
and Engineering*

MDPI

Editorial

Integrated Ecosystem Assessments for Fisheries Management in the Yellow Sea, the East China Sea, and the East/Japan Sea

Sang Heon Lee [1],* and Seok-Hyun Youn [2]

[1] Department of Oceanography and Marine Research Institute, Pusan National University, Busan 46241, Republic of Korea
[2] Oceanic Climate & Ecology Research Division, National Institute of Fisheries Science, Busan 46083, Republic of Korea; younsh@korea.kr
* Correspondence: sanglee@pusan.ac.kr; Tel.: +82-51-510-2256

Citation: Lee, S.H.; Youn, S.-H. Integrated Ecosystem Assessments for Fisheries Management in the Yellow Sea, the East China Sea, and the East/Japan Sea. *J. Mar. Sci. Eng.* **2023**, *11*, 845. https://doi.org/10.3390/jmse11040845

Received: 3 April 2023
Accepted: 11 April 2023
Published: 17 April 2023

Marine environmental conditions are highly distinct in the Yellow Sea, the East China Sea, and the East/Japan Sea, with characteristics such as the shallow and turbid conditions of the Yellow Sea, relatively warm subtropical conditions of the East China Sea, and deep and semi-enclosed nature of the East/Japan Sea. Physico-chemical properties and, subsequently, biological characteristics are different among the three seas. In recent decades, dramatic changes in the physical structure and vertical distribution of chemical properties have been reported in the Yellow Sea, the East China Sea, and the East/Japan Sea. These recent environmental changes have greatly affected the physiological status, community structure, and bloom pattern of phytoplankton and, thus, consequently altered the seasonal distributions and nutritional status of higher trophic levels such as zooplankton, fish, and marine mammals. However, to date, we do not know much about the current status of the marine ecosystems in these three distinct seas. Since 2018, the integrated ecosystem assessment for ecosystem-based fisheries management has been implemented in the Yellow Sea, the East China Sea, and the East/Japan Sea by the National Institute of Fisheries Science, Korea.

This Special Issue will provide basic information for the current status of the aforementioned marine ecosystems and will form an important basis for future monitoring of marine ecosystem response to ongoing climate changes in the Yellow Sea, the East China Sea, and the East/Japan Sea.

In this Special Issue, we present a total of 12 articles covering a wide range of topics for water column dynamics derived by the Kuroshio Current and various ecosystem components in the Yellow Sea, the East China Sea, and the East/Japan Sea. A brief overview of all the articles follows.

Lee et al. (2022a) [1] investigate the water column dynamics of the East Korea Warm Current in the western East Sea responding to spatiotemporal variability in the Kuroshio Current. This article expands our understanding of the potential mechanisms for climate change and its effects on the oceanic environmental conditions and subsequent response of fishery resources in the western East Sea.

Kim et al. (2022a) [2] present morphological characteristics of the brown seaweed *Sargassum thundergii* adapting to local environmental conditions along the Korean Coast. Their results, based on biota-environment matching analysis, prove that geographically different morphological characteristics of the brown seaweed *Sargassum thundergii* result from different local environmental factors, especially tidal condition.

Park et al. (2022) [3] examine the summer phytoplankton community structure as a consequence of recent marine environment changes in the Northern East China Sea, finding a significant change in key dominant species from micro-sized diatoms and dinoflagellates in the 2000s to nano- and pico-sized flagellates in the period 2016–2020. This change in the dominant phytoplankton community structure is likely due to the recent low-nutrient conditions after the construction of the Three Gorges Dam in Changjiang River.

Kim et al. (2022b) [4] identify size-fractionated phytoplankton communities seasonally in the Yellow Sea based on the HPLC results. This study provides important information on the contribution of seasonal variations in small-sized diatom to the phytoplankton community in the Yellow Sea, which has been overlooked to date.

Jang et al. (2021) [5] report the recent primary productions in the Yellow Sea, the South Sea of Korea, and the East/Japan Sea and potential reasons for the lower current productions relating to major environmental changes in each sea. Moreover, the authors recognize that small-sized phytoplankton have significantly negative impacts on the primary productions, which should be further investigated for their ecological roles under rapid warming condition.

Kang et al. (2022) [6] assess a new algorithm based on a deep learning model suitable for the estimation of phytoplankton size classes (micro, nano, and pico size) in Korean waters. This algorithm is expected to be useful for understanding long-term variations in phytoplankton size structure using satellite ocean color data.

Shin et al. (2022) [7] examine seasonal water mass effects on the spatiotemporal distribution characteristics of copepods in the Northeastern East China Sea. The authors classify major copepod species for different regional water masses based on cluster analysis. These major species can be useful indicators for future environmental changes in the Northeastern East China Sea.

Kim et al. (2022c) [8] present a sound scattering layer distribution and density estimation of *Euphausia pacifica* in the central Yellow Sea Bottom Cold Water by hydroacoustic and net surveys. This article contributes toward improving our understanding of the structure of the marine ecosystems in the Yellow Sea Bottom Cold Water.

Kim et al. (2022d) [9] investigate stable isotopes and fatty acid compositions in four dominant zooplankton in relation to their potential diets in the East/Japan Sea. Given the results from stable isotopes and fatty acid compositions, the authors confirm that the change in the dietary intake of zooplankton is generally dependent on phytoplankton conditions.

Choi et al. (2022) [10] implement a DNA barcoding approach to investigate pelagic marine fish eggs as an indicator of spawning and intrusive species in the Ulleung Basin of the East/Japan Sea. The study discovers the eggs of *Trachipterus trachypterus* and *Trachipterus jacksonensis* for the first time in the Northwestern Pacific Ocean, which broadens their potential spawning ecology and geographical distribution.

Kim et al. (2022e) [11] investigate seasonal dietary changes in wild *Hippcampus haema* inhabiting Geoje Hansan Bay in South Korea using the metabarcoding technique. The authors find a survival strategy of wild *H. haema*, which are adapting to the dynamically changing coastal environment by consuming prey suitable for its mouth size.

Lee et al. (2022b) [12] report the first estimate for the Dokdo sea lion (*Zalophus japonicas*) population reconstructed from a discrete time stage-structured population model, showing a rapid decline in their population number with human hunting pressure and a 70% decline in the initial population after 10 years of hunting. The authors propose that there exists the potential for the extermination of a large local population of marine mammals due to human over-hunting.

This Special Issue covers topics for physical response and various major components in marine ecosystems such as the brown seaweed *Sargassum thundergii*, phytoplankton, zooplankton, fish eggs, wild Korean seahorse (*Hippcampus haema*), and an extinct marine mammal. This volume broadens our existing knowledge on the marine ecosystems in the Yellow Sea, the East China Sea, and the East/Japan Sea. Lastly, we hope that this volume provides a valuable foundation for understanding the current status of marine ecosystems in the Yellow Sea, the East China Sea, and the East/Japan Sea and their future response to ongoing climate change.

Funding: This research was supported by the grant (R2023056) from the National Institute of Fisheries Science (NIFS), funded by the Ministry of Oceans and Fisheries, Republic of Korea.

Conflicts of Interest: The authors declare no conflict of interest.

References

1. Lee, C.-I.; Jung, Y.-W.; Jung, H.-K. Response of Spatial and Temporal Variations in the Kuroshio Current to Water Column Structure in the Western Part of the East Sea. *J. Mar. Sci. Eng.* **2022**, *10*, 1703. [CrossRef]
2. Kim, S.; Choi, S.K.; Van, S.; Kim, S.T.; Kang, Y.H.; Park, S.R. Geographic Differentiation of Morphological Characteristics in the Brown Seaweed *Sargassum thunbergii* along the Korean Coast: A Response to Local Environmental Conditions. *J. Mar. Sci. Eng.* **2022**, *10*, 549. [CrossRef]
3. Park, K.-W.; Oh, H.-J.; Moon, S.-Y.; Yoo, M.-H.; Youn, S.-H. Effects of Miniaturization of the Summer Phytoplankton Community on the Marine Ecosystem in the Northern East China Sea. *J. Mar. Sci. Eng.* **2022**, *10*, 315. [CrossRef]
4. Kim, Y.; Youn, S.-H.; Oh, H.-J.; Joo, H.; Jang, H.-K.; Kang, J.-J.; Lee, D.B.; Jo, N.; Kim, K.; Park, S.; et al. Seasonal Compositions of Size-Fractionated Surface Phytoplankton Communities in the Yellow Sea. *J. Mar. Sci. Eng.* **2022**, *10*, 1087. [CrossRef]
5. Jang, H.-K.; Youn, S.-H.; Joo, H.; Kim, Y.; Kang, J.-J.; Lee, D.; Jo, N.; Kim, K.; Kim, M.-J.; Kim, S.; et al. First concurrent measurement of primary production in the Yellow Sea, the South Sea of Korea, and the East/Japan Sea, 2018. *J. Mar. Sci. Eng.* **2021**, *9*, 1237. [CrossRef]
6. Kang, J.J.; Oh, H.J.; Youn, S.-H.; Park, Y.; Kim, E.; Joo, H.T.; Hwang, J.D. Estimation of Phytoplankton Size Classes in the Littoral Sea of Korea Using a New Algorithm Based on Deep Learning. *J. Mar. Sci. Eng.* **2022**, *10*, 1450. [CrossRef]
7. Shin, S.S.; Choi, S.Y.; Seo, M.H.; Lee, S.J.; Soh, H.Y.; Youn, S.H. Spatiotemporal Distribution Characteristics of Copepods in the Water Masses of the Northeastern East China Sea. *J. Mar. Sci. Eng.* **2022**, *10*, 754. [CrossRef]
8. Kim, H.; Kim, G.; Kim, M.; Kang, D. Spatial Distribution of Sound Scattering Layer and Density Estimation of *Euphausia pacifica* in the Center of the Yellow Sea Bottom Cold Water Determined by Hydroacoustic Surveying. *J. Mar. Sci. Eng.* **2022**, *10*, 56. [CrossRef]
9. Kim, J.; Yun, H.-Y.; Won, E.J.; Choi, H.; Youn, S.-H.; Shin, K.-H. Influences of Seasonal Variability and Potential Diets on Stable Isotopes and Fatty Acid Compositions in Dominant Zooplankton in the East Sea, Korea. *J. Mar. Sci. Eng.* **2022**, *10*, 1768. [CrossRef]
10. Choi, H.-Y.; Choi, H.-C.; Kim, S.; Oh, H.-J.; Youn, S.-H. Discovery of pelagic eggs of two species from the rare mesopelagic fish genus *Trachipterus* (Lampriformes: Trachipteridae). *J. Mar. Sci. Eng.* **2022**, *10*, 637. [CrossRef]
11. Kim, M.-J.; Kim, H.-W.; Lee, S.-R.; Kim, N.-Y.; Lee, Y.-J.; Joo, H.-T.; Kwak, S.-N.; Lee, S.-H. Feeding Strategy of the Wild Korean Seahorse (*Hippocampus haema*). *J. Mar. Sci. Eng.* **2022**, *10*, 357. [CrossRef]
12. Lee, Y.-J.; Cho, G.; Kim, S.; Hwang, I.; Im, S.-O.; Park, H.-M.; Kim, N.-Y.; Kim, M.-J.; Lee, D.; Kwak, S.-N.; et al. The First Population Simulation for the *Zalophus japonicus* (Otariidae: Sea Lions) on Dokdo, Korea. *J. Mar. Sci. Eng.* **2022**, *10*, 271. [CrossRef]

Journal of
*Marine Science
and Engineering*

Response of Spatial and Temporal Variations in the Kuroshio Current to Water Column Structure in the Western Part of the East Sea

Chung-Il Lee [1], Yong-Woo Jung [2] and Hae-Kun Jung [3,*]

[1]   Department of Marine Bioscience, Gangneung-Wonju National University, Gangneung 25457, Korea
[2]   Department of Oceanography, Chonnam National University, Gwangju 61186, Korea
[3]   Fisheries Resources and Environment Research Division, East Sea Fisheries Research Institute, National Institute of Fisheries Science, Gangneung 25435, Korea
*   Correspondence: hkjung85@korea.kr; Tel.: +82-33-660-8533

Abstract: Using geographic sea surface current data, long-term changes in spatial and temporal variations in the Kuroshio Current 1993–2021 were analyzed, and the relationship between the Kuroshio Current and oceanic conditions, such as water column structure and intensity of East Korea Warm Current (EKWC) in the western part of the East Sea (WES), was investigated. Long-term changes in the Kuroshio Current intensity were positively correlated with the Pacific Decadal Oscillation and East Asian Winter Monsoon indices. When the Kuroshio Current was strong, its main axis passing around the Ryukyu Islands moved eastward, and the intensity of EKWC separated from the Kuroshio Current and flowed into the WES, indicating weakened conditions. When the intensity of the EKWC was weakened, its main axis moved away from the inshore area of the WES. As a result, the vertical distribution range of the cold and low saline water mass located in the bottom layer extended to shallower depths in the inshore area of the WES with increasing chlorophyll-*a*.

Keywords: western part of the East Sea; Kuroshio Current; East Korea Warm Current; Pacific decadal oscillation

Citation: Lee, C.-I.; Jung, Y.-W.; Jung, H.-K. Response of Spatial and Temporal Variations in the Kuroshio Current to Water Column Structure in the Western Part of the East Sea. *J. Mar. Sci. Eng.* **2022**, *10*, 1703. https://doi.org/10.3390/jmse10111703

Academic Editor: Rafael J. Bergillos

Received: 10 October 2022
Accepted: 8 November 2022
Published: 9 November 2022

Publisher's Note: MDPI stays neutral with regard to jurisdictional claims in published maps and institutional affiliations.

1. Introduction

The East Sea is a continental sea in the Northwest Pacific that connects to open oceans such as the North Pacific and East China Seas (ECS). Hence, changes in oceanic conditions in the East Sea are influenced by both the direct and teleconnection effects of atmospheric and oceanic circulation in the North Pacific [1,2]. The Kuroshio Current, which originates in the equatorial current system, plays a key role in transmitting heat energy from the equatorial region into the East Sea, and its intensity affects oceanic and biological conditions in the East Sea [1,3–6]. The Tsushima Warm Current, which passes into the western channel of the Korea Strait separated from the Kuroshio Current, is divided into two branches, one of which is the East Korea Warm Current (EKWC), which flows into the western part of the East Sea (WES) [7,8]. Thus, the EKWC originating from the Kuroshio Current is a key factor for the variations in the upper layer and deep-water circulation in the East Sea [8–10]. However, the composition of the water mass comprising the TWS and EKWC indicated distinct seasonal patterns.

The Tsushima Warm Current that flows into the WES primarily comprises two types of water masses: one that originates in the Taiwan Current and another a branch of the Kuroshio Current [11–14]. In the summer, the composition of the warm water mass moving through the Korea Strait has a similar ratio between the Taiwan Current and Kuroshio Current. In the winter, the composition of the Kuroshio Current exceeds 80% of the total ratio [15]. Therefore, the Kuroshio Current accounts for a large proportion of the warm water mass flowing into the WES via the Korea Strait during winter, and changes in the intensity of the Kuroshio Current may be critical for controlling the volume transport of warm water masses flowing into the WES [1,2,7,8,10].

Variations in EKWC volume transport are an essential factor for the changes in oceanic conditions in the WES, such as water temperature and salinity [1,16], and the intensity of the EKWC fluctuated similar to that of the Kuroshio Current [3]. The changes in oceanic conditions in the WES caused by the changes in the intensity of the Kuroshio Current were also closely associated with biological conditions, such as the spatial and temporal distribution of fishery resources in the WES [4,17]. During the strong EKWC period, the distribution range of warm water species, such as common squid, and their biomass expanded in the WES [3]. In addition, the EKWC and Tsushima Warm Current, separated from the Kuroshio Current, contain the nutrients transported by the Changjiang River entering the ECS. The intensity of the Kuroshio Current contributes to the variations in surface nutrients in the WES [18]. However, variations in oceanic conditions associated with the current system are simultaneously affected by current intensity and spatial changes [19,20]. The intensity and spatial changes in the Kuroshio Current influence the changes in the water column structure and chlorophyll-*a* (chl-*a*) in southern Japan and the Kuroshio Extension region in the eastern Pacific [21,22]. Previous studies have primarily focused on the intensity of the Kuroshio Current to explain oceanographic and biological conditions in the WES [1], and volume transport in the Korea Strait and Kuroshio Current has similar fluctuation patterns [3]. Therefore, the intensity of the Kuroshio Current influences the changes in oceanic conditions, such as the intensity of the Tsushima Warm Current and EKWC. However, spatial and temporal variations in the location of the main stream of currents can be one of the fundamental drivers of the changes in physical and biological environments [23,24]. Spatial changes in the main axis of the Kuroshio Current are also major factors influencing the changes in warm-water mass inflow in the WES, such as the Tsushima Warm Current and EKWC [1–3]. In addition, to understand the relationship between the Kuroshio Current and oceanic conditions in the WES, we must focus on both changes in the upper layer and water column structure in the WES.

The aim of this study was to elucidate the response mechanism of oceanic conditions in the WES to the variations in the intensity and spatial distribution of the Kuroshio Current. In particular, we addressed the following questions: when does the intensity of the Kuroshio Current strengthen, and does the volume transport of EKWC in the WES increase? In addition, we assessed the effect of change in the intensity of the Kuroshio Current on water column structure and chl-*a* in the WES, and its relationship with atmospheric and oceanic circulation in the North Pacific. To this end, we selected oceanographic data from eight fixed stations including the inshore and offshore areas near 37° N in the WES that passed the main axis of the EKWC, and then focused on the winter season with the highest proportion of the Kuroshio Current in the EKWC. We analyzed (1) long-term changes in the volume transport of warm water mass passing into the WES associated with the variations in the intensity and spatial distribution of the Kuroshio Current, (2) the effect of changes in the Kuroshio Current on changes in vertical structure and chl-*a* in the WES, and (3) the mechanisms underlying the response of oceanic conditions in the WES to the changes in the Kuroshio Current, focusing on the oceanic circulation system in the North Pacific.

2. Data and Methods

2.1. Oceanic Conditions in the WES

Oceanographic data were collected using conductivity, temperature, and depth probes provided by the National Institute of Fisheries Science in Korea. We used water temperature and salinity data in February between 1993–2021 from eight fixed stations near 37° N that passed the main axis of the EKWC (Figure 1). To analyze the long-term changes in oceanic conditions associated with the intensity and spatial variations in the EKWC, salinity in the upper 100 m layer was used, and the intensity of stratification (stratification index) was calculated using the water temperature difference between the upper 10 m and 100 m layers (Figure 1).

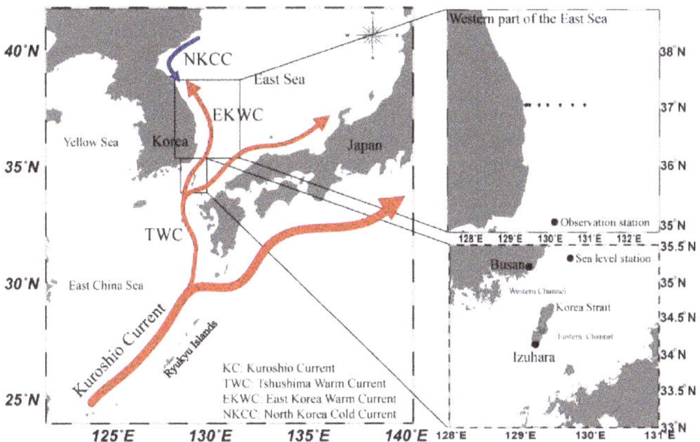

Figure 1. Study area for oceanic conditions such as water temperature and salinity in the western part of the East Sea and volume transport in the western channel of the Korea Strait.

2.2. Kuroshio Current Intensity and Longitudinal Shift of the Main Axis

The intensity of the Kuroshio Current and its longitudinal shift (west–east) were analyzed using geostrophic sea surface current data obtained from the Archiving, Validation, and Interpretation of Satellite Oceanographic database archived monthly 0.25° grids from 1993 to 2021. The Kuroshio Current intensity (KCI) was estimated using the average geostrophic sea surface current around 28° N and 125–129° E (Figure 2). Considering the time lag effect of change in the Kuroshio Current in the ECS on oceanic conditions in the WES [25–27], sea surface current data in January were used. The longitudinal shift in the main axis of the Kuroshio Current in January was calculated as the longitudinal position of the strongest current velocity around 125–129° E at the same latitude (28° N) (Figure 2).

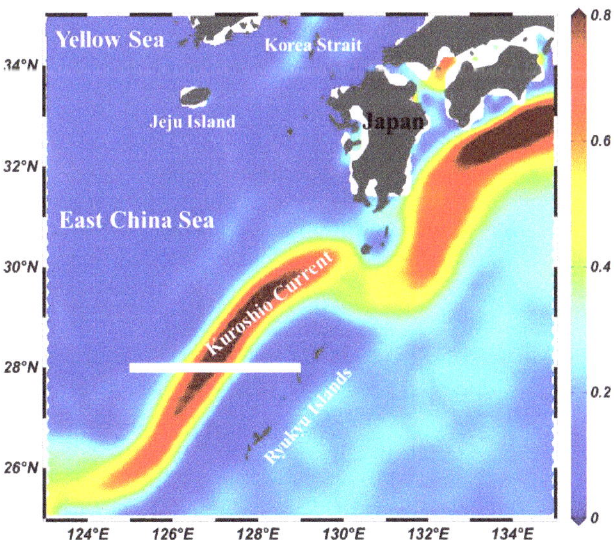

Figure 2. Mean geostrophic sea surface current velocity (m/s) from satellite in January during 1993–2021. The white bar indicates the area where the intensity and longitudinal shift of Kuroshio Current was studied.

2.3. Intensity of the East Korea Warm Current

The intensity of EKWC was defined as the volume transport of the western channel in the Korea Strait. The volume transport of the western channel in the Korea Strait during winter (January–March) from 1993 to 2019 was calculated using sea level data from Izuhara, Japan, from the Japan Meteorological Agency (JMA) and Busan, South Korea from the Korea Hydrographic and Oceanographic Agency (KHOA).

$$V = 1/f/\rho * \Delta p/\Delta x \tag{1}$$

where V is the volume transport (hm^3/s), f is the Coriolis force, ρ is the density of seawater (kg/m^3), Δp is the pressure difference between each tidal station (hPa), and Δx is the distance between each tidal station (51.17 km) [1,28].

2.4. Chlorophyll-a

To analyze the long-term variations in chl-*a* in the WES (35–40° N, 127–132° E) during February, the merged satellite data from the Copernicus Marine Environmental Monitoring Service (CMEMS) was used [29]. CMEMS GlobColour-merged chl-*a* product relies on the following sensors: SeaWiFS (1997–2010), MERIS (2002–2012), MODIS Aqua (2002–present), VIIRS-NPP (2012–present) and OLCI-S3A (2016–present) [29]. The monthly Level-4 product chl-*a* for the global ocean from 1998 to 2021 was obtained from the Copernicus website (https://marine.copernicus.eu/access-data (accessed on 31 August 2022)). Level 4 data were re-mapped to a standard projection with a spatial resolution of 4 km [29].

2.5. Climate Indices

The Pacific Decadal Oscillation Index (PDOI) and East Asian Winter Monsoon Index (EAWMI) were used to elucidate the relationship between the intensity of the Kuroshio Current and the atmospheric and oceanic circulation systems in the North Pacific. The EAWMI is defined as the difference between the two regions (27.5–37.5° N, 110–170° E, 50–60° N, 80–140° E) in area-averaged zonal wind speed at 300 hectopascal [30].

$$EAWMI = U_{300}\ (27.5\text{-}37.5°\ N,\ 110\text{-}170°\ E) - U_{300}\ (50\text{-}60°\ N,\ 80\text{-}140°\ E)$$

The PDOI is defined as the leading empirical orthogonal function of monthly sea surface temperature anomalies in the Pacific poleward of 20° N [31].

2.6. Statistical Analysis

The sequential *t*-test analysis of regime shifts (STARS) developed by Rodionov (2006) was applied to determine the time scale of the regime and magnitudes of the intensity of Kuroshio Current and volume transport in the western channel of the Korea Strait. Sequential *t*-test analysis of regime shifts was designed for sequential data processing and can identify statistically significant departures from the mean.

3. Results

3.1. Long-Term Changes in the Kuroshio Current and Their Relationship with Climate Effects

The KCI showed a positive anomaly (strong velocity) before 2010 and a rapidly decreasing current velocity pattern after 2010 (Figure 3). The STAR analysis results also showed that step change in the KCI was detected in 2010 (Figure 3). These fluctuations in the KCI were related to the longitudinal shift in the main axis of the Kuroshio Current ($r = 0.55$, $p < 0.01$). The analysis of the correlation coefficient between the longitudinal shift in the main axis of the Kuroshio Current and surface current velocity around the ECS revealed that the surface current velocity around the Ryukyu Islands, which are located in the main stream of the Kuroshio Current, has a positive relationship with the longitudinal shift in the main axis of the Kuroshio Current (Figure 4). In contrast, the western part of the Kuroshio Current mainstream had a negative relationship with the longitudinal shift in the main axis of the Kuroshio Current (Figure 4), implying that when the Kuroshio velocity was high, the main axis of the Kuroshio Current migrated eastward. However, when the Kuroshio velocity was low, it moved westward.

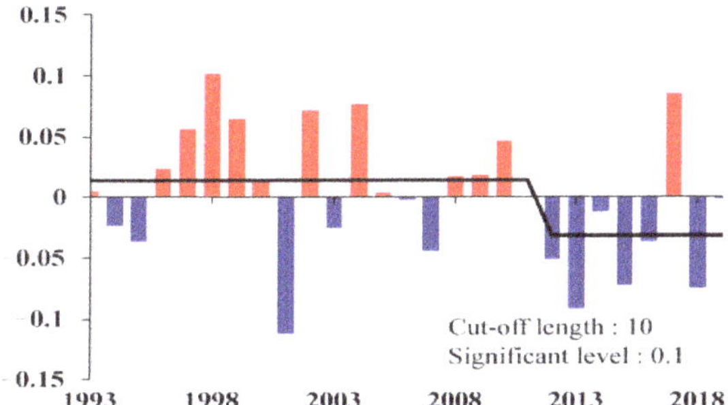

Figure 3. Long-term variations in the Kuroshio Current intensity in January. Black line represents step changes estimated using sequential *t*-test analysis of regime shifts analysis.

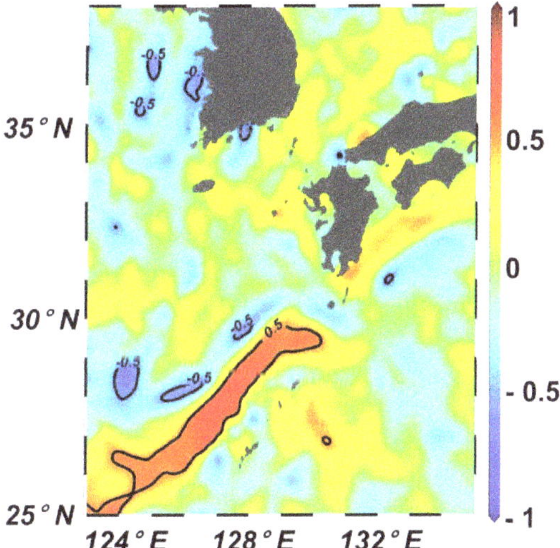

Figure 4. Horizontal distribution of the correlation coefficient between the geostrophic surface current velocity in January and the longitudinal shift of the Kuroshio Current Index in January. Black lines represent significant levels: $p < 0.01$.

Long-term changes in KCI were related to atmospheric and oceanic conditions in the North Pacific. The EAWMI and PDOI are key climate indices for the intensity of the Northwest Pacific winter monsoon and fluctuations in sea surface temperature in the North Pacific associated with atmospheric and oceanic circulations. The correlation coefficients between the surface current velocity around the ECS and EAWMI and PDOI showed that both climate indices had a positive relationship with surface current velocity around the Ryukyu Islands (Figure 5). In particular, a highly significant coefficient of variation was observed with a two-month lag phase than with a phase with no lag (Figure 5).

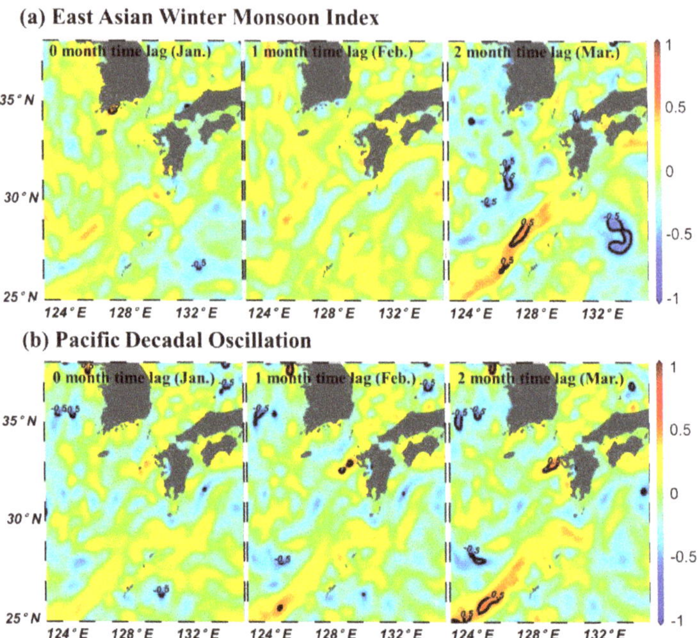

Figure 5. Horizontal distribution of the correlation coefficient between the geostrophic surface current velocity in January: (**a**) East Asian Winter Monsoon Index (January–March), and (**b**) Pacific Decadal Oscillation Index (January–March). Black lines represent significant levels: $p < 0.01$.

3.2. Long-Term Changes in the East Korea Warm Current

Long-term changes in the EKWC have a negative correlation with the KCI (Table 1). After 2010, the intensity of the EKWC exhibited a rapidly increasing pattern, whereas the KCI exhibited a decreasing pattern (Figures 3 and 6). The EKWC and KCI had a one-month lag phase (leading KCI) and had positive correlation coefficients (Table 1). The KCI in January had a more significant relationship with volume transport in February than it did with that in January and March (Table 1).

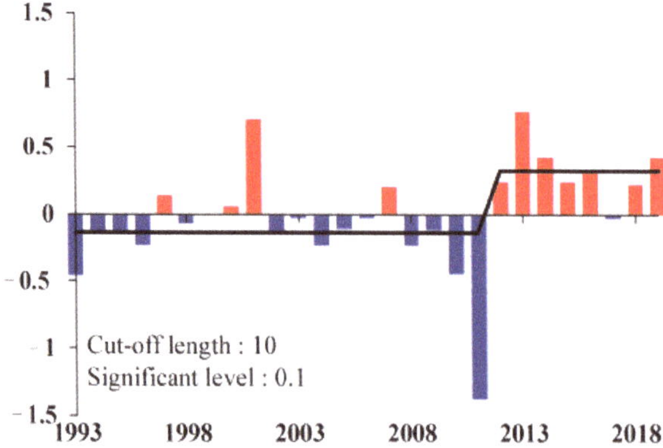

Figure 6. Long-term changes in East Korea Warm Current intensity in February. Black line represents step changes estimated using sequential *t*-test analysis of regime shifts analysis.

Table 1. Correlation coefficient between volume transport in western channel of Korea Strait and Kuroshio Current intensity.

	Volume Transport in Western Channel of Korea Strait			
	January	February	March	Mean (January–March)
Kuroshio	$-0.26^{\,N}$	$-0.47^{\,*}$	$-0.27^{\,N}$	$-0.36^{\,N}$

*: $p < 0.01$, N: $p \geq 0.01$.

3.3. Variations in Oceanic Conditions in the Western Part of the East Sea Associated with the East Korea Warm Current

Long-term changes in salinity at a depth of 100 m in the WES near 37° E exhibited a distinct fluctuation pattern during weakened EKWC periods from the late 1990s to 2010; the salinity at the depth of 100 m around the inshore area $\leq 129.79°$ E was lower than 34.1 (Figure 7a). However, after 2010, the salinity at a depth of 100 m around the inshore area increased (Figure 7a), thereby indicating strong a EKWC (Figure 6). Furthermore, the stratification index was related to the intensity of the EKWC (Figure 7b). The stratification index increased after the late 1990s, which weakened EKWC. After the early 2010s, the stratification index was gradually reduced with a strong EKWC (Figure 7b). In summary, a low-saline water mass in the upper 100 m layer with strong stratification formed during the period of weakened EKWC. In contrast, during the strengthened EKWC period, highly saline water with weak stratification was observed (Figure 7). After 2019, the salinity and stratification index were slightly decreased (≤ 34.01) and increased (≥ 8), respectively; however, due to the absence of EKWC intensity data, the relationship between water column structure and EKWC intensity could not be analyzed in this study.

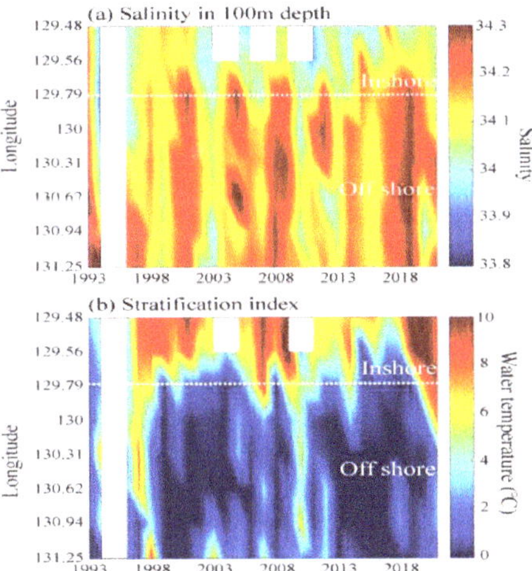

Figure 7. Long-term changes in (**a**) salinity at 100 m depth and (**b**) stratification index.

3.4. Long-Term Changes in Chlorophyll-a in the Inshore Area of the Western Part of the East Sea

Long-term fluctuations in the intensity of stratification in February were related to chl-*a* in February along the inshore area of the WES (Figure 8). The chl-*a* in inshore area of the WES had a more significant positive relationship than those in other regions, according to the analyzed correlation coefficients between the surface amount of Chl-*a* in February around the WES and the intensity of stratification in the inshore area (Figure 8). This suggests that the chl-*a* in the inshore area increased during winter owing to strengthened stratification (Figure 8).

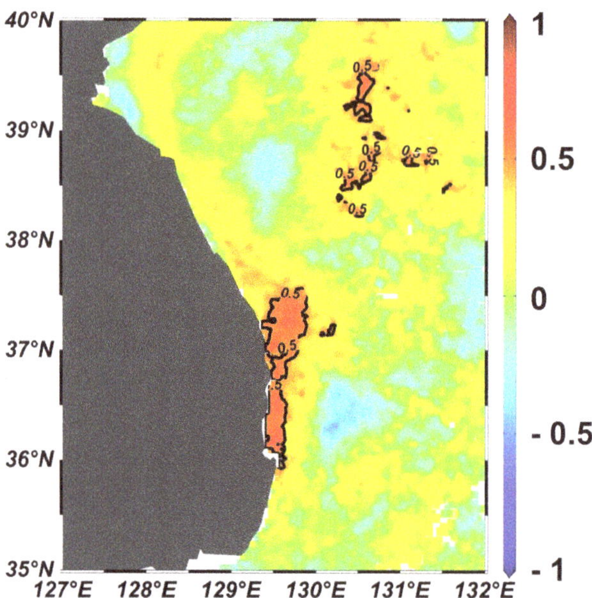

Figure 8. Horizontal correlation coefficient distribution between chlorophyll-*a* in February and stratification index in February. Black lines represent significant levels: $p < 0.01$.

4. Discussion

Long-term changes in the oceanic conditions of the WES are affected by the fluctuations in both the thermal energy of the atmosphere and the current system in North Pacific [1,2,4]. In particular, the EAWMI and PDOI, which show atmospheric and oceanic circulations in the NP, are important factors influencing the changes in oceanic conditions in the WES [1,2,4]. In this study, the EAWMI and PDOI exhibited positive relationship with the KCI (Figure 9). In previous studies, the relationship between the Kuroshio Current and Pacific Decadal Oscillation has been reported. During the positive mode of the Pacific Decadal Oscillation, the intensity of the Kuroshio Current was increased by strong Aleutian low-pressure [31,32]. With strengthened Aleutian low-pressure, the pressure gradient between the high- and middle-latitude areas of the North Pacific as well as the westerlies around the middle-latitude area increased [31,32]. Consequently, the sea surface temperature in the central North Pacific exhibited a decreasing trend owing to strong winds, heat loss, and vertical mixing; however, the intensity of subtropical gyres such as the Kuroshio Current was enhanced by strong Ekman transport [33–35]. Additionally, variation in Aleutian low-pressure intensity is one of the primary factors controlling the EAWM by interacting with the Siberian High pressure [36]. The pressure gradient between the enhanced Aleutian low-pressure and Siberian High pressure, as well as the intensity of the winter monsoon over the WES, increased [1,2]. Consequently, during periods of strong Aleutian low-pressure, northwesterly winter winds, and cold, dry air masses from Siberia were intensified near the WES [1,2].

Fluctuations in wind stress curl associated with Aleutian low-pressure and Pacific Decadal Oscillation intensity variations have a positive relationship with the barotropic transport of the Kuroshio Current [37]; however, in this study, the response of geostrophic velocity in the ECS indicates regional differences. The surface current velocity around the ECS, which are located in the main stream of Kuroshio Current, had a positive relationship with the longitudinal shift in the main axis of the Kuroshio Current (Figure 9). This implies that when the KCI was enhanced, the main axis of the Kuroshio Current migrated eastward. The geostrophic velocity in the mainstream region of the Kuroshio Current and the western section of the Ryukyu Islands indicates different responses to the changes in atmospheric and oceanic circulations in the North Pacific [38,39]. Previous studies have explained that in the period with strong Kuroshio Current and enhanced Aleutian low-pressure, the effect of Ekman transport on the flow of the Kuroshio Current moving through the ECS was amplified, and the main axis of the Kuroshio Current migrated eastward with strong velocity [1,16]. Consequently, the warm water mass separated from the Kuroshio Current, such as the Tsushima Warm Current, as well as the volume transport in the Korea Strait was weakened [1,16]. Furthermore, atmospheric conditions near the Korea Strait have a critical role in controlling volume transport in the Korea Strait. During winter, volume transport in the Korea Strait was reduced with increased northwesterly winds related to a strong EAWM [16,40]. In this study, similar results were obtained. When the intensity of the EKWC was enhanced, the KCI showed a decreasing pattern with strong EAWMI. Consequently, the response of oceanic conditions to the KCI revealed regional differences in the Northwest Pacific [1,2,41]. In previous studies, oceanic conditions, such as sea level and water temperature in the WES, driven by branch currents of the Kuroshio Current including the Tsushima Warm Current and EKWC indicated different fluctuation patterns with the Kuroshio Current region, such as the ECS and Kuroshio Extension in North Pacific [1,2,41].

Figure 9 Schematic of the relationship between intensity of Kuroshio Current and East Korea Warm Current. Abbreviations: ECS, East China Seas; EAWM, East Asian Winter Monsoon; PDO, Pacific Decadal Oscillation.

The changes in the intensity of the EKWC flowing through the Korea Strait into the WES influence oceanic and biological conditions of the WES [42,43]. For the enhanced EKWC period, the main axis of the EKWC was adjacent to the inshore area but separated from the inshore area during the periods of reduced EKWC [44]. This spatial distribution along the main axis of the EKWC was related to the changes in the vertical structure of the water column in the inshore area [2,9]. During enhanced EKWC, the vertical volume of the warm water mass increased, and the vertical structure in the inshore area indicated mixed conditions. However, as the EKWC weakened and the main axis moved away from the inshore area, the North Korea Cold Current extended to the inshore area. Consequently, the vertical distribution range of the North Korea Cold Current, which was cold with less saline (33.9–34.1‰), and nutrient rich water mass with high oxygen content [9,45,46], extended to shallower depths in the inshore area of the WES [42]. Furthermore, vertical mixing became more active owing to the strong winter monsoon in the winter [2,43], and a replete condition of nutrient replenishment below the upper layer was supplied to the upper layer [43]. In February, chl-*a* in the inshore area of the WES increased.

The results of this study provide novel insights into the possible mechanisms underlying the effect of climate change on oceanic conditions in the WES. Especially, these results provide major information that can be used to better understand the effects of the intensity changes in the Kuroshio current related to climate change on the change of EKWC and water column structure in the WES. Furthermore, by explaining the mechanism underlying the changes in oceanic conditions in habitats ground, this study expands our understanding of the changes in distribution and biomass of fisheries resources. However, to accurately investigate the changes in low trophic level, such as primary production, understanding the nutrient circulation cycle, supplement, and demand is necessary. In this study, we did not explain the process involving nutrient source. Therefore, in future studies, we will assess the possible mechanisms underlying the response of climate change to physical, chemical, and biological conditions in the WES, including nutrient, primary production, and fisheries resources.

5. Conclusions

Our study elucidated the impact of the intensity of the Kuroshio Current on the variations in the vertical water column structure in the inshore area of the WES. Long-term fluctuations in WES oceanic conditions are affected by both atmospheric and oceanic circulation. The intensity of the Kuroshio Current was related to atmospheric conditions, such as the intensity of the winter monsoon, as well as oceanic conditions, such as the volume transport of warm water mass into the WES. The interaction between atmospheric and oceanic conditions can play a major role in controlling the changes in the water column structure and primary production. During periods of increased KCI, the severity of the winter monsoon increased; however, the warm water mass passing into the WES separated from the Kuroshio Current weakened. Consequently, the North Korea Cold Current range expanded, and cold-water masses were distributed under the upper layer in the inshore area of the WES. The water column structure is a possible primary cause of the high chl-*a* observed in the area.

Author Contributions: Conceptualization, Methodology, Validation, Formal analysis, and Writing—Original Draft, H.-K.J.; Supervision, Project Administration and Funding Acquisition, and Writing—Review and Editing, C.-I.L.; Data Curation and Visualization, Y.-W.J. All authors have read and agreed to the published version of the manuscript.

Funding: This research was supported by the National Institute of Fisheries Science, Ministry of Oceans and Fisheries, Korea (R2022035) and Korea Institute of Marine Science & Technology Promotion (KIMST) funded by the Ministry of Oceans and Fisheries (20220558).

Data Availability Statement: All data supporting the results of this study are provided in this manuscript.

Acknowledgments: We would like to thank the members of the Fisheries Resources and Environment Research Division in National Institute of Fisheries Science for their assistance in field observations and data analyses.

Conflicts of Interest: The authors declare no conflict of interest.

References

1. Jung, H.K.; Rahman, S.M.; Kang, C.K.; Park, S.Y.; Lee, S.H.; Park, H.J.; Lee, C.I. The influence of climate regime shifts on the marine environment and ecosystems in the East Asian Marginal Seas and their mechanisms. *Deep Sea Res. Part II Top. Stud. Oceanogr.* **2017**, *143*, 110–120. [CrossRef]
2. Jung, H.K.; Rahman, S.M.; Choi, H.C.; Park, J.M.; Lee, C.I. Recent trends in oceanic conditions in the western part of East/Japan Sea: An analysis of climate regime shift That occurred after the late 1990s. *J. Mar. Sci. Eng.* **2021**, *9*, 1225. [CrossRef]
3. Zhang, C.I.; Lee, J.B.; Kim, S.; Oh, J.H. Climatic regime shifts and their impacts on marine ecosystem and fisheries resources in Korean waters. *Prog. Oceanogr.* **2000**, *47*, 171–190. [CrossRef]
4. Tian, Y.; Kidokoro, H.; Watanabe, T.; Iguchi, N. The late 1980s regime shift in the ecosystem of Tsushima warm current in the Japan/East Sea: Evidence from historical data and possible mechanisms. *Prog. Oceanogr.* **2008**, *77*, 127–145. [CrossRef]
5. Tang, T.Y.; Tai, J.H.; Yang, Y.J. The flow pattern north of Taiwan and the migration of the Kuroshio. *Cont. Shelf Res.* **2000**, *20*, 349–371. [CrossRef]
6. Hsin, Y.C.; Wu, C.R.; Shaw, P.T. Spatial and temporal variations of the Kuroshio east of Taiwan, 1982–2005: A numerical study. *J. Geophys. Res. Oceans* **2008**, *113*, C04002. [CrossRef]
7. Cho, Y.K.; Kim, K. Branching mechanism of the Tsushima Current in the Korea Strait. *J. Phys. Oceanogr.* **2000**, *30*, 2788–2797. [CrossRef]

8. Lee, D.K.; Niiler, P.P.; Lee, S.R.; Kim, K.; Lie, H.J. Energetics of the surface circulation of the Japan/East Sea. *J. Geophys. Res. Oceans* **2000**, *105*, 19561–19573. [CrossRef]
9. Cho, Y.K.; Kim, K. Seasonal variation of the East Korea Warm Current and its relation with the cold water. *La Mer* **1996**, *34*, 172–182.
10. Kim, Y.Y.; Cho, Y.K.; Kim, Y.H. Role of cold water and beta-effect in the formation of the East Korean Warm Current in the East/Japan Sea: A numerical experiment. *Ocean Dyn.* **2018**, *68*, 1013–1023. [CrossRef]
11. Beardsley, R.C.; Limeburner, R.; Yu, H.; Cannon, G.A. Discharge of the Changjiang (Yangtze River) into the East China sea. *Cont. Shelf Res.* **1985**, *4*, 57–76. [CrossRef]
12. Fang, G.H.; Zhao, B.R.; Zhu, Y.H. Water volume transport through the Taiwan Strait and the continental shelf of the East China Sea measured with current meters. In *Oceanography of Asian Marginal Seas*; Takano, K., Ed.; Elsevier: New York, NY, USA, 1991; pp. 345–358. [CrossRef]
13. Nitani, H. Beginning of the Kuroshio. In *Kuroshio, Physical Aspect of the Japan Current*; Yoshida, S.K., Ed.; University Washington Press: Seattle, WA, USA, 1972; Volume 129, p. 163.
14. Lie, H.J.; Cho, C.H. On the origin of the Tsushima Warm Current. *J. Geophys. Res. Oceans* **1994**, *99*, 25081–25091. [CrossRef]
15. Guo, X.; Miyazawa, Y.; Yamagata, T. The Kuroshio onshore intrusion along the shelf break of the East China Sea: The origin of the Tsushima Warm Current. *J. Phys. Oceanogr.* **2006**, *36*, 2205–2231. [CrossRef]
16. Jung, Y.; Park, J.H.; Hirose, N.; Yeh, S.W.; Kim, K.J.; Ha, H.K. Remote impacts of 2009 and 2015 El Niño on oceanic and biological processes in a marginal sea of the Northwestern Pacific. *Sci. Rep.* **2022**, *12*, 741. [CrossRef]
17. Tian, Y.; Kidokoro, H.; Watanabe, T. Long-term changes in the fish community structure from the Tsushima warm current region of the Japan/East Sea with an emphasis on the impacts of fishing and climate regime shift over the last four decades. *Prog. Oceanogr.* **2006**, *68*, 217–237. [CrossRef]
18. Shim, M.J.; Yoon, Y.Y. Long-term variation of nitrate in the East Sea, Korea. *Environ. Monit. Assess.* **2021**, *193*, 720. [CrossRef] [PubMed]
19. Kawabe, M. Variations of current path, velocity, and volume transport of the Kuroshio in relation with the large meander. *J. Phys. Oceanogr.* **1995**, *25*, 3103–3117. [CrossRef]
20. Douglass, E.M.; Jayne, S.R.; Bryan, F.O.; Peacock, S.; Maltrud, M. Kuroshio pathways in a climatologically forced model. *J. Oceanogr.* **2012**, *68*, 625–639. [CrossRef]
21. Wang, Y.; Tang, R.; Yu, Y.; Ji, F. Variability in the sea surface temperature gradient and its impacts on chlorophyll-a concentration in the Kuroshio extension. *Remote Sens.* **2021**, *13*, 888. [CrossRef]
22. Chow, C.H.; Lin, Y.C.; Cheah, W.; Tai, J.H. Injection of high chlorophyll-a waters by a branch of Kuroshio Current into the nutrient-poor North Pacific subtropical gyre. *Remote Sens.* **2022**, *14*, 1531. [CrossRef]
23. Chang, Y.L.K.; Miyazawa, Y.; Miller, M.J.; Tsukamoto, K. Influence of ocean circulation and the Kuroshio large meander on the 2018 Japanese eel recruitment season. *PLoS ONE* **2019**, *14*, e0223262. [CrossRef] [PubMed]
24. Liu, S.; Liu, Y.; Li, J.; Cao, C.; Tian, H.; Li, W.; Tian, Y.; Watanabe, Y.; Lin, L.; Li, Y. Effects of oceanographic environment on the distribution and migration of Pacific saury (*Cololabis saira*) during main fishing season. *Sci. Rep.* **2022**, *12*, 13585. [CrossRef] [PubMed]
25. Baek, S.H.; Kim, Y.; Lee, M.; Ahn, C.Y.; Cho, K.H.; Park, B.S. Potential cause of decrease in bloom events of the harmful dinoflagellate Cochlodinium polykrikoides in southern Korean coastal waters in 2016. *Toxins* **2020**, *12*, 390. [CrossRef] [PubMed]
26. Kim, H.C.; Yamaguchi, H.; Yoo, S.; Zhu, J.; Okamura, K.; Kiyomoto, Y.; Katsuhisa, T.; Kim, S.-W.; Park, T.; Oh, I.S.; et al. Distribution of Changjiang diluted water detected by satellite chlorophyll-a and its interannual variation during 1998–2007. *J. Oceanogr.* **2009**, *65*, 129–135. [CrossRef]
27. Bai, Y.; He, X.; Pan, D.; Chen, C.T.A.; Kang, Y.; Chen, X.; Cai, W.J. Summertime Changjiang River plume variation during 1998–2010. *J. Geophys. Res.* **2014**, *119*, 6238–6257. [CrossRef]
28. Takikawa, T.; Yoon, J.H.; Cho, K.D. The Tsushima warm current through Tsushima Straits estimated from ferryboat ADCP data. *J. Phys. Oceanogr.* **2005**, *35*, 1154–1168. [CrossRef]
29. Garnesson, P.; Mangin, A.; Fanton d'Andon, O.; Demaria, J.; Bretagnon, M. The CMEMS GlobColour chlorophyll a product based on satellite observation: Multi-sensor merging and flagging strategies. *Ocean Sci.* **2019**, *15*, 819–830. [CrossRef]
30. Jhun, J.G.; Lee, E.J. A new East Asian winter monsoon index and associated characteristics of the winter monsoon. *J. Clim.* **2004**, *17*, 711–726. [CrossRef]
31. Mantua, N.J.; Hare, S.R.; Zhang, Y.; Wallace, J.M.; Francis, R.C. A Pacific interdecadal climate oscillation with impacts on salmon production. *Bull. Am. Meteorol. Soc.* **1997**, *78*, 1069–1080. [CrossRef]
32. Sugimoto, S.; Hanawa, K. Decadal and interdecadal variations of the Aleutian Low activity and their relation to upper oceanic variations over the North Pacific. *J. Meteorol. Soc. Jpn Ser. II* **2009**, *87*, 601–614. [CrossRef]
33. Mantua, N.J.; Hare, S.R. The Pacific Decadal Oscillation. *J. Oceanogr.* **2002**, *58*, 35–44. [CrossRef]
34. Miller, R.L.; Perlwitz, J.P.; Tegen, I. Modeling Arabian dust mobilization during the Asian summer monsoon: The effect of prescribed versus alculated SST. *Geophys. Res. Lett.* **2004**, *30*, L22214. [CrossRef]
35. Schneider, N.; Cornuelle, B.D. The forcing of the Pacific decadal oscillation. *J. Clim.* **2005**, *18*, 4355–4373. [CrossRef]
36. Gong, D.Y.; Ho, C.H. The Siberian High and climate change over middle to high latitude Asia. *Theor. Appl. Climatol.* **2002**, *72*, 1–9. [CrossRef]

37. Andres, M.; Kwon, Y.O.; Yang, J. Observations of the Kuroshio's barotropic and baroclinic responses to basin-wide wind forcing. *J. Geophys. Res. Oceans* **2011**, *116*, C04011. [CrossRef]
38. Li, R.; Zhang, Z.; Wu, L. High-resolution modeling study of the Kuroshio path variations south of Japan. *Adv. Atmos. Sci.* **2014**, *31*, 1233–1244. [CrossRef]
39. Wu, C.R.; Wang, Y.L.; Chao, S.Y. Disassociation of the Kuroshio Current with the Pacific Decadal Oscillation Since 1999. *Remote Sens.* **2019**, *11*, 276. [CrossRef]
40. Ma, C.; Wu, D.; Lin, X.; Yang, J.; Ju, X. On the mechanism of seasonal variation of the Tsushima Warm Current. *Cont. Shelf Res.* **2012**, *48*, 1–7. [CrossRef]
41. Gordon, A.L.; Giulivi, C.F. Pacific decadal oscillation and sea level in the Japan/East Sea. *Deep-Sea Res. I Oceanogr. Res. Pap.* **2004**, *51*, 653–663. [CrossRef]
42. Lee, J.Y.; Kang, D.J.; Kim, I.N.; Rho, T.; Lee, T.; Kang, C.K.; Kim, K.R. Spatial and temporal variability in the pelagic ecosystem of the East Sea (Sea of Japan): A review. *J. Mar. Syst.* **2009**, *78*, 288–300. [CrossRef]
43. Park, K.A.; Park, J.E.; Kang, C.K. Satellite-observed chlorophyll-a concentration variability in the East Sea (Japan Sea): Seasonal cycle, long-term trend, and response to climate index. *Front. Mar. Sci.* **2022**, *9*, 807570. [CrossRef]
44. Pak, G.; Kim, Y.H.; Park, Y.G. Lagrangian approach for a new separation index of the East Korea Warm Current. *Ocean Sci. J.* **2019**, *54*, 29–38. [CrossRef]
45. Mooers, C.N.K.; Bang, I.; Sandoval, F.J. Comparison between observation and numerical simulations of Japan (East) Sea flow and mass fields in 1999 through 2001. *Deep Sea Res. II Top. Stud. Oceanogr.* **2005**, *52*, 1639–1661. [CrossRef]
46. Ashjian, C.J.; Davis, C.S.; Gallager, S.M.; Alatalo, P. Characterization of the zooplankton community, size composition, and distribution in relation to hydrography in the Japan/East Sea. *Deep Sea Res. II Top. Stud. Oceanogr.* **2005**, *52*, 1363–1392. [CrossRef]

Journal of
*Marine Science
and Engineering*

Article

Geographic Differentiation of Morphological Characteristics in the Brown Seaweed *Sargassum thunbergii* along the Korean Coast: A Response to Local Environmental Conditions

Sangil Kim [1], Sun Kyeong Choi [2], Seohyeon Van [1], Seong Taek Kim [3], Yun Hee Kang [4,*] and Sang Rul Park [2,*]

[1] Ocean Climate and Ecology Research Division, National Institute of Fisheries Science, Busan 46083, Korea; sikim10@korea.kr (S.K.); shvan@korea.kr (S.V.)
[2] Estuarine and Coastal Ecology Laboratory, Department of Marine Life Sciences, Jeju National University, Jeju 63243, Korea; choisk@jejunu.ac.kr
[3] B&G Ecotech Co., Ltd., Jinju 52818, Korea; kst73@naver.com
[4] Department of Earth and Marine Sciences, Jeju National University, Jeju 63243, Korea
* Correspondence: unirang78@gmail.com (Y.H.K.); srpark@jejunu.ac.kr (S.R.P.); Tel.: +82-64-754-3425 (S.R.P.)

Abstract: Intraspecific variation in morphology is widespread among seaweed species in different habitats. We examined the morphological variation in *Sargassum thunbergii* involving diverse environmental factors. We quantified 16 morphological characteristics on 15 rocky intertidal shores in Korea. A cluster analysis based on morphology identified three groups. Group M1 comprised populations on the northern part of the east coast, where the thalli was short and thick, with large leaf and air-vesicle. Group M3 consisted of populations on the west coast exclusively separated from other populations, with short, slender and sparsely branched thalli. Group M2 comprised populations on the southern part of the east coast and on the south coast (including Jeju Island), with longest thalli and lateral branches. Principal coordinate analyses showed that group M1 and M3 were mostly influenced by strong wave action and large tidal amplitudes, respectively. Group M2 were under the influence of warm temperatures and high irradiance. Biota-environment matching analysis showed that the morphology is affected by combinations of different local environmental factors and also that tidal condition is important as a single variable, suggesting that morphology of *S. thunbergii* reflects and adapts to local environmental conditions.

Keywords: *Sargassum thunbergii*; morphological variability; seaweed morphology; multiple environmental factors; intertidal zone

Citation: Kim, S.; Choi, S.K.; Van, S.; Kim, S.T.; Kang, Y.H.; Park, S.R. Geographic Differentiation of Morphological Characteristics in the Brown Seaweed *Sargassum thunbergii* along the Korean Coast: A Response to Local Environmental Conditions. *J. Mar. Sci. Eng.* **2022**, 10, 549. https://doi.org/10.3390/jmse10040549

Academic Editor: Azizur Rahman

Received: 25 March 2022
Accepted: 14 April 2022
Published: 16 April 2022

Publisher's Note: MDPI stays neutral with regard to jurisdictional claims in published maps and institutional affiliations.

1. Introduction

The populations of plant species in different habitat show alternative morphological forms. The degree of intraspecies morphological differences is dependent on dispersal distances and gene flow [1]. Considering that their life history traits were closely related with morphology, morphological variability has important ecological and physiological implications for the plant themselves and the organisms associated with them [2–5]. Seaweeds having broad habitat ranges, from intertidal to subtidal zones, usually show high intraspecific morphological variations [6]. Such morphological variability is believed to extend selective advantages to individuals inhabiting various environmental conditions [7–10]. For example, large brown seaweeds such as *Laminaria longicruis* and *Eisenia arborea* have narrow and strap-like blades to enhance their survival and fitness under wave-exposed condition [8,11]. Thus, the existing morphological characteristics in seaweeds would reflect local environmental conditions where they inhabit.

Seaweed morphology is influenced by many environmental factors, including season [12], depth [13,14], nutrient availability [2,15], wave action [16], irradiance [17,18], temperature and salinity [19–21]. For example, *Ulva fenestrate* shows the optimal thallus growth at high irradiance and low temperature [17] and seaweeds growing in wave-exposed sites (e.g., *Codium fragile*, *Fucus* spp. and *Ecklonia radiata*) are often much smaller than those growing in sheltered waters [16,22–24]. Similarly, *Saccharina latissima* has narrow blade in hyposaline condition [19]. Due to the relationships between morphological development and extrinsic conditions, some seaweed species can be used as environmental indicators [25,26]. Hence, improved understanding of the impacts of environmental conditions on seaweed morphology can be an important element in investigations of coastal ecological health [27,28].

Morphological variability in seaweed results from the interactions of the mutifactors [17,18]. Wave action is well known as a major physical factor that influence thallus size and intraspecific morphological variability in seaweed because many studies have focused on the effects of wave action [29]. Relatively few have investigated the influence of other abiotic factors, such as salinity, desiccation, nutrient concentration and temperature, which affect morphology indirectly by altering physiological responses to stressors [2,21,30,31]. Furthermore, biotic factors, biomass losses and defenses induced by grazing pressure can affect morphological variability [27,30]. Although morphological variation is affected by the complex interactions of multiple stressors [23,32], many investigations have regarded only one or two factors. A better understanding of the relationships between morphology and environment requires a more comprehensive approach that considers multiple factors simultaneously.

The genus *Sargassum* C. Agardh is one of the most species-rich genera among the Fucales, Phaeophyceae [33]. *Sargassum thunbergii*, an ecologically and economically important brown seaweed [34], is widely distributed from middle to lower intertidal zones in the coastal areas of Northeast Asia including Korea, China and Japan [35–37]. This species has high morphological variability and adjusts its growth form in accordance with varying levels of local environmental conditions [34,38–40]. *S. thunbergii* provides habitat, shelter and nursery areas for a wide variety of flora and fauna [41]. This species is especially an important feed source for abalone and the rapidly developing holothurian aquaculture. Additionally, *S. thunbergii* is a major source of bioactive components such as alginate and pharmaceutical products and widely used in biosorptions of trace metal ions [42–44]. Lately, due to large biomass and high productivity, this species has been proposed as an appropriate candidate species to artificially restore large seaweed forests in the intertidal zone in China [45]. The technology for commercial cultivation of *S. thunbergii* has been developed to meet various types of demand [34]. However, ecological and physiological information available for cultivation, conservation and restoration of *S. thunbergii* is still lacking [41,46,47], even though several studies have been conducted on phylogenetic or population genetic diversity [48–50]. In particular, there have been almost no attempts to identify environmental factors affecting intraspecific morphological variation of *S. thunbergii*.

Understanding how morphological traits of seaweed respond to varying environmental factors is important in making management policy for its conservation, restoration and aquaculture. In this study, we hypothesized that morphological differences among *S. thunbergii* populations occur along the Korean coast depending on local environmental conditions. The objectives of this study were (1) to identify the morphological differences among *S. thunbergii* populations grown on Korean coasts and (2) to examine the determinant of their morphology through quantifying the relationship between morphology of *S. thunbergii* and local environmental factors.

2. Materials and Methods

2.1. Study Area

The present study was conducted at 15 study sites along the coasts of South Korea including Jeju Island (Figure 1a). The eastern coast of Korea is exposed to high wave action and its coastline is simple and monotonous. The tidal amplitude is very small (<30 cm) and seaweed habitats are well developed. The southern part on the eastern coast is affected by the East Korean Warm Current (EKWC), which originates in the Tsushima Warm Current, while the northern part is influenced by the North Korea Cold Current. The western coast of Korea, which faces the Yellow Sea, is a typical drowned valley shoreline (muddy flats with high turbidity) mostly dominated by macrotidal (tidal amplitudes reaching ca. 9 m). As the West Korea Coastal Current flows around the western coast of Korea, water temperature and salinity are relatively low. The southern coast of Korea has many semi-enclosed bays with moderate wave exposure and is mesotidal, with semi-diurnal tidal range of 1–3 m. Tsushima Warm Current, a branch of the Kuroshio Current, transports warm waters to the southern sea of Korea and is divided to the Yellow Sea Warm Current and the East Korean Warm Current, which affects the southwestern and the southeastern coast of Korea, respectively. Jeju Island, located in the southern sea of Korea, is the largest island in Korea. The climate of Jeju Island is strongly influenced by the Tsushima Warm Current flowing around the island. The tidal regime is similar to that of the southern coast of Korea.

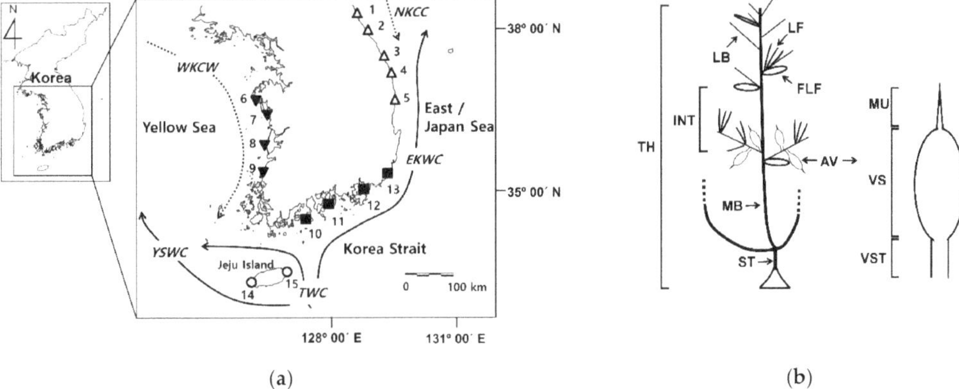

Figure 1. Study sites (**a**) and schematic diagram of *Sargassum thunbergii* (**b**); (**a**) Study sites and oceanic currents around Korea. Eastern coast (△): 1. Sokcho (SO), 2. Jumunjin (JM), 3. Samcheok (SC), 4. Jukbyeon (JB), 5. Yeongdeok (YD); western coast (▼): 6. Taean (TA), 7. Daecheon (DC), 8. Sinsido Island (SI), 9. Yeonggwang (YG); southern coast (■): 10. Goheung (GH), 11. Namhae (NH), 12. Geojaedo Island (GJ), 13. Busan (BS); Jeju Island (○): 14. Gosan (GS), 15. Seongsan (SS). Solid lines indicate warm currents. The Yellow Sea Warm Current (YSWC) and the East Korean Warm Current (EKWC) stem from the Tsushima Warm Current (TWC), which originates in the Kuroshio Warm Current. Dashed lines indicate cold currents: the West Korea Cold Water (WKCW) and the North Korea Cold Current (NKCC) (modified from Gong et al. [51]); (**b**) Schematic diagram describing the morphology of *S. thunbergii*. Thallus height (TH); main branch (MB); stipe (ST); lateral branch (LB); internode interval (INT); leaf (LF); fulcrant leaf (FLF); air-vesicle (AV); mucro (MU); vesicle (VS); stalk of vesicle (VST).

Fifteen sites where *S. thunbergii* dominated in the rocky shores were selected according to [40]; the distance among sampling sites was more than 40 km (Figure 1a). The eastern sites (site 1–5) were on steep shores subjected to strong wave action. Most *S. thunbergii* populations were permanently submerged even during low tide. The northern and the southern parts on the eastern coasts are affected by cold (NKCC) and warm (EKWC) waters, respectively. Site 3 is simultaneously affected by both NKCC and EKWC. The western sites (site 6–9) have very turbid waters and broad expanses of flat and rocky substratum. During low tide periods, *S. thunbergii* populations are exposed to protracted aerial exposure. The southern sites (site 10–13) and two sites (site 14 and 15) on Jeju Island have moderate wave exposure and tidal ranges. *S. thunbergii* populations in these sites are distributed from middle to lower rocky intertidal habitats with gentle slopes and thus, the populations on the southern coasts are periodically exposed to air during low tide.

2.2. Environmental Parameters

Air temperature and sea surface temperature ($^\circ$C), maximum wave height (m), daily incident photon irradiance (mol photons m^{-2} d^{-1}) and salinity data were obtained from the Korean Ocean Observing and Forecasting System (KOOFS, sms.khoa.go.kr/koofs) and the National Climate Data Service System (NCDSS, sts.kma.go.kr), where daily data were assembled from tidal stations, marine buoys and meteorological observatories at each study site for the last 20 years (1992–2011). Air temperature, sea surface temperature and salinity were measured every 30 or 60 min depending on those observation systems and then averaged daily. Maximum wave height averaged daily the highest wave heights which were calculated at 30 or 60 min intervals from data logged every 0.5 or 1 s. Accumulated daily solar irradiance (MJ m^{-2}) obtained from NCSS converted to the daily incident photon flux density (mol photons m^{-2}d^{-1}) [52]. However, we could not obtain a full set of environmental data at sites 1, 4, 7, 10 and 14. These sites were excluded data analyses for the relationship between morphological variation and environments. The magnitude of tidal range were classified into three categories of dominant water oscillation: category 1, microtidal system (the east coast; generally mixed diurnal or semi-diurnal tides); category 2, mesotidal system (the south coast and Jeju Island; generally semi-diurnal tides); category 3, macrotidal system (the west coast; generally semi-diurnal tides).

2.3. Sampling Design and Morphological Measurements

Sargassum thunbergii was identified based on morphological characters described by previous literature and their growth form was adjusted in accordance with various local environmental conditions [38–40]. The species also has a high seasonal variation in morphology. In particular, juveniles and subadult plants are morphologically immature during winter, for example, incomplete development of lateral branch and modular organs (leaf, air-vesicle and receptacle) because they are not yet fully grown. For summer and fall, they shed their modular organs and branches due to thallus decay after sexual reproduction and therefore, their morphology is complete and intact during spring season. To measure clearly morphological responses to local conditions and prevent taxonomic misidentification, we haphazardly collected 20 fully grown individuals with all modular organs at each site during spring period, late April–early June 2011 (Figure A1) and 6–16 individuals were used in analysis. The samples were fixed immediately in a 5–10% formalin-seawater solution in the field, transported to the laboratory and processed immediately. Thallus height, main branch diameter, the length and diameter of stipe and longest lateral branch, internode interval, the length, width and thickness of leaf, the length, width and volume of air-vesicle and the length of stalk and mucro were measured to the nearest 0.1 mm using digital calipers as shown in Figure 1b.

2.4. Data Analysis

Hierarchical cluster analysis (with group-average linking) and Principal Coordinates Analysis (PCoA) were carried out to identify morphological dissimilarity among the sites; the PCoA ordination plot included morphological vectors. These were based on Euclidean distance with square root-transformed and normalized data because each of the morphological measurements were not on the same scale. Significant differences among the groups formed by cluster analysis were tested by Similarity Profile (SIMPROF). To examine the relationship between morphological characteristics and environmental parameters, BIO-ENV (biota-environment) procedure was applied: this procedure explains the best subset of environmental variables correlated with biological variables [53]. For example, Umanzor et al. [54] demonstrated that the attenuation of irradiance and desiccation stress among the several experimental treatments they performed were best factors to be correlated with the abundance of macroinvertebrates. To use BIO-ENV, the two matrices for environment and morphology must exactly refer to a common set of samples (sampling site). We did not have, however, environmental data of all study sites as mentioned in Section 2.2. BIO-ENV were carried out using the morphological data of the study sites corresponding to where environmental variables were obtained and PCoA ordination plot was also modified accordingly. Here, as superimposing environmental vectors on the modified PCoA ordination plot, we explained correlations among environmental variables, morphological groups and the modified PCoA axes. Multivariate analyses were performed with PRIMER v6 and PERMANOVA+ software (PRIMER-e, Auckland, New Zealand) [53,55].

Significant differences in each of the morphological measurements among the group formed by cluster analysis were tested using a one-way ANOVA. Morphological data were tested for normality and homogeneity of variance to the assumptions of parametric statistics using Shapiro–Wilk's and Levene's tests, respectively. If these assumptions were not met, data were log (x + 1) transformed prior to analysis. When a significant difference was observed among variables, the means were examined using Student–Newman–Keuls (SNK) tests to determine where the significant difference occurred. Statistical significance was set at $\alpha < 0.05$. All values are presented as means $\pm$ SE (standard error) and ANOVA were conducted using SPSS software (version 20.0, IBM SPSS Statistics, New York, NY, USA).

3. Results

3.1. Environmental Parameters

Environmental conditions varied with 10 sampling sites where full sets of data were obtained (Figure 2). The overall air and water temperatures were higher at sampling sites located at southern coasts (including Jeju Island) and lower at sites located at western and eastern coasts (Figure 2a,b). The surface photon irradiance was slightly higher at the southern sites than at other sites (Figure 2c). The wave heights were apparently higher at sites on the east coast than in other regions (Figure 2d). The salinity was lower off the west coast than at other sites (Figure 2e).

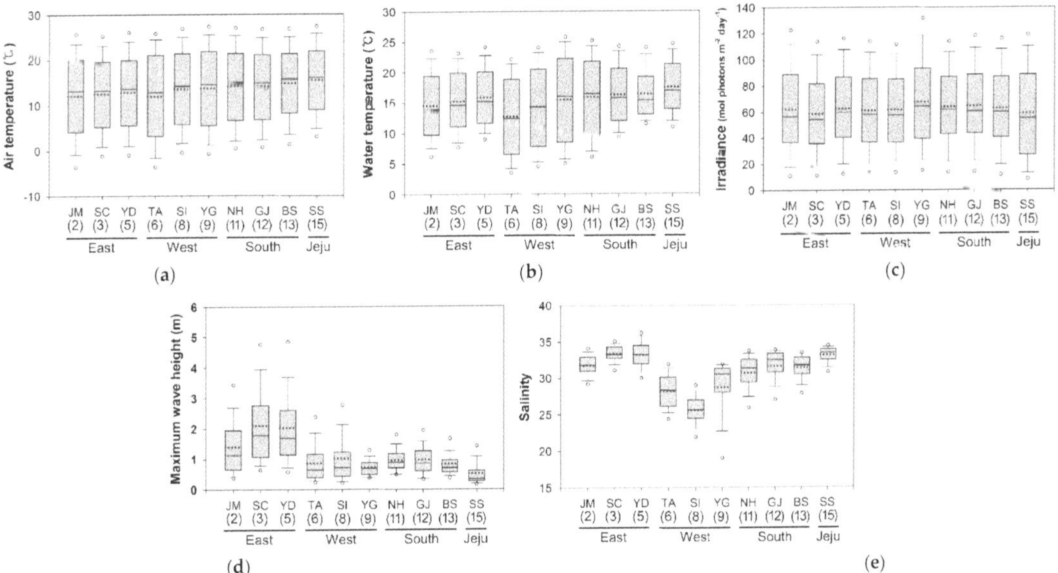

Figure 2. Environmental parameters in the study sites: (**a**) Air temperature; (**b**) water temperature; (**c**) incident irradiance; (**d**) maximum wave height; (**e**) salinity. Open circles are 5th and 95th percentile; solid line and dotted line indicate median value and mean value, respectively.

3.2. Morphological Variations

Morphological characteristics of *Sargassum thunbergii* were divided into five significant groups by the hierarchical cluster analysis with SIMPROF test ($\pi = 0.949$, $p = 0.001$,

Figure 3a). Site 1 and 2, the northern part on the eastern coasts, were independently separated from other sites (we denominated them group M1). The western sites also formed an exclusive group although they were divided into two groups, site 7 and site 6–9. When SIMPROF test was independently carried out on a subset of the western sites only, they (site 7 and site 6–9) did not show significant difference ($\pi = 0.043$, $p = 0.89$): it would be reasonable to consider the western sites as virtually a single group (M3). The southern sites including Jeju Island and the remains of eastern coasts were significantly split into two groups. SIMPROF test with a subset of sites of these two groups, however, showed nonsignificant differences ($\pi = 0.223$, $p = 0.7$) and therefore, they totally would be a single group (M2). Consequently, morphological characteristics can be grouped into three clusters by an arbitrary distance level of 5.5 (Figure 3b).

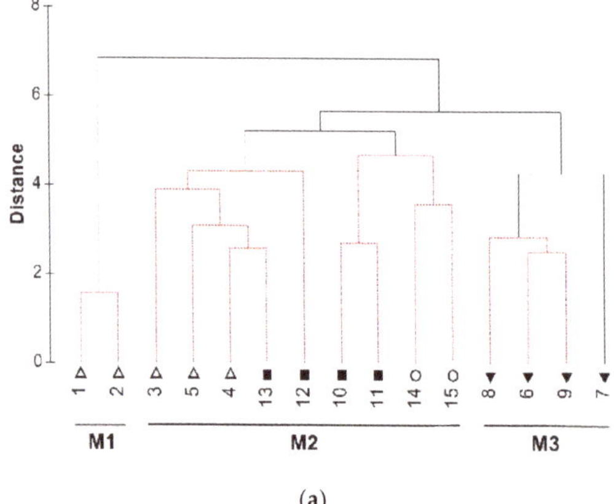

(a)

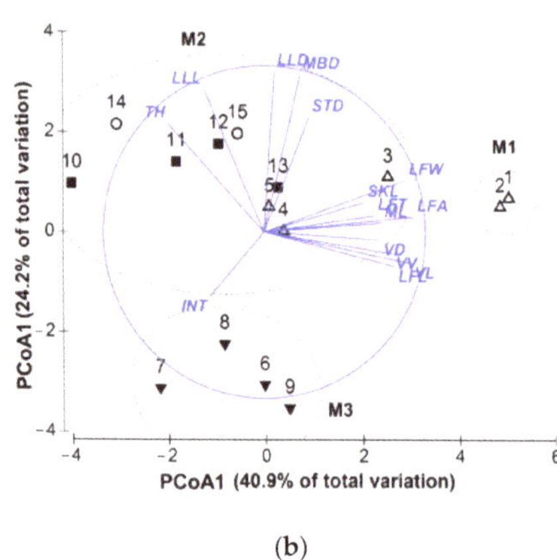

(b)

Figure 3. Hierarchical cluster dendrogram (**a**) and PCoA ordination plot (**b**) based on Euclidean distance: (**a**) Red dotted lines represent no significant differences between sites by SIMPROF test; (**b**) PCoA ordination plot showed that the study sites can be grouped into three clusters (M1–M3) at an arbitrary distance level of 5.5. Correlations between morphological variables and PCoA axes are represented by superimposed vectors (Pearson correlation, $r > 0.5$) in (**b**). Sampling sites are indicated by symbols and numbers, which are provided in Figure 1a. STD (stipe diameter); TH (thallus height); LLL (length of longest lateral branch); LLD (diameter of longest lateral branch); MBD (main branch diameter); INT (internode interval); LFL (leaf length); LFW (leaf width); LFT (leaf thickness); LFA (leaf area); VL (vesicle length); VD (vesicle diameter); VV (vesicle volume); SKL (stalk length of vesicle); MU (mucro length).

Morphological characteristics were significantly different among groups M1, M2 and M3 (Table 1). Thallus height and longest lateral branch length were significantly longer in group M2 than in the other groups. The smallest diameters of main and lateral branch occurred in group M3. Stipe diameters were larger in groups M1 and M2 than in group M3. Stipe diameters were significantly different among the groups, but stipe lengths were not. The internode interval was the longest in group M3, meaning sparsely branched. Leaf- and vesicle-related measurements were significantly larger in group M1 than in the other groups. The PCoA plot summarized these findings well (Figure 3b). The first two axes of the PCoA explained 65.1% of the total variation (PCoA1: 40.9%; PCoA2: 24.2%). The axis of PCoA1 was positively correlated with leaf- (LFL: $r = 0.810$, LFW: $r = 0.878$, LFT: $r = 0.684$, LFA: $r = 0.929$) and vesicle-related traits (VL: $r = 0.912$, VD: $r = 0.718$, VV: $r = 0.794$, SKL: $r = 0.620$, ML: $r = 0.727$). Group M1 (with large values for leaf and vesicle) was strongly correlated with vectors related to leaf and vesicle and separated from the others along the axis of PCoA1. The axis of PCoA2 was positively correlated with thallus height ($r = 0.652$), longest branch length ($r = 0.857$), branch diameter (MDB: $r = 0.948$, LLD: $r = 0.966$) and stipe diameter ($r = 0.692$). Group M2 (with large values for thallus height and branch, but small leaf and vesicle) was strongly positively correlated with vectors related to thallus height and lateral branch traits. Group M3 (with sparse branching and slender and small thalli) was positively correlated with internode interval.

Table 1. One-way ANOVA results in morphological measurements of *Sargassum thunbergii* grouped by hierarchical clustering. Data are mean $\pm$ SE (n = 12–93).

Morphological Characteristics	Code	Group M1	Group M2	Group M3
Thallus				
Thallus height (cm) ***	TH	27.8 ± 1.1 [c]	68.3 ± 2.7 [a]	35.2 ± 1.3 [b]
Main branch diameter (mm) ***	MBD	2.0 ± 0.1 [a]	2.0 ± 0.1 [a]	1.2 ± 0.1 [b]
Stipe length (mm) [NS]	STL	4.1 ± 0.2 [a]	3.7 ± 0.2 [a]	3.5 ± 0.2 [a]
Stipe diameter (mm) ***	STD	6.6 ± 0.3 [a]	4.9 ± 0.2 [b]	3.2 ± 0.1 [c]
Longest lateral branch length (cm) ***	LLL	4.4 ± 0.4 [b]	9.8 ± 0.5 [a]	3.2 ± 0.1 [c]
Longest lateral branch diameter (mm) ***	LLD	1.1 ± 0.0 [b]	1.2 ± 0.0 [a]	0.4 ± 0.0 [c]
Internode interval (mm) ***	INT	2.0 ± 0.0 [c]	4.2 ± 0.2 [b]	4.9 ± 0.2 [a]
Leaf				
Leaf length (mm) ***	LFL	9.2 ± 0.2 [a]	6.0 ± 0.2 [c]	6.7 ± 0.2 [b]
Leaf width (mm) ***	LFW	0.8 ± 0.0 [a]	0.5 ± 0.0 [b]	0.4 ± 0.0 [c]
Leaf thickness (mm) **	LFT	0.41 ± 0.0 [a]	0.36 ± 0.0 [b]	0.34 ± 0.0 [b]
Leaf area (mm^2) ***	LFA	7.2 ± 0.2 [a]	3.1 ± 0.2 [b]	2.9 ± 0.1 [b]
Vesicle				
Air-vesicle length (mm) ***	VL	2.9 ± 0.1 [a]	2.2 ± 0.0 [b]	2.4 ± 0.1 [b]
Air-vesicle diameter (mm) *	VD	1.0 ± 0.0 [a]	0.9 ± 0.0 [b]	0.9 ± 0.0 [b]
Air-vesicle volume (mm^3) ***	VV	6.7 ± 0.4 [a]	4.1 ± 0.2 [b]	4.6 ± 0.4 [b]
Stalk length (mm) ***	SKL	3.1 ± 0.1 [a]	2.4 ± 0.1 [b·]	2.2 ± 0.1 [b]
Mucro length (mm) ***	ML	1.3 ± 0.1 [a]	0.9 ± 0.0 [b]	1.0 ± 0.1 [b]

Superscript of same small letter indicates no significant difference in morphological parameters among groups; *: $p < 0.05$; **: $p < 0.01$; ***: $p < 0.001$; [NS]: no significance.

3.3. Relationships between Morphological Variations and Environments

We found that the morphological characteristics of *S. thunbergii* have been highly correlated with environmental variables (Figure 4 and Table 2). The locations of morphological groups in the modified PCoA plot which were analyzed using the study sites with environmental variables corresponded to the locations of morphological characteristics in Figure 3b (Figure 4). The study sites that clustered in groups R1, R2 and R3 corresponded to those of groups M1, M2 and M3, but site 3 did not fit this pattern. Site 3 belonged to group R1 corresponding to group M1, rather than to group R2 corresponding to group M2. The grouping clearly corresponded to the direction of the overlain vectors for environmental variables (Figure 4). The vector for wave height had a moderately positive correlation with group R1 ($r = 0.500$), which separated this group from the others. The vectors for air temperature, water temperature and irradiance were strongly positively correlated with group R2 ($r = 0.772$, $r = 0.785$ and $r = 0.583$, respectively). The vector for tidal regime and salinity were strongly positively ($r = 0.878$) and negatively ($r = 0.832$) correlated with group R3, respectively.

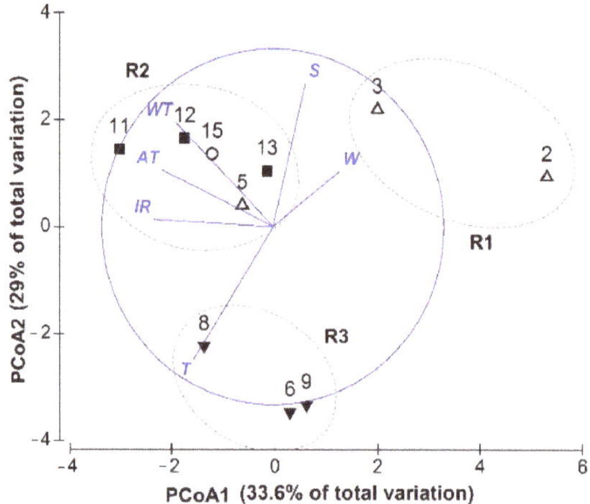

Figure 4. PCoA ordination plot based on Euclidean distances of square root-transformed morphological parameters. Correlations between environmental variables and PCoA axes are represented by vectors superimposed on the PCoA plot (Pearson correlation, *r* > 0.5). Dashed ellipses are groups distinguished at a distance level of 5.5 in the hierarchical clustering analysis. Group R is also a morphological group and we denominated the name R as illustrating the relationship between morphology and environmental variables. AT (air temperature); WT (water temperature); IR (irradiance); W (wave height); S (salinity); T (tidal regime).

Table 2. The best results acquired from BIO-ENV analysis of combined environmental variables matching morphology of *Sargassum thunbergii* (Spearman correlation coefficient, ρ).

Number of Variables	Best Combinations of Environmental Variables
1	T (0.342), IR (0.339), AT (0.233), WT (0.182), S (0.099), W (0.002)
2	**IR+T** (0.464) [5], AT+T (0.397), AT+IR (0.393), WT+T (0.390)
3	**AT+IR+T** (0.519) [2], **WT+IR+T** (0.519) [2], AT+WT+T (0.440)
4	**AT+WT+IR+T** (0.556) [1], AT+S+IR+T (0.446), WT+S+IR+T (0.443)
5	**AT+WT+S+IR+T** (0.475) [4], AT+WT+IR+W+T (0.354)

AT (Air temperature); WT (Water temperature); W (Wave height); IR (Irradiance); S (Salinity); T (Tidal regime). Superscript is five best results in numerical order.

The best permutations of combined environmental variables (BIO-ENV analysis) matching to the morphology of *S. thunbergii* are listed in Table 2. The single variable which best explains the site grouping, in manner consistent with the morphological patterns, is tidal regime (ρ = 0.342). Tidal regime was always involved in the best five combinations, indicating it is important to shape the morphology of *S. thunbergii* as a single variable. On the other hand, the successful matches of the best five combinations included two or more environmental variables at least. Spearman correlation coefficients were always greater for combinations of multiple variables than for each single variable. Their Spearman correlation coefficients did not show dramatic differences across the permutations. This result indicates that the morphology of *S. thunbergii* is also influenced by combinations of multiple environmental variables simultaneously rather than by a single dominant.

4. Discussion

4.1. Relationships between Morphological Variation and Environments

The morphology of *Sargassum thunbergii* markedly changed around the Korean coastline. We found the changes were highly correlated with different environmental conditions on the different shores. Seaweed morphologies are often highly plastic. Thalli in wave-exposed areas are frequently much smaller and tougher than those growing in wave-protected areas [16,22–24]. Additionally, many abiotic and biotic factors, including wave action [24,29], light intensity [17,18], temperature [21], salinity [19,20] and grazing [56,57], directly or indirectly affect the growth and morphology of seaweeds. Although these factors interactively affect algal morphology, few studies have examined the relationships between seaweed morphology and multiple environmental factors [58]. Our results showed that determining the morphology of *S. thunbergii* was linked in a complex manner with different local environmental conditions.

Group R3 (corresponding to group M3) occurred exclusively on the Korean west coast and their morphological traits were obviously distinct from the other groups. Thalli in this group had short and slender branches and long internode intervals (Figure 3b and Table 1); hence, they were small, soft in texture and sparsely branched thalli. Large tidal amplitude was a best factor grouping the sampling sites as a single variable, which exclusively separated group R3 from other groups (Figure 4 and Table 2). The west coast of Korea has calm waters, but the large tidal amplitudes and a very gentle slope prolong the aerial exposure times of intertidal creatures. During prolonged atmospheric exposure, thalli are exposed to intense stresses, including desiccation, strong solar radiation, mineral nutrient starvation and extreme fluctuations in salinity and temperature. Yu et al. [45] found that photosynthetic activity of the thalli was reduced by desiccation, heat stress and prolonged aerial exposure, as shown previously for other seaweed species [59–61]. Mueller et al. [58] recently emphasized the role of tidal regime-related stressors in the morphological expression of intertidal seaweeds. The distinctive morphology of group R3 in our study can be explained by the extremely stressful conditions of the macrotidal regime. In addition, many seaweeds produce diminutive thalli in low-salinity waters [62]. As an example, short fucoid thalli with narrow stipes are frequently observed in low-salinity intertidal zones [20,31]. In the present study, the morphological characteristics of the members of group R3 (also M3) were negatively correlated with salinity (Figure 4). Thus, our results support that physiological responses to the harsh stresses associated with macrotidal regime likely affected the growth and morphogenesis of *S. thunbergii*.

The morphologies of group R2 (corresponding to M2) were strongly correlated with temperatures and irradiance and weakly correlated with wave height (Figure 4). The study sites in this group ranged from Jeju Island to the southern part of the mainland east coast, where the influences of two warm currents prevail (the TWC and EKWC; see Figure 1b). The intertidal zone of the south coast of Korea is particularly sheltered by the ria configuration of the coastline. Thalli in this group were larger than any other groups: they have long and thick branches and thick stipes (Figure 3b and Table 1). Among all study sites, biomass per thallus should be highest in these southern locations. The thick stipes are able to help supporting heavy thalli [63]. Large forms of *S. thunbergii* often occur in warm and/or sheltered locations [35,38,40], but to date, there has been no direct evidence that the size is related to temperature or shelter. Here, we provide more robust evidence to support this postulation. Many species of seaweeds reach larger sizes or biomass in wave-sheltered waters than in areas exposed to strong hydrodynamic forces (e.g., [11,22,24,64,65]). Growth and survival of seaweed need optimal range of temperature, water motion to allow for nutrient uptake, but not too strong to remove algal thallus, and sufficient light of high levels without photo-inhibition. As identifying the above-mentioned optimal conditions for growth of *S. thunbergii* is not the focus of this study, we do not cover that in detail here. Nevertheless, given our results and distribution of large forms in warm and/or sheltered areas, morphological characteristics of *S. thunbergii* in the southern locations likely result from proper combinations of those environmental factors. Actually, growth of *S. thunbergii* in marine farms increases with temperature and the thallus length reached about 2 m [66]; note the fact that most marine farms are located in sheltered and environmentally suitable places for the maximum growth and survival of the subject organism. These results indicate that strong light, warm temperatures and shelter may promote the formation of large thalli in *S. thunbergii*.

The morphologies of group R1 (corresponding to M1) were correlated with wave height (Figure 4). Specimens of group M1, on the northern part of the east coast, had large leaves and vesicles and their thalli and lateral branch were very short, thick and bushy (Figure 3b and Table 1). Such a morphological form would likely be advantageous in the wave-exposed environments of the northern part of the east coast. For example, thick and flexible stipe and small and thick thallus are very effective in preventing dislodgement and/or breakage from the strong hydrodynamic forces [8,16]. *S. thunbergii* on the east coast have been mostly submerged even during the low tide due to very small tidal amplitude. Thalli on this coast were very dense, which may have caused self-shading. Large air vesicles likely have functional significance in competition for light by providing increased buoyancy to raise thalli towards the surface. Large leaves can also enhance light interception due to maximized leaf surface. Therefore, morphological characteristics of group M1 would be a typical seaweed form in wave-exposed areas.

4.2. Environmental Factors in the Intertidal Zone

Although the sites on the east coast of Korea are subjected to strong wave exposure (Figure 2d and see 2.1 study site), wave height did not completely separate the five sites (site 1–5) from other sites. At first glance, it can make a hasty prediction that morphologies on the east coast (site 1–5) would strongly correlate with wave height. However, the morphologies on this coast fell into two groups, northern part (M1 and R1) and southern part (M2 and R2) according to correlations with different environmental variables (Figures 3b and 4). The site 3 is geographically located between the sites 1–2 and 4–5 on the eastern coast. Interestingly, the same pattern also appeared on the PCoA plots and site 3 belonged to both morphological groups M2 and R1 (Figures 3b and 4). These mean that morphologies of specimens on site 3 have an intermediate form resulting from a combined effect of different environmental variables. Actually, morphological characteristics of both group M1 and M2, or intermediate forms between them were observed at the specimens from site 3. There are two different water masses near site 3, where two currents with different temperatures meet (Figure 1a). Thalli at site 3 were very likely exposed to a complexity of environmental factors, including wave heights, warm and cold water temperatures and fluctuating irradiance. These indicate that combined effects of local environmental variables influenced the morphological grouping although the whole of the Korean east coast is affected by high waves. Our findings are in line with previous studies that demonstrated seaweed morphology in response to local environmental conditions when the degree of wave exposures were similar [23,32].

Many studies have suggested that the degree of wave exposure is the most important determinant of morphology and survival in seaweed (reviewed by Hurd [29]). Water turbulence is likely a principal factor in the subtidal zone, where environmental conditions are stable compared to the intertidal zone, but this is just one of many interacting factors (e.g., irradiance, temperature, nutrients) in the intertidal zone that vary dramatically in space and time [29]. In this study, wave height was only a partially dominant factor in the northern part of the east coast due to locally different environmental variations (Figure 4). Rather, as a single variable, tidal regime was more important (Table 2); it separated exclusively the western coast sites as a whole from other sites (Figure 4). This is probably because tide-induced stresses are greater than wave-induced impacts in intertidal zone [58]. Therefore, tidal condition would be a more important factor shaping morphology of intertidal seaweeds than wave forces at regional scale.

Meanwhile, there are different root causes in the morphological differentiation of seaweeds. For example, dwarf forms in seaweeds have several different causes in response to the harsh environments such as high on the shore [67,68] and tide-induced stress [58] as well as high wave action [16,24]. In this study, although thallus height was significantly shorter in group M1 and M3 than in group M2 (Table 1), the underlying causal factors were different. The short size of thalli in group M1 is linked to strong wave action, while that of thalli in group M3 was linked to tide-induced stress. Members of group M2 had the largest thallus due to favorable conditions such as temperatures, irradiance and less wave impact. Thus, the morphological differentiation in intertidal seaweeds is controlled by combinations of various local environmental factors rather than by a single dominant effect.

5. Conclusions

This study examined morphological variability in intertidal brown seaweed *Sargassum thunbergii* and their relationships with diverse environmental variables at regional scale. Tidal-induced stresses were important as a single variable at regional scale while wave force was a lesser environmental factor which should be considered together with local conditions in intertidal zone. Simultaneously, our results showed morphological differentiation adjusting to locally different environmental conditions, highlighting consideration of multiple environmental variables to examine biological and ecological processes of intertidal seaweeds. Diverse morphologies of *S. thunbergii* resulted from adaptation to different habitat conditions and could provide information about the environments where they inhabit. This study provides valuable information regarding coastal ecological heath, restoration and aquaculture using *S. thunbergii*. There are, of course, other factors affecting seaweed morphology (e.g., nutrients, season and grazing) which were not involved in this study. Further studies incorporating them can improve understanding morphological variability of intertidal seaweeds.

Author Contributions: Conceptualization, S.K., Y.H.K. and S.R.P.; methodology, S.K. and S.K.C.; validation, S.K., S.K.C. and S.T.K.; formal analysis, S.K. and S.V.; investigation, S.K., S.K.C., S.V. and S.T.K.; data curation, S.K.; writing—original draft preparation, S.K.; writing—review and editing, Y.H.K. and S.R.P.; visualization, S.K. and S.V.; supervision, S.R.P.; project administration, Y.H.K. and S.R.P.; funding acquisition, S.R.P. All authors have read and agreed to the published version of the manuscript.

Funding: This research was supported by National Institute of Fisheries Science (NIFS), Korea (project title: Research on the characteristics of marine ecosystem and oceanographic observation in Korean waters, No. R2022055). This research was also supported by Basin Science Research Program through the National Research Foundation of Korea (NRF) funded by the Ministry of Education (2019R1I1A1A01057599) to S.R.P.

Institutional Review Board Statement: Not applicable.

Informed Consent Statement: Not applicable.

Data Availability Statement: Not applicable.

Acknowledgments: We would like to thank S.W. Jung, and Y.S. Oh for field assistance and laboratory support. We also appreciate the productive suggestions from editors and anonymous reviewers and would like to give our thanks to them.

Conflicts of Interest: The authors declare no conflict of interest.

Appendix A

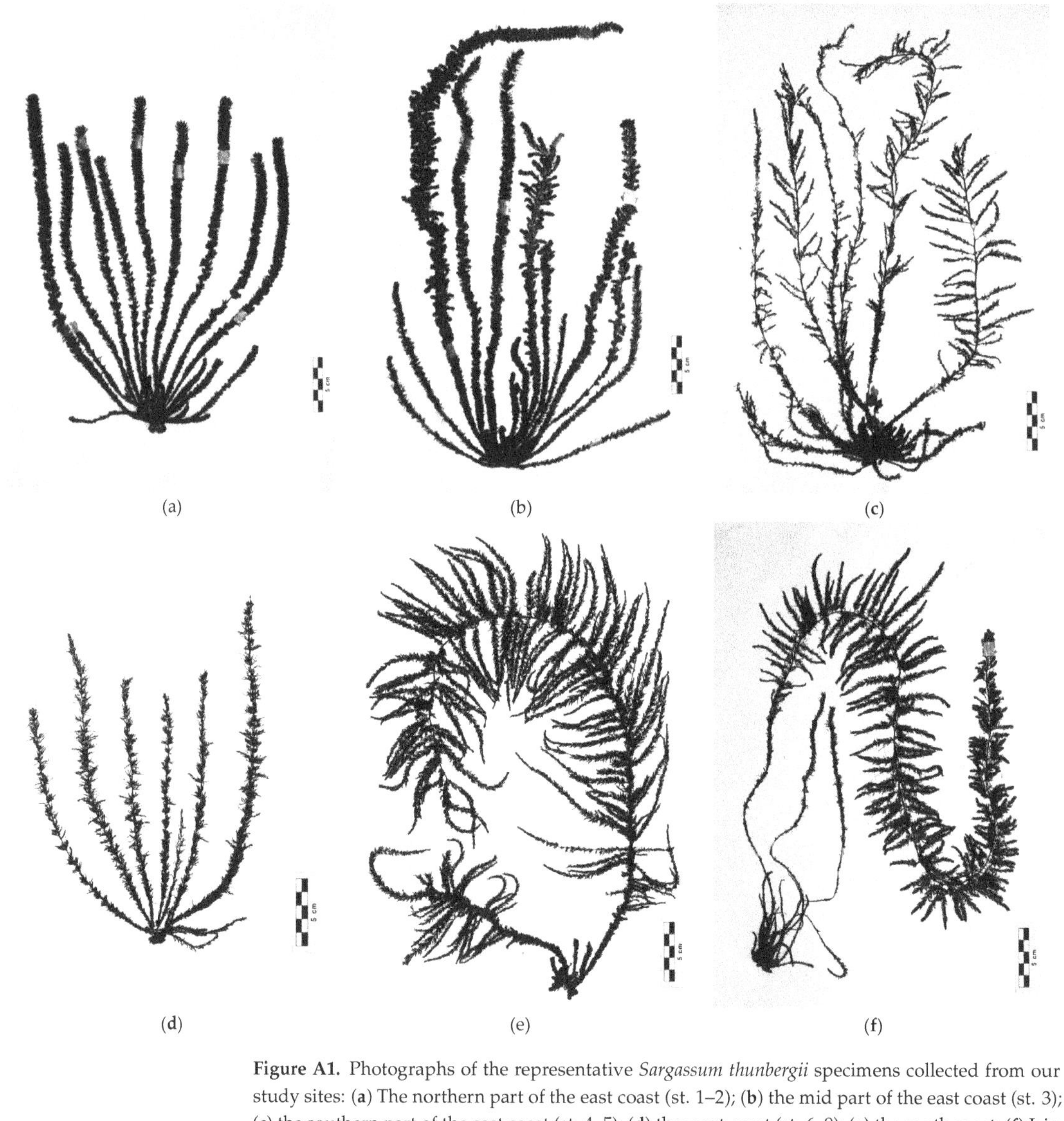

(a)

(b)

(c)

(d)

(e)

(f)

Figure A1. Photographs of the representative *Sargassum thunbergii* specimens collected from our study sites: (**a**) The northern part of the east coast (st. 1–2); (**b**) the mid part of the east coast (st. 3); (**c**) the southern part of the east coast (st. 4–5); (**d**) the west coast (st. 6–9); (**e**) the south coast; (**f**) Jeju Island. Specimens (**c**,**e**,**f**) are just a part of a whole individual because almost all the large branches have been cut off to make a specimen on the limited paper. Scale bar indicates 5 cm.

References

1. Slatkin, M. Gene flow and the geographic structure of natural populations. *Science* **1987**, *236*, 787–792. [CrossRef] [PubMed]
2. Blanchette, C.A.; Miner, B.G.; Gaines, S.D. Geographic variability in form, size and survival of *Egregia menziesii* around Point Conception, California. *Mar. Ecol. Prog. Ser.* **2002**, *239*, 69–82. [CrossRef]
3. Fowler-Walker, M.J.; Gillanders, B.M.; Connell, S.D.; Irving, A.D. Patterns of association between canopy-morphology and understorey assemblages across temperate Australia. *Estuar. Coast. Shelf. Sci.* **2005**, *63*, 133–141. [CrossRef]

4. Cacabelos, E.; Olabarria, C.; Incera, M.; Troncoso, J.S. Effects of habitat structure and tidal height on epifaunal assemblages associated with macroalgae. *Estuar. Coast. Shelf. Sci.* **2010**, *89*, 43–52. [CrossRef]

5. Wernberg, T.; Vanderklift, M.A. Contribution of temporal and spatial components to morphological variations in the kelp *Ecklonia* (Laminariales). *J. Phycol.* **2010**, *46*, 153–161. [CrossRef]

6. Littler, M.M.; Littler, D.S. The evolution of thallus form and survival strategies in benthic marine macroalgae: Field and laboratory tests of a functional form model. *Am. Nat.* **1980**, *116*, 25–44. [CrossRef]

7. Duggins, D.O.; Eckman, J.E.; Siddon, C.E.; Klinger, T. Population, morphometric and biomechanical studies of three understory kelps along a hydrodynamic gradient. *Mar. Ecol. Prog. Ser.* **2003**, *265*, 57–76. [CrossRef]

8. Roberson, M.L.; Coyer, J.A. Variation in blade morphology of the kelp *Eisenia arborea*: Incipient speciation due to local water motion? *Mar. Ecol. Prog. Ser.* **2004**, *282*, 115–128. [CrossRef]

9. Benedetti-Cecchi, L.; Bertocci, I.; Vaselli, S.; Maggi, E. Morphological plasticity and variable spatial patterns in different populations of the red alga *Rissoella verrucosa*. *Mar. Ecol. Prog. Ser.* **2006**, *315*, 87–98. [CrossRef]

10. Charrier, B.; Le Bail, A.; de Reviers, B. Plant Proteus: Brown algal morphological plasticity and underlying developmental mechanisms. *Trends Plant. Sci.* **2012**, *17*, 468–477. [CrossRef]

11. Gerard, V.A.; Mann, K.H. Growth and production of *Laminaria longicruris* (Phaeophyta) populations exposed to different intensities of water movement. *J. Phycol.* **1979**, *15*, 33–41. [CrossRef]

12. Falace, A.; Bressan, G. Seasonal variations of *Cystoseira barbata* (Stackhouse) C. Agardh frond architecture. *Hydrobiologia* **2006**, *555*, 193–206. [CrossRef]

13. Engelen, A.H.; Åberg, P.; Olsen, J.L.; Stam, W.T.; Breeman, A.M. Effects of wave exposure and depth on biomass, density and fertility of the fucoid seaweed *Sargassum polyceratium* (Phaeophyta, Sargassaceae). *Eur. J. Phycol.* **2005**, *40*, 149–158. [CrossRef]

14. Stewart, H. Morphological variation and phenotypic plasticity of buoyancy in the macroalga *Turbinaria ornata* across a barrier reef. *Mar. Biol.* **2006**, *149*, 721–730. [CrossRef]

15. Malta, E.-J.; Ferreira, D.G.; Vergara, J.J.; Pérez-Lloréns, L. Nitrogen load and irradiance affect morphology, photosynthesis and growth of *Caulerpa prolifera* (Bryopsidales: Chlorophyta). *Mar. Ecol. Prog. Ser.* **2005**, *298*, 101–114. [CrossRef]

16. D'Amours, O.; Scheibling, R.E. Effect of wave exposure on morphology, attachment strength and survival of the invasive green alga *Codium fragile* ssp. tomentosoides. *J. Exp. Mar. Biol. Ecol.* **2007**, *351*, 129–142. [CrossRef]

17. Toth, G.B.; Harrysson, H.; Wahlström, N.; Olsson, J.; Oerbekke, A.; Steinhagen, S.; Kinnby, A.; White, J.; Albers, E.; Edlund, U.; et al. Effects of irradiance, temperature, nutrients, and pCO_2 on the growth and biochemical composition of cultivated *Ulva fenestrata*. *J. Appl. Phycol.* **2020**, *32*, 3243–3254. [CrossRef]

18. Monro, K.; Poore, A.G.B.; Brooks, R. Multivariate selection shapes environment-dependent variation in the clonal morphology of a red seaweed. *Evol. Ecol.* **2007**, *21*, 765–782. [CrossRef]

19. Vettori, D.; Nikora, V.; Biggs, H. Implications of hyposaline stress for seaweed morphology and biomechanics. *Aquat. Bot.* **2020**, *162*, 103188. [CrossRef]

20. Kalvas, A.; Kautsky, L. Morphological variation in *Fucus vesiculosus* populations along temperature and salinity gradients in Iceland. *J. Mar. Biol. Assoc. U. K.* **1998**, *78*, 985–1001. [CrossRef]

21. Kübler, J.E.; Dudgeon, S.R. Temperature dependent change in the complexity of form of *Chondrus crispus* fronds. *J. Exp. Mar. Biol. Ecol.* **1996**, *207*, 15–24. [CrossRef]

22. Blanchette, C.A. Size and survival of intertidal plants in response to wave action: A case study with *Fucus gardneri*. *Ecology* **1997**, *78*, 1563–1578. [CrossRef]

23. Ruuskanen, A.; Bäck, S.; Reitalu, T. A comparison of two cartographic exposure methods using *Fucus vesiculosus* as an indicator. *Mar. Biol.* **1999**, *134*, 139–145. [CrossRef]

24. Fowler-Walker, M.J.; Wernberg, T.; Connell, S.D. Differences in kelp morphology between wave sheltered and exposed localities: Morphologically plastic or fixed traits? *Mar. Biol.* **2006**, *148*, 755–767. [CrossRef]

25. Kilar, J.A.; McLachlan, J. Branching morphology as an indicator of environmental disturbance: Testing the vegetative fragmentation of *Acanthophora spicifera* and the turf morphology of *Laurencia papillosa*. *Aquat. Bot.* **1986**, *24*, 115–130. [CrossRef]

26. Yñiguez, A.T.; McManus, J.W.; Collado-Vides, L. Capturing the dynamics in benthic structures: Environmental effects on morphology in the macroalgal genera *Halimeda* and *Dictyota*. *Mar. Ecol. Prog. Ser.* **2010**, *411*, 17–32. [CrossRef]

27. Miner, B.G.; Sultan, S.E.; Morgan, S.G.; Padilla, D.K.; Relyea, R.A. Ecological consequences of phenotypic plasticity. *Trends Ecol. Evol.* **2005**, *20*, 685–692. [CrossRef]

28. Balata, D.; Piazzi, L.; Rindi, F. Testing a new classification of morphological functional groups of marine macroalgae for the detection of responses to stress. *Mar. Biol.* **2011**, *158*, 2459–2469. [CrossRef]

29. Hurd, C.L. Water motion, marine macroalgal physiology, and production. *J. Phycol.* **2000**, *36*, 453–472. [CrossRef]

30. Hay, M.E. The functional morphology of turf-forming seaweeds: Persistence in stressful marine habitats. *Ecology* **1981**, *62*, 739–750. [CrossRef]

31. Gylle, A.M.; Nygård, C.A.; Ekelund, N.G.A. Desiccation and salinity effects on marine and brackish *Fucus vesiculosus* L. (Phaeophyceae). *Phycologia* **2009**, *48*, 156–164. [CrossRef]

32. Wernberg, T.; Thomsen, M.S. The effect of wave exposure on the morphology of *Ecklonia radiata*. *Aquat. Bot.* **2005**, *83*, 61–70. [CrossRef]

33. Boundouresque, C.F. Taxonomy and Phylogeny of Unicellular Eukaryotes. In *Environmental Microbiology: Fundamentals and Applications*; Bertrand, J.C., Caumette, P., Lebaron, P., Matheron, R., Normand, P., Sime-Ngando, T., Eds.; Springer: Dordrecht, The Netherlands, 2015; pp. 191–257. [CrossRef]
34. Liu, F.-L.; Li, J.-J.; Liang, Z.-R.; Zhang, Q.-S.; Zhao, F.-J.; Jueterbock, A.; Critchley, A.T.; Morrell, S.L.; Assis, J.; Tang, Y.-Z.; et al. A concise review of the brown seaweed *Sargassum thunbergia*—A knowledge base to inform large-scale cultivation efforts. *J. Appl. Phycol.* **2021**, *33*, 3469–3482. [CrossRef]
35. Umezaki, I. Ecological studies of *Sargassum thunbergii* (Mertens) O. Kuntze in Maizuru Bay, Japan Sea. *Bot. Mag. Tokyo* **1974**, *87*, 285–292. [CrossRef]
36. Zhang, Q.S.; Li, W.; Liu, S.; Pan, J.H. Size-dependence of reproductive allocation of *Sargassum thunbergii* (Sargassaceae, Phaeophyta) in Bohai Bay, China. *Aquat. Bot.* **2009**, *91*, 194–198. [CrossRef]
37. Cho, S.M.; Lee, S.M.; Ko, Y.D.; Mattio, L.; Boo, S.M. Molecular systematic reassessment of *Sargassum* (Fucales, Phaeophyceae) in Korea using four gene regions. *Bot. Mar.* **2012**, *55*, 473–484. [CrossRef]
38. Yendo, K. The Fucaceae of Japan. *J. Coll. Sci. Tokyo Imp. Univ.* **1907**, *21*, 1–174.
39. Okamura, K. *Icones of Japanese Algae*; Kazamashobo: Tokyo, Japan, 1923; Volume V, pp. 6–8.
40. Oak, J.H.; Lee, I.K. Taxonomy of the genus *Sargassum* (Fucales, Phaeophyceae) from Korea I. Subgenus *Bactrophycus* Section *Teretia*. *Algae* **2005**, *20*, 77–90. [CrossRef]
41. Chu, S.H.; Zhang, Q.S.; Liu, S.K.; Tang, Y.Z.; Zhang, S.B.; Lu, Z.C.; Yu, Y.Q. Tolerance of *Sargassum Thunbergii* germlings to thermal, osmotic and desiccation stress. *Aquat. Bot.* **2012**, *96*, 1–6. [CrossRef]
42. Kang, J.Y.; Khan, M.N.A.; Park, N.H.; Cho, J.Y.; Lee, M.C.; Fujii, H.; Hong, Y.K. Antipyretic, analgesic, and anti-inflammatory activities of the seaweed *Sargassum fulvellum* and *Sargassum thunbergii* in mice. *J. Ethnopharmacol.* **2008**, *116*, 187–190. [CrossRef]
43. Kim, J.-A.; Karadeniz, F.; Ahn, B.-N.; Kwon, M.S.; Mun, O.-J.; Bae, M.J.; Seo, Y.; Kim, M.; Lee, S.-H.; Kim, Y.Y.; et al. Bioactive quinone derivatives from the marine brown alga *Sargassum thunbergii* induce anti-adipogenic and pro-osteoblastogenic activities. *J. Sci. Food Agric.* **2015**, *96*, 783–790. [CrossRef] [PubMed]
44. Pan, Y.; Wernberg, T.; de Bettignies, T.; Holmer, M.; Li, K.; Wu, J.; Lin, F.; Yu, Y.; Xu, J.; Zhou, C.; et al. Screening of seaweeds in the East China Sea as potential bio-monitors of heavy metals. *Environ. Sci. Pollut. Res.* **2018**, *25*, 16640–16651. [CrossRef] [PubMed]
45. Yu, Y.Q.; Zhang, Q.S.; Tang, Y.Z.; Zhang, S.B.; Lu, Z.C.; Chu, S.H.; Tang, X.X. Establishment of intertidal seaweed beds of *Sargassum thunbergii* through habitat creation and germling seeding. *Ecol. Eng.* **2012**, *44*, 10–17. [CrossRef]
46. Li, X.M.; Zhang, Q.S.; Tang, Y.Z.; Yu, Y.Q.; Liu, H.L.; Li, L.X. Highly efficient photoprotective responses to high light stress in *Sargassum thunbergii* germlings, a representative brown macroalga of intertidal zone. *J. Sea Res.* **2014**, *85*, 491–498. [CrossRef]
47. Liang, Z.; Wang, F.; Sun, X.; Wang, W.; Liu, F. Reproductive biology of *Sargassum thunbergii* (Fucales, Phaeophyceae). *Am. J. Plant. Sci.* **2014**, *5*, 2574–2581. [CrossRef]
48. Stiger, V.; Horiguchi, T.; Yoshida, T.; Coleman, A.W.; Masuda, M. Phylogenetic relationships within the genus *Sargassum* (Fucales, Phaeophyceae), inferred from ITS-2 nrDNA, with an emphasis on the taxonomic subdivision of the genus. *Phycol. Res.* **2003**, *51*, 1–10. [CrossRef]
49. Zhao, F.; Wang, X.; Liu, J.; Duan, D. Population genetic structure of *Sargassum thunbergii* (Fucales, Phaeophyta) detected by RAPD and ISSR markers. *J. Appl. Phycol.* **2007**, *19*, 409–416. [CrossRef]
50. Li, J.-J.; Hu, Z.-M.; Gao, X.; Sun, Z.-M.; Choi, H.-G.; Duan, D.-L.; Endo, H. Oceanic currents drove population genetic connectivity of the brown alga *Sargassum thunbergii* in the north-west Pacific. *J. Biogeogr.* **2017**, *44*, 230–242. [CrossRef]
51. Gong, Y.; Suh, Y.-S.; Seong, K.-T.; Han, I.-S. *Climate Change and Marine Ecosystem*; Academy Book: Seoul, Korea, 2010; pp. 39–64.
52. Carruthers, T.J.B.; Longstaff, B.J.; Dennison, W.C.; Abal, E.G.; Aioi, K. Measurement of Light Penetration in Relation to Seagrass. In *Global Seagrass Research Motheds*; Short, F.T., Coles, R.G., Short, C.A., Eds.; Elsevier Science: Amsterdam, The Netherlands, 2001; pp. 369–392.
53. Clarke, K.R.; Gorley, R.N. *PRIMER Version 6: User Manual/Tutorial*; Primer-E Ltd.: Plymouth, UK, 2006.
54. Umanzor, S.; Ladah, L.; Calderon-Aguilera, L.E.; Zertuche-González, J.A. Testing the relative importance of intertidal seaweeds as ecosystem engineers across tidal heights. *J. Exp. Mar. Biol. Ecol.* **2019**, *511*, 100–107. [CrossRef]
55. Anderson, M.J.; Gorley, R.N.; Clarke, K.R. *PERMANOVA+ for PRIMER: Guide to Software and Statistical Methods*; Primer-E Ltd.: Plymouth, UK, 2008.
56. Taylor, R.B.; Sotka, E.; Hay, M.E. Tissue-specific induction of herbivore resistance: Seaweed response to amphipod grazing. *Oecologia* **2002**, *132*, 68–76. [CrossRef]
57. Diaz-Pulido, G.; Villamil, L.; Almanza, V. Herbivory effects on the morphology of the brown alga *Padina boergesenii* (Phaeophyta). *Phycologia* **2007**, *46*, 131–136. [CrossRef]
58. Mueller, R.; Fischer, A.M.; Bolch, C.J.; Wright, J.T. Environmental correlates of phenotypic variation: Do variable tidal regimes influence morphology in intertidal seaweeds? *J. Phycol.* **2015**, *51*, 859–871. [CrossRef] [PubMed]
59. Pearson, G.A.; Davison, I.R. Freezing rate and duration determine the physiological response of intertidal fucoids to freezing. *Mar. Biol.* **1993**, *115*, 353–362. [CrossRef]
60. Dudgeon, S.R.; Kübler, J.E.; Vadas, R.L.; Davison, I.R. Physiological responses to environmental variation in intertidal red algae: Does thallus morphology matter? *Mar. Ecol. Prog. Ser.* **1995**, *117*, 193–206. [CrossRef]
61. Liu, F.; Pang, S.J. Stress tolerance and antioxidant enzymatic activities in the metabolisms of the reactive oxygen species in two intertidal red algae *Grateloupia turuturu* and *Palmaria palmata*. *J. Exp. Mar. Biol. Ecol.* **2010**, *382*, 82–87. [CrossRef]

62. Norton, T.A.; Mathieson, A.C.; Neushul, M. Morphology and Environment. In *The Biology of Seaweeds*; Lobban, C.S., Wynne, M.J., Eds.; University of California Press: Oakland, CA, USA, 1981; pp. 421–451.
63. Sjøtun, K.; Fredriksen, S.; Rueness, J. Effect of canopy biomass and wave exposure on growth in *Laminaria hyperborea* (Laminariaceae: Phaeophyta). *Eur. J. Phycol.* **1998**, *33*, 337–343. [CrossRef]
64. De Paula, E.J.; de Oliveira F°, E.C. Wave exposure and ecotypical differentiation in *Sargassum cymosum* (Phaeophyta-Fucales). *Phycologia* **1982**, *21*, 145–153. [CrossRef]
65. Viejo, R.M.; Arrontes, J.; Andrew, N.L. An experimental evaluation of the effect of wave action on the distribution of *Sargassum muticum* in northern Spain. *Bot. Marina* **1995**, *38*, 437–442. [CrossRef]
66. Chen, J.; Zhang, J.H.; Li, J.Q.; Zhang, Y.T.; Sui, H.D.; Wu, W.G.; Niu, Y.L.; Gao, Z.K. The growth characteristics of long-line cultured seaweed *Sargassum thunbergii* in the Sanggou Bay. *Prog. Fish Sci.* **2016**, *37*, 120–126.
67. Sideman, E.J.; Mathieson, A.C. Morphological variation within and between natural populations of non-tide pool *Fucus distichus* (Phaeophyta) in New England. *J. Phycol.* **1985**, *21*, 250–257. [CrossRef]
68. Stengel, D.B.; Dring, M.J. Morphology and in situ growth rates of plants of *Ascophyllum nodosum* (Phaeophyta) from different shore levels and responses of plants to vertical transplantation. *Eur. J. Phycol.* **1997**, *35*, 193–202. [CrossRef]

Journal of
*Marine Science
and Engineering*

Article

Effects of Miniaturization of the Summer Phytoplankton Community on the Marine Ecosystem in the Northern East China Sea

Kyung-Woo Park [1], Hyun-Ju Oh [1], Su-Yeon Moon [1], Man-Ho Yoo [2] and Seok-Hyun Youn [1,*]

[1] Oceanic Climate & Ecology Research Division, National Institute of Fisheries Science, Busan 46083, Korea; areshan12@naver.com (K.-W.P.); hyunjuoh@korea.kr (H.-J.O.); suyeonmoon1119@gmail.com (S.-Y.M.)
[2] HAERANG Technology and Policy Research Institute, Suwon 16229, Korea; ryu10005@hanmail.net
* Correspondence: younsh@korea.kr; Tel.: +82-51-720-2233

Abstract: After the construction of the Three Gorges Dam (Changjiang River), the northern East China Sea has been exposed to major environmental changes in the summer due to climate change and freshwater control. However, little is known regarding phytoplankton in this area. Here, we investigated differences in the summer phytoplankton-community structure as a consequence of marine-environment changes from 2016 to 2020. In the 2000s, the key dominant species in the summer phytoplankton community in the northern East China Sea were diatoms and dinoflagellates. In this study, however, nanoflagellates of ≤ 20 μm were identified as the dominant species throughout the survey period, with abundances ranging from 43.1 to 69.7%. This change in the phytoplankton-community structure may be ascribed to low nutrient concentrations in the area, especially phosphate, which was below the detection limit, seriously hampering phytoplankton growth. The relative contribution of picophytoplankton to the total chlorophyll a biomass was highest in the surface mixed layer with low nutrient concentrations. Spatially, higher percentages were observed along the east-side stations than the west-side stations, where nutrient concentrations were relatively high. Conclusively, decreased nutrients led to phytoplankton miniaturization. Accordingly, as the dominance of picophytoplankton increases, energy transfer is expected to decrease at the upper trophic level.

Keywords: northern East China Sea; Changjiang diluted water; phytoplankton community; chl *a* size fraction; picophytoplankton; phosphate restriction

Citation: Park, K.-W.; Oh, H.-J.; Moon, S.-Y.; Yoo, M.-H.; Youn, S.-H. Effects of Miniaturization of the Summer Phytoplankton Community on the Marine Ecosystem in the Northern East China Sea. *J. Mar. Sci. Eng.* **2022**, *10*, 315. https://doi.org/10.3390/jmse10030315

Academic Editor: Azizur Rahman

Received: 25 January 2022
Accepted: 20 February 2022
Published: 23 February 2022

Publisher's Note: MDPI stays neutral with regard to jurisdictional claims in published maps and institutional affiliations.

1. Introduction

The northern East China Sea is exposed to various currents depending on the season and exhibits singular seasonal fluctuations [1]. It is known to be a highly valuable fishing ground because of its high primary productivity [2–4]. In summer, in particular, it shows diverse water mass characteristics, with the surface layer affected by the freshwater flowing in from the coastal areas of China, the bottom layer affected by the cold deep water of the Yellow Sea, and the eastern part affected by the high temperature and salinity of the Kuroshio water. The fronts formed at the boundaries of water masses with different characteristics show unique biological and chemical patterns and have significant effects on the distribution of zooplankton and fish along the food chain [5]. The inflow of freshwater due to Changjiang diluted water shows marked seasonal variability, with the minimum discharge in the winter and the maximum discharge in the summer, which greatly affects the seasonal salinity distribution in the northern East China Sea [6,7]. Moreover, as a major nutrient source [8], the coastal waters of China that are adjacent to the Changjiang are rich in nutrients, with high primary productivity [9–11] and characteristics of an estuary dominated by Chinese coastal waters. Geographically, a large difference in water depths is observed between the eastern and western waters, and complex topographical characteristics and diverse water masses show complex environmental interactions with seasonally variable intensity. Thus, the East China Sea, which is characterized by high productivity, is used jointly by Korea, Japan, and China as the largest fishing ground for migratory fish species [3]. Studies on phytoplankton distributions in the East China Sea were mainly conducted by Chinese and Japanese researchers in the 1980s and 1990s [12–18]. The focus areas of phytoplankton abundances and chlorophyll a (chl-*a*) concentrations were high in the waters affected by the Changjiang dilution water in spring and autumn [4], changes in phytoplankton standing stocks due to the stratification of the bottom layer structure in summer [19], the relationship between phytoplankton-community structures and water masses in summer [20,21], zooplankton (phytoplankton predators) [22–25], and changes in chl-*a* concentrations related to nutrients and suspended matter along with changes in phytoplankton-community structures [26]. In Korea, studies have been conducted on the changes in phytoplankton-community distributions and structures [27–30] since the 1990s. Several studies have been conducted on different topics, including the distribution characteristics of chl-*a* according to summer nutrients and suspended substances [31,32] and long-term changes in surface water temperatures and chl-*a* biomasses [33]. It was also reported that the primary productivity decreased by approximately 86% in waters adjacent to the Changjiang River after the construction of the Three Gorges Dam and that the phytoplankton-community structure in the East China Sea changed [34,35]. However, there appears to be a lack of recent data on this subject matter. Phytoplankton communities in marine ecosystems are sensitive to environmental changes, resulting in noticeable changes in community compositions and standing stocks following changes in physical- and chemical-environmental factors. Therefore, these changes in the phytoplankton-community structure can be used as an indicator of changes in the marine ecosystem. Moreover, in order to properly understand the structure and function of a marine ecosystem, it is necessary to understand temporal and spatial changes that occur in the phytoplankton-community structure in response to environmental factors [36]. The results of this study and those of previous studies were compared to confirm the change in the marine environment in the northern East China Sea due to the effect of constructing the Three Gorges Dam and the effects on the pelagic ecosystems according to the changes in the phytoplankton community structure were investigated.

2. Materials and Methods

2.1. Cruises and Sampling

To determine the distribution profiles of summer phytoplankton communities in the northern East China Sea, we conducted five field surveys from 2016 to 2020 (23 August–6 September 2016; 27 August–6 September 2017; 2–10 August 2018; 20–30 August 2019; and 3–15 August 2020) at the study site. The study site covered three lines and 15 sampling stations, and the surveys were conducted using the ocean research vessel, Tamgu 3 (797 tons; National Institute of Fisheries Sciences), as indicated in Figure 1 and Table 1. For nutrient analysis and quantitative analysis of phytoplankton, we collected samples at seven maximum water depths (0, 10, 20, 30, 50, 75, and 100 m) using Niskin bottles (8 L polyvinyl chloride) attached to a CTD/rosette sampler. The vertical distributions of water temperatures and salinities were measured using a calibrated SBE 9/11 CTD instrument (Sea-Bird Electronics, Bellevue, WA, USA), where descending data were used for the CTD data analysis.

2.2. Dissolved Inorganic Nutrients

For nutrient analysis, we filtered seawater samples (10 mL) through 0.45 μm disposable membrane filter units (Advantec, Tokyo, Japan), placed the samples in conical tubes (15 mL), washed them with hydrochloric acid (HCl, 10%), and immediately stored them at $-20\,°C$. After thawing the samples at room temperature ($20 \pm 2\,°C$), we analyzed the ammonium (NH_4), nitrite (NO_2), nitrate (NO_3), phosphate (PO_4), and silicon dioxide (SiO_2)

concentrations with an automatic nutrient analyzer (Quaatro, Seal Analytical, Norderstedt, Germany). The sum of the NH_4, NO_2, and NO_3 concentrations was calculated as the dissolved inorganic nitrogen (DIN).

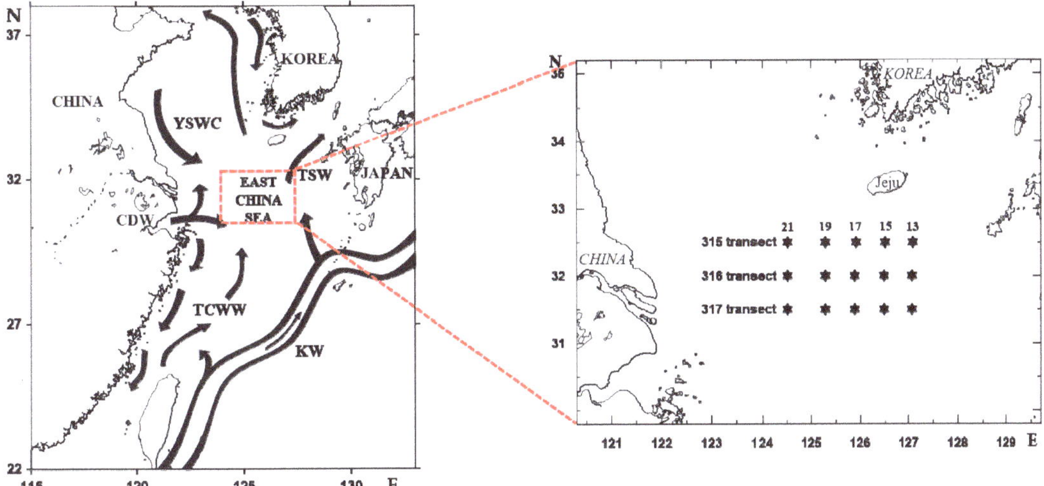

Figure 1. Sampling station in the northern East China Sea, 2016–2020. CDW: Changjiang diluted water; YSWC: Yellow Sea cold water; KW: Kuroshio water; TCWW: Taiwan current warm water; TSW: Tsushima surface water.

Table 1. Description of the sampling sites in the northern East China Sea for the cruise period from 2016 to 2020.

Station	Latitude	Longitude	Bottom Depth (m)
315-13	32.5	127.0	125
315-15	32.5	126.5	107
315-17	32.5	125.9	91
315-19	32.5	125.3	68
315-21	32.5	124.5	46
316-13	32.0	127.0	119
316-15	32.0	126.5	99
316-17	32.0	125.9	76
316-19	32.0	125.3	55
316-21	32.0	124.5	39
317-13	31.5	127.0	105
317-15	31.5	126.5	88
317-17	31.5	125.9	67
317-19	31.5	125.3	54
317-21	31.5	124.5	46

2.3. Phytoplankton Abundances and Dominant Species

For quantitative analysis of phytoplankton, each sample was collected in a 1 L square PE bottle at a standard water depth at each sampling station, fixed with Lugol's solution (diluted to a final concentration of 1%), and transported to our laboratory. After 48 h decantation, the lower layer (200 mL) containing the sedimented algae was put in a small measuring cylinder. After being allowed 48 h for decantation, the lower layer (20 mL) containing the sedimented algae was put in a glass vial and stored in a dark box. The concentrated samples were observed under an optical microscope (Nikon eclipse, Ni-U, Nikon Imaging Japan Inc., Tokyo, Japan) at $100\times$ to $1000\times$ magnification, using a Sedwick–Rafter Chamber for species identification and counting [37–40]. The resulting data were converted to the number of cells·L^{-1}, and species accounting for >5% of the total standing stock were classified as dominant species.

2.4. Picophytoplankton Abundances

To prepare picophytoplankton samples, collected seawater was filtered through a 3 μm polycarbonate membrane filter (Whatman, Florham Park, NJ, USA), split into 5 mL aliquots, and placed in cryogenic tubes. Glutaraldehyde was added to each tube at a final concentration of 1% and allowed to settle for 15 min at room temperature. The samples were then stored at −80 °C. To prepare the samples for picophytoplankton counting, they were thawed shortly before analysis and mixed with yellow-green fluorescent microspheres (0.5 μm diameter beads; Polysciences, Inc., Warrington, PA, USA), an internal reference material for standardizing scattering and fluorescence. Counting was conducted using a flow cytometer equipped with a 488 nm (1 W) argon ion laser (NovoCyte 2060R, ACEA Biosciences Inc., San Diego, CA, USA; BD Accuri^{TM} C6 Plus, BD Biosciences Inc., Franklin Lakes, NJ, USA) (Figure 2), and each picophytoplankton group was distinguished based on the characteristic side scattering of red fluorescence at a 90°-angle, the chl-*a* concentration, and the orange fluorescence from phycoerythrin (Figure 2). The flow cytometry data were analyzed using NovoExpress (Ver. 1.2.5, ACEA Biosciences Inc., San Diego, CA, USA) and BD Accuri^{TM} C6 Plus Analysis Software (Ver. 1.0.23.1, BD Biosciences Inc., Franklin Lakes, NJ, USA).

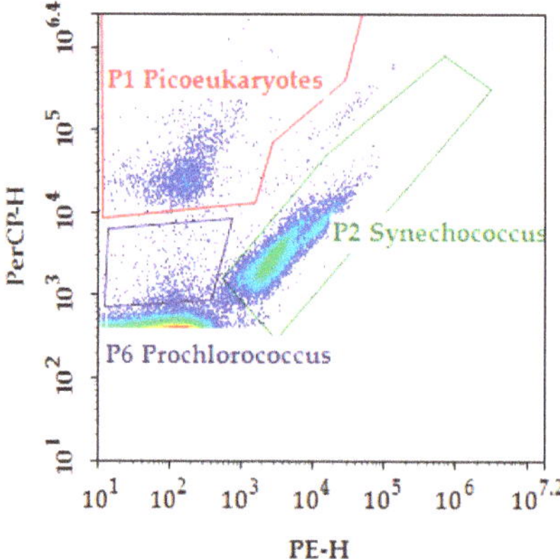

Figure 2. Flow-cytometric analysis of a picophytoplankton sample. The signatures of each picoplankton group were discriminated based on their orange and red fluorescence intensities. PerCP-H: red-fluorescence, PE-H: orange-fluorescence, P1: picoeukaryote count, P2: *Synechococcus* count, P6: *Prochlorococcus* count.

2.5. Chl-a Size Fractionation

Chl-*a* concentrations were calculated as previously described [41]. To measure chl-*a* concentrations according to the phytoplankton size (>20 μm: micro; 20 μm ≥ chl-*a* > 3 μm: nano; ≤3 μm: pico), each phytoplankton sample (0.5 L) was sequentially filtered through a 20 μm membrane (Polycarbonate Track Etched Membrane disk, 47 mm diameter, GVS, Sanford, ME, USA), a 3 μm polycarbonate membrane filter (47 mm diameter, Whatman, Florham Park, NJ, USA), and a 0.45 μm membrane filter (47 mm diameter, ADVANTEC, Tokyo, Japan) and mounted on the filter holders. After measuring the chl-*a* concentrations in the microphytoplankton, nanophytoplankton, and picophytoplankton, the total chl-*a* concentration was calculated by addition. All filters were transferred to our lab in frozen storage (−80 °C), and chl-*a* was extracted after solvation in 90% acetone and settling in a dark and cool chamber for 24 h. The extract was then filtered through a 0.45 μm syringe filter (PTFE, Advantec, Florham Park, NJ, USA) to remove particulate matter. Finally, the absorbance values were measured using a 10-Au field fluorometer (Tuner Designs, San Jose, CA, USA) calibrated with standard chl-*a* (Sigma-Aldrich, Darmstadt, Germany).

2.6. Data Analyses

The R statistical program (Ver. 4.0.3) was used to analyze statistical correlations between environmental factors and phytoplankton groups collected in the northern East China Sea. First, the decorana function of the vegan package of R was used to examine the distributions of biological parameters, which revealed that the DCA1 axis length (1.6993) was less than 3. Accordingly, a redundancy analysis (RDA) was performed.

3. Results

3.1. Physical Environments

Figures 3 and 4 show plots of the surface and vertical profiles of the mean water temperature and salinity in the northern East China Sea. The average surface water temperature in August ranged from 27.3 to 28.9 °C, with an overall average of 28.0 ± 0.5 °C and no distinctive spatial-distribution profile (Figure 3, left). The distribution range of surface salinity was 29.1–32.1, with an average of 30.2 ± 0.9 and a low-west and high-east distribution profile (Figure 3, right). The average vertical water temperature ranged from 15.1 to 28.9 °C, with an overall average of 22.6 ± 4.8 °C, and revealed that a strong thermocline formed at a depth of 20–30 m (Figure 4, top). The average salinity ranged from 29.1 to 34.5 psu, with an overall average of 32.0 ± 1.6 psu (Figure 4, bottom). The western part of the sea had a low salinity (≤31.0 psu from the surface layer to a depth of 20 m), reflecting the influence of the Changjiang dilution water. The deep layer of the eastern part of the sea had a high salinity of ≥34.0 psu, reflecting the influence of the Kuroshio water. In August, the northern East China Sea formed a complex water mass structure under the influence of various water masses, depending on the water depths between the sampling stations.

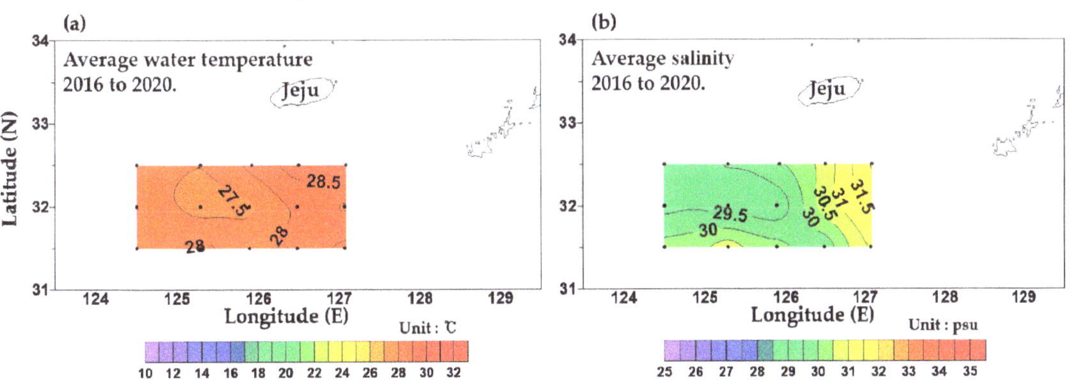

Figure 3. Spatial distribution of the average surface temperature (**a**, °C) and average surface salinity (**b**, psu) in the northern East China Sea from 2016 to 2020.

3.2. Dissolved Inorganic Nutrient Concentrations

The average surface-layer DIN, PO_4, and SiO_2 concentrations in the northern East China Sea in August were 3.5–10.3 µM, <0.1–0.1 µM, and 2.2–5.5 µM, respectively (Figure 5). The DIN concentration increased from north to south, while the PO_4 and SiO_2 concentrations showed a spatial-distribution profile with a gradual decrease from north to south. The average vertical DIN, PO_4, and SiO_2 concentration ranges were 2.5–19.5 µM, <0.1–0.8 µM, and 2.5–19.5 µM, respectively (Figure 6). Throughout the survey period, the DIN tended to increase from the surface to the bottom layer, and the PO_4 concentration was extremely low in the surface mixed layer, at the detection limit of ≤0.1 µM.

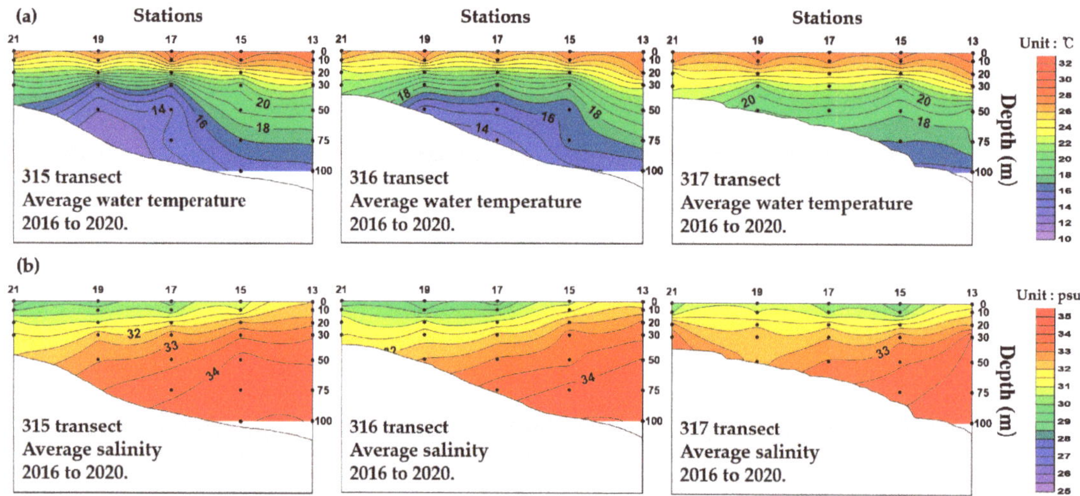

Figure 4. Vertical distribution of average surface temperatures (**a**, °C) and average surface salinities (**b**, psu) in the northern East China Sea from 2016 to 2020.

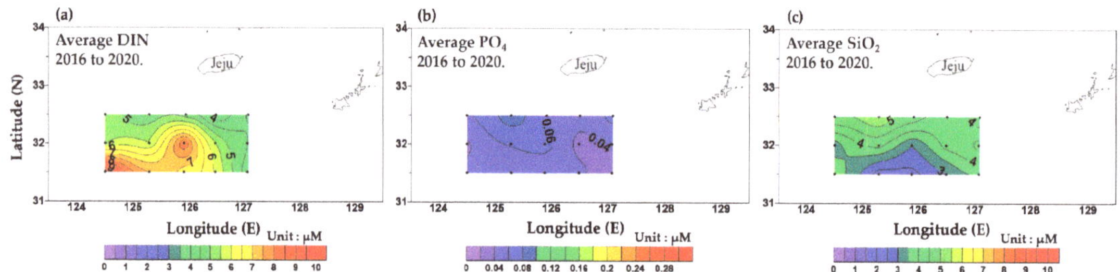

Figure 5. Spatial distributions of average surface dissolved inorganic nitrogen (**a**, DIN), PO$_4$ (**b**), and SiO$_2$ (**c**) concentrations (in µM) in the northern East China Sea from 2016 to 2020.

3.3. Phytoplankton Abundance and Dominant Species

In the summer, the average phytoplankton standing stock at the depth of each station ranged from 7955 to 1,090,202 cells·L^{-1}, with an overall average count of 149,656 ± 192, 162 cells·L^{-1}. Spatially, in the western part of the sea (affected by the Changjiang River dilution water), the standing stock was approximately three times higher than that in the eastern part of the sea (affected by offshore waters). Due to strong vertical stratification in the summer, over 90% of the total standing stock appeared in the surface mixed layer (water depth: 0–30 m), demonstrating a remarkable difference between the surface and bottom layers. Nanoflagellates (≤20 µm) appeared to be the dominant species throughout the survey period. Moreover, with *Scrippsiella acuminata* (which inhabits the coastal environment) reaching a high abundance in the western part of the sea, flagellates predominated in summer. In addition, at specific times during the survey, diatoms such as *Chaetoceros lorenzianus*, *Guinardia flaccida*, *Paralia sulcate*, and *Thalassiosira* spp., and dinoflagellates such as *Alexandrium* spp. appeared as the dominant species (Table 2).

3.4. Relative Contributions of Size-Fractionated Chl-a to the Overall Chl-a Concentration

The average surface-layer concentration of chl-*a* in the northern East China Sea in summer ranged from 0.19 to 1.25 µg·L^{-1}, with an overall average of 0.57 ± 0.33 µg·L^{-1} that gradually decreased going from west to east (Figure 7). The vertical distribution of chl-*a* ranged from 0.04 to 1.99 $\mu g \cdot L^{-1}$ with an overall average of 0.52 ± 0.40 $\mu g \cdot L^{-1}$, where the distribution was higher in the surface mixed layer along the thermocline and gradually decreased going toward the bottom. In addition, the subsurface chl-*a* maximum layer developed at a depth of 10 to 20 m in the western part of the sea, and at 20 to 30 m in the eastern part, depending on the locations of the sampling stations (Figure 8).

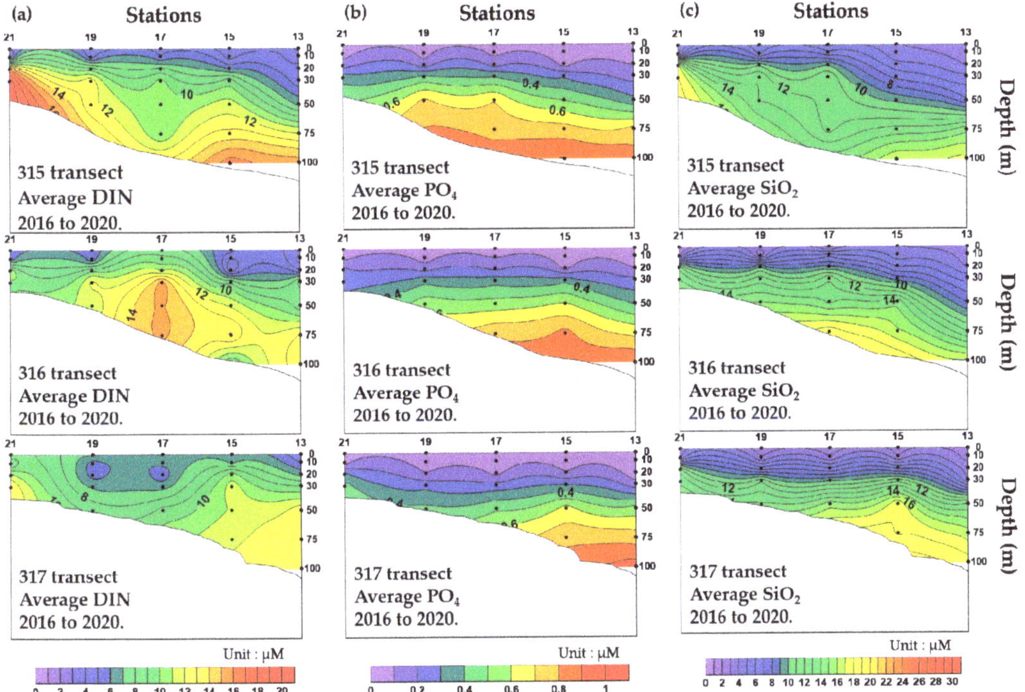

Figure 6. Vertical distribution of the average DIN (**a**), PO$_4$ (**b**), and SiO$_2$ (**c**) concentrations (in μM) in the northern East China Sea from 2016 to 2020.

Table 2. Dominant phytoplankton taxa in the northern East China Sea from 2016 to 2020.

Dominance	2016	2017	2018	2019	2020
1st	Nanoflagellates (<20 µm; 50.5%)	Nanoflagellates (<20 µm; 50.1%)	Nanoflagellates (<20 µm; 43.1%)	Nanoflagellates (<20 µm; 68.6%)	Nanoflagellates (<20 µm; 69.7%)
2nd	*Scrippsiella acuminata* (15.2%)	*Chaetoceros lorenzianus* (14.3%)	*Paralia sulcata* (10.2%)	*Alexandrium* spp. (8.4%)	*Paralia sulcata* (6.5%)
3rd	*Guinardia flaccida* (9.4%)	*Diploneis* spp. (11.0%)	*Chaetoceros* spp. (6.0%)		

The average relative contributions of microphytoplankton, nano phytoplankton, and picophytoplankton to the chl-*a* concentrations were 2.9–54.1% (23.2 ± 11.9%), 13.7–59.9% (28.7 ± 12.6%), and 16.4–770% (48.1 ± 16.2%), respectively. Spatially, microphytoplankton contributed more to the chl-*a* concentration in the western part of the sea, and picophytoplankton contributed more in the eastern part of the sea, showing opposite spatial distributions. Vertically, the relative contributions of microphytoplankton and nanophytoplankton increased from the surface layer to the bottom layer, whereas that of the picophytoplankton tended to decrease toward the bottom layer (Table 3).

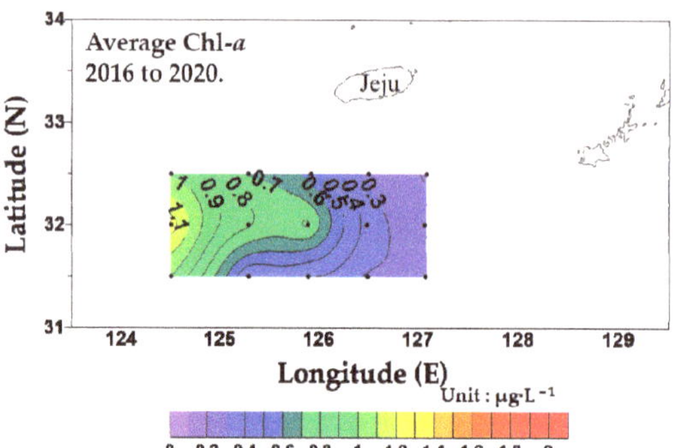

Figure 7. Spatial distributions of the average surface chl-*a* concentrations (µg·L^{-1}) in the northern East China Sea from 2016 to 2020.

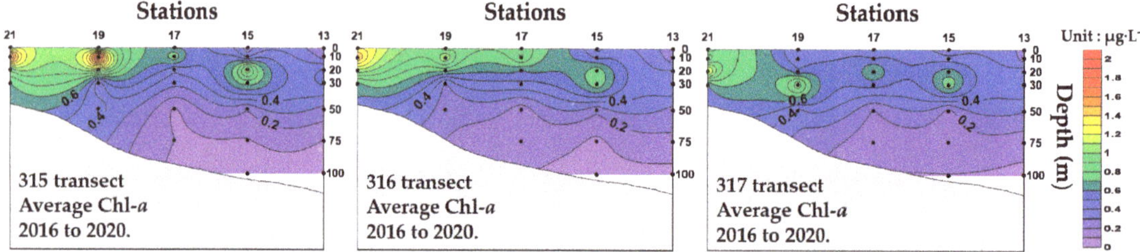

Figure 8. Vertical distributions of the average chl-*a* concentrations (µg·L^{-1}) in the northern East China Sea from 2016 to 2020.

Table 3. Vertical variation of the chl-*a* size composition in the northern East China Sea from 2016 to 2020.

315 Line		Chl-*a* Size Composition (%)			316 Line		Chl-*a* Size Composition (%)			317 Line		Chl-*a* Size Composition (%)		
St.*	Depth (m)	M	N	P	St.	Depth (m)	M	N	P	St.	Depth (m)	M	N	P
	0	12.0	18.4	69.6		0	12.6	14.0	73.4		0	20.2	15.0	64.8
	10	10.7	17.4	71.9		10	9.3	13.7	77.0		10	16.8	17.0	66.2
	20	5.2	18.6	76.2		20	17.0	16.1	66.9		20	16.6	18.9	64.5
13	30	10.5	21.2	68.3	13	30	10.6	28.8	60.6	13	30	12.6	17.0	70.4
	50	13.2	21.9	64.9		50	11.3	19.8	68.9		50	17.8	25.3	56.9
	75	18.4	31.5	50.1		75	13.7	27.6	58.7		75	18.6	30.3	51.1
	100	18.2	44.5	37.3		100	16.7	42.0	41.3		100	23.9	42.3	33.8
	0	14.3	32.6	53.1		0	17.9	22.3	59.8		0	12.4	22.9	64.7
	10	8.6	23.9	67.5		10	11.9	21.0	67.1		10	24.3	17.3	58.4
	20	14.2	18.5	67.3		20	2.9	22.2	74.9		20	20.5	21.7	57.8
15	30	25.8	21.0	53.2	15	30	15.0	29.5	55.5	15	30	23.2	27.5	49.3
	50	12.1	22.6	65.3		50	19.9	47.2	32.9		50	13.5	41.7	44.8
	75	10.0	33.9	56.1		75	13.0	58.2	28.8		75	23.1	50.9	26.0
	100	12.7	52.5	34.8		100	17.7	58.9	23.4			□	□	□
	0	17.5	20.0	62.5		0	40.8	18.5	40.7		0	30.6	14.5	54.9
	10	19.8	21.1	59.1		10	32.5	17.3	50.2		10	26.0	18.3	55.7
	20	10.0	29.6	60.4		20	24.1	20.9	55		20	40.7	16.7	42.6
17	30	13.7	35.8	50.5	17	30	27.3	30.9	41.8	17	30	40.7	22.3	37.0
	50	9.9	47.8	42.3		50	24.8	51.2	24.0		50	24.3	45.6	30.1
	75	26.9	52.8	20.3		75	23.7	59.9	16.4			□	□	□

Table 3. *Cont.*

315 Line		Chl-*a* Size Composition (%)			316 Line		Chl-*a* Size Composition (%)			317 Line		Chl-*a* Size Composition (%)		
St.*	Depth (m)	M	N	P	St.	Depth (m)	M	N	P	St.	Depth (m)	M	N	P
19	0	23.8	19.4	56.8	19	0	37.2	23.1	39.7	19	0	30.9	21.2	47.9
	10	20.7	21.4	57.9		10	34.7	21.2	44.1		10	27.3	18.3	54.4
	20	23.4	27.5	49.1		20	27.1	28.3	44.6		20	28.8	19.9	51.3
	30	8.9	40.6	50.5		30	37.1	40.3	22.6		30	44.2	30.5	25.3
	50	20.8	58.5	20.7		50	25.6	55.7	18.7		50	19.6	47.0	33.4
21	0	29.9	15.8	54.3	21	0	53.6	18.1	28.3	21	0	34.4	29.6	36.0
	10	41.9	19.7	38.4		10	45.8	17.7	36.5		10	30.6	31.9	37.5
	20	40.4	32.3	27.3		20	47.1	23.1	29.8		20	54.1	16.8	29.1
	30	51.4	29.6	19.0		30	35.5	34.5	30.0		30	42.8	25.0	32.2
	Mean	18.8	29.3	51.9		Mean	24.4	30.4	45.2		Mean	26.6	26.1	47.3
	SD	10.9	12.1	16.0		SD	12.9	14.7	18.3		SD	10.6	10.7	13.7

*St.: station; M: micro (>20 μm); N: nano (20 μm $\geq$ chl-*a* > 3 μm); P: pico ($\leq$3 μm).

3.5. Picophytoplankton Cell Abundances

Figure 9 shows plots of the picophytoplankton standing stocks in three groups, as determined by flow cytometry. *Synechococcus* and Picoeukaryotes cells did not show spatially distinct distribution profiles, whereas *Prochlorococcus* cells appeared to be limited at some east-side sampling stations (13, 15, 17 on each line) and abundant in the west-side sampling stations. Regarding the vertical distribution of picophytoplankton in summer, all three genera showed the highest standing stocks at the depth where the thermocline formed (rather than at the surface layer) and gradually decreased below the thermocline.

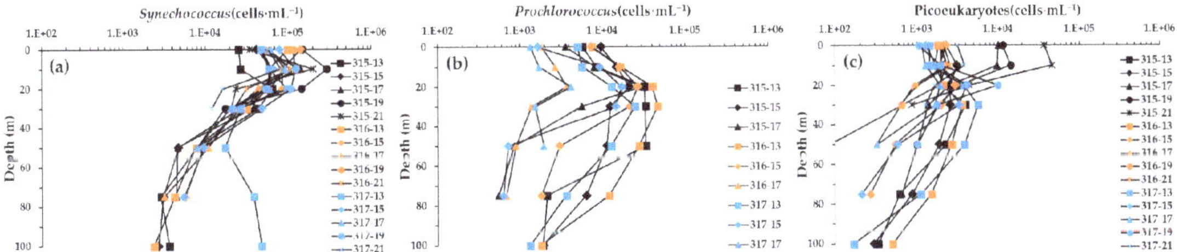

Figure 9. Vertical distributions of *Synechococcus* (**a**), *Prochlorococcus* (**b**), and picoeukaryote (**c**) cell abundances (cells·mL^{-1}) found at each station in the northern East China Sea from 2016 to 2020.

4. Discussion

4.1. Changes in Phytoplankton-Community Structures in Response to Changes in the Summer Marine Environment

To determine the changes in the phytoplankton-community structure associated with changes in the marine environment of the northern East China Sea in the summer, we compared the data obtained from the same sampling stations (Lines 315 and 316) at the same depth in a previous study and in this study (Table 4). The comparison revealed decreased nutrients (especially phosphate) in the surface layer as a characteristic difference in the marine environment. In terms of the community structure, diatoms and dinoflagellates were identified as the dominant species in the previous study conducted by Oh [42], and nanoflagellates ($\leq$20 μm) were identified as the dominant species in this study. This difference may be ascribed to decreased nutrients, especially in terms of phosphate. Among the nutrients flowing into the East China Sea, the DIN concentration in the estuary of the Changjiang River doubled from 1985 to 1998, and the DIN flux increased by 1.3-fold [43]. As a result, in the estuary and coastal waters of China, the area containing red tides increased from 2000 km^2 in the 1980s to over 7000 km^2 in the 2000s [44] and contributed to a 21-fold increase in the number of red tide occurrences. The frequent occurrence of red tides in the Changjiang River estuary consumes a large amount of phosphorus, resulting in a phosphorus deficiency (rather than excessive nitrogen) [11,45]. The Changjiang dilution water affects the northern East China Sea. However, the transport volume decreased by 9–18% after the construction of the Three Gorges Dam (2003–2009), and the amount of suspended solids decreased by up to 56% [46]. The decrease in the inflow of the Changjiang dilution water also decreased the amount of nutrients, which caused a change in the community structure of phytoplankton from being dominated by diatoms to being dominated by flagellates [47]. Furthermore, the main source of phosphorus in the East China Sea is the Kuroshio water mass, although its influence has recently decreased in the northern East China Sea. In this regard, it was reported that the phosphate inflow decreased as the influence of the Taiwan warm current (which is caused by upwelling of the Kuroshio water mass) also weakened [48]. The decrease in phosphate in the East China Sea caused a decrease in the proportion of diatoms from 85% in 1984 to 60% in 2000 [49]. These results were observed in both the East China Sea and the Yellow Sea. In this respect, Lin et al. [43] reported that the standing stock of diatoms markedly decreased from the 1980s to the 1990s because of the increased surface-water temperature of the Yellow Sea and decreased silicate and phosphate concentrations. Likewise, Lee [50] reported extremely low phosphate concentrations, even below the detection limit, in a study of the central part of the Yellow Sea in summer, which resulted in a phytoplankton transition due to phosphate restriction, in which flagellates ($\leq$10 µm) were the dominant species at all sampling stations, except for some bottom layer-dominant species, such as species in the *Navicula* spp. and *Skeleonema* spp. It was also reported that the restriction of nutrients greatly affects the growth of phytoplankton when their concentrations fall below the minimum levels of DIN 1.0 µM, phosphate 0.2 µM, and silicate 2.0 µM [51]. Phosphate can also act as an important factor that hampers the growth of diatoms rather than flagellates [49], and if phosphate is deficient, then even if nitrate is abundant, phytoplankton cannot use nitrate, which hinders growth [52]. In this survey, the surface mixed layer, which accounted for 90% of the total standing stock, maintained sufficient nitrate and silicate concentrations, but had an extremely low phosphate concentration, even lower than 0.1 µM. The fact that nanoflagellates with relatively low nutrient requirements could dominate over diatoms in all sampling stations can be explained by the phosphate concentration acting as a limiting factor.

Table 4. Comparison of environmental variables and dominant species in the northern East China Sea between August 1998 [42] and August 2016 to 2020 (this study).

Parameters		August 1998	August 2016–2020
Temperature (°C)	Surface	27.4–29.4 (28.7 ± 0.5)	25.4–31.3 (28.0 ± 1.4)
	Vertical	12.7–29.4 (21.5 ± 5.9)	11.8–31.3 (23.1 ± 5.5)
Salinity	Surface	26.7–29.7 (28.3 ± 0.9)	26.2–34.0 (30.1 ± 1.9)
	Vertical	27.5–34.7 (31.2 ± 2.5)	26.2–34.6 (31.7 ± 1.9)
Nitrite (µM)	Surface	0.07–0.36 (0.21 ± 0.08)	0.01–0.16 (0.05 ± 0.04)
	Vertical	0.07–0.57 (0.19 ± 0.14)	0.01–1.61 (0.15 ± 0.26)
Nitrate (µM)	Surface	1.26–3.54 (2.51 ± 0.66)	0.12–11.37 (2.77 ± 2.70)
	Vertical	0.87–3.54 (2.05 ± 0.74)	0.09–35.40 (5.87 ± 5.54)
Phosphate (µM)	Surface	0.16–0.45 (0.25 ± 0.09)	ND–0.16 (0.05 ± 0.04)
	Vertical	0.16–1.03 (0.39 ± 0.23)	ND–0.82 (0.21 ± 0.24)
Silicate (µM)	Surface	8.95–13.77 (9.69 ± 1.45)	0.01–11.46 (4.23 ± 3.10)
	Vertical	8.95–13.77 (9.59 ± 0.88)	0.01–40.17 (7.87 ± 5.71)
Dominant species		*Pseudonitzschia pungens* *Prorocentrum dentatum* *Skeleonema costatum*	Nanoflagellates (<20 µm) *Scrippsiella acuminata* *Paralia sulcata*

Values in parentheses are means.

4.2. Changes in Phytoplankton Community Sizes and Structures in the Northern East China Sea

The results of this survey study indicated that the picophytoplankton contributed to >60% of the chl-*a* concentration in summer standing stock; in particular, the surface mixed layer (from the surface to a depth of 20 m), which had low nutrient concentrations, had the highest relative chl-*a* concentration. Spatially, higher chl-*a* percentages were observed at the east-side stations than at the west-side stations, where nutrient concentrations were relatively high (Table 3). These tendencies were verified using statistical analysis. RDA of environmental factors and the chl-*a* size revealed that the nutrient concentrations increased with increasing water depths, which was indicative of an environment where the nutrient availability became increasingly limited from the bottom layer to the surface layer due to strong stratification caused by the rise in the surface water temperature in summer. Regarding the relationship between the phytoplankton size and environmental factors, the fraction of nanophytoplankton increased toward the bottom layer, where nutrient concentrations were relatively high, whereas picophytoplankton were inversely related to all nutrients, showing the highest fraction at the surface layer, where nutrient concentrations were low. In particular, picophytoplankton had the strongest inverse correlation with phosphate, which was associated with extreme phosphate depletion in the surface mixed layer (Figure 10). Similar results have been reported in other sea areas in recent years. Size-fractionation analysis that integrated satellite-data analysis results and field observations in the Mediterranean Sea confirmed that picophytoplankton < 2 μm accounted for 31–92% of the total phytoplankton biomass with seasonal variations, primarily due to low nitrogen and phosphorus availability [53,54]. Agawin et al. [55] reported that the relative contribution of picophytoplankton to the total chl-*a* biomass was over 50% in Blanes Bay in the Mediterranean Sea, which was attributed primarily to high temperatures and nutrient-poor waters. The composition ratio of picophytoplankton was 44–90% and 54–64%, respectively, in the Mediterranean Tyrrhenian Sea and Levantine Basin waters [56,57]. An analysis of summer satellite data of the Yamato Basin and Japan Basin showed that the percentage of picophytoplankton was more than 50% [58]. In addition, the composition ratio of nanophytoplankton and picophytoplankton in many waters was reported to be over 60% [53,59–62] (Table 5). In summer surveys of the East Sea of Korea, the relative contribution of picophytoplankton to the total chl-*a* biomass ranged from 35 to 63% [58,63], whereas those of nanophytoplankton and picophytoplankton (≤20 μm) were 74% on average in the surface layer of the eastern Yellow Sea, in a study focused on the dominance of small phytoplankton [50]. Similarly, Furuya et al. [64] showed that picophytoplankton made a high relative contribution (up to 80%) due to the oligotrophic environment of the surface layer in a summer survey of the northern East China Sea. Son et al. [33] also reported that their analysis of long-term satellite observations of the chl-*a* biomass led to the finding that the contribution of microphytoplankton to the chl-*a* concentration sharply decreased in the northern East China Sea, whereas those of nanophytoplankton and picophytoplankton increased. These results have a common denominator in that they were derived from oligotrophic environments, where smaller phytoplankton cells had larger surface areas per unit volume, making small phytoplankton become dominant, owing to their capacity for rapid nutrient exchange through the cell surface [65–67].

Miniaturization of the phytoplankton size can result in decreased primary productivity. Indeed, analysis of primary productivity data derived from the Moderate Resolution Imaging Spectroradiometer Aqua led to a report showing that the annual primary production in the East Sea is decreasing by 13% per decade, as a result of an increase in the relative contribution of picophytoplankton (≤2 μm) to the total biomass [68]. Moreover, Lee et al. [63] verified a strong inverse correlation between the dominance rate of phytoplankton of ≤5 μm and the total primary production in the Amundsen Sea and attributed this to the low rate of carbon uptake by small phytoplankton. The miniaturization of phytoplankton also affects the food web, and it is projected that the increase in picophytoplankton abundance would lead to the dominance of microzooplankton (<200 μm) and gelatinous zooplankton (salps, doliolids, and ctenophores) and a decrease in biomass [69].

Moreover, a previous study showed that mesozooplankton (<200 μm), e.g., copepods, did not directly feed on picophytoplankton, but rather indirectly fed on ciliates that feed on picophytoplankton [70]. Accordingly, as the dominance of picophytoplankton increases, the biomass of mesozooplankton that do not directly feed on them will likely decrease. Consequently, the marine food chain will likely change from a simple diatom-based web with high primary productivity to a complex microbial food web centering on small phytoplankton with low productivity, negatively affecting overall marine productivity and reducing the efficiency of carbon transfer to consumers at higher trophic levels (Figure 11).

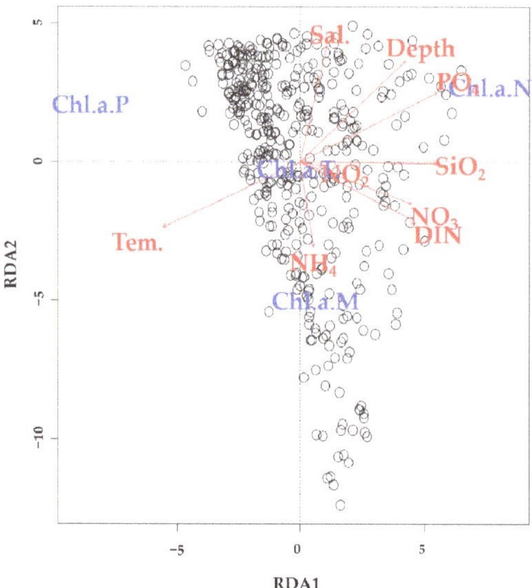

Figure 10. Redundancy analysis (RDA) ordination plots showing relationships between environmental and biological conditions in the northern East China Sea.

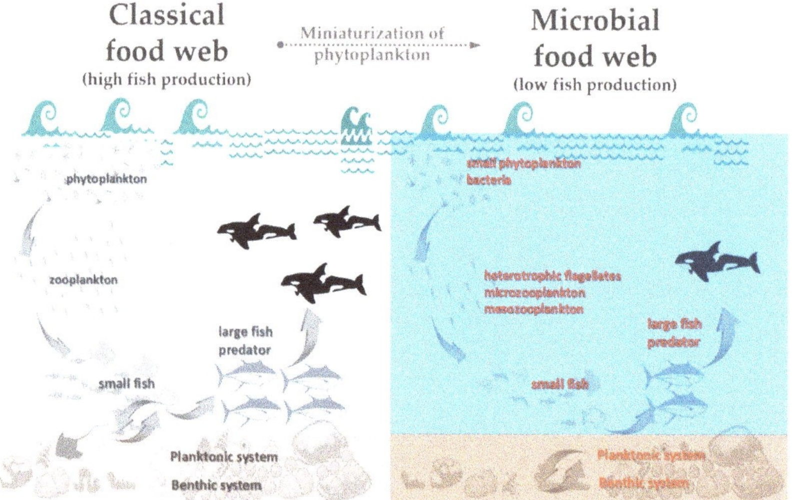

Figure 11. Changes in the food chain due to succession of the phytoplankton community in the northern East China Sea.

Table 5. Size fractionation of phytoplankton observed in the different coastal waters of the northern East China Sea and global waters.

Area	Date	Relative Ratio (%)			References
		Pico Size	Nano Size	Micro Size	
Northen East China Sea	2018–2020/ seasonally	45.6	31.2	23.2	This study
Blanes Bay	1997/summer	>50			[55]
Levantine Basin	March 1992	54.3–64.2			[56]
Tyrrhenian Sea (South)	July 2005	44–81			[57]
	December 2005	76–90			
Western Subarctic Pacific	23–29 June 2010	63			
Japan Basin	5–11 July 2010	56			[58]
Yamato Basin	18–20 July 2010	56			
Ulleung Basin	22–24 July 2010	38			
Mediterranean Sea		31–92			[53]
Adriatic Sea (North)	August 1986 and 1988; July 1987			10–23	[59]
Atlantic Meridional Transect (Oligortophic)	April, October 1996 April, October 1997	80	16	4	[60]
Algerian Basin	October 1996	42–62	38–58		[61]
South China Sea	summer 1998	63	22	16	[62]
	winter 1998	51	14	36	

Author Contributions: Conceptualization, K.-W.P. and S.-H.Y.; methodology K.-W.P. and M.-H.Y.; validation, K.-W.P. and S.-H.Y.; formal analysis, K.-W.P., S.-Y.M. and M.-H.Y.; investigation, K.-W.P., H.-J.O., S.-Y.M., M.-H.Y. and S.-H.Y.; data curation, K.-W.P.; writing—original draft preparation, K.-W.P.; writing—review and editing, S.-H.Y.; visualization, K.-W.P.; supervision, S.-H.Y.; project administration, H.-J.O. and S.-H.Y.; funding acquisition, H.-J.O. and S.-H.Y. All authors have read and agreed to the published version of the manuscript.

Funding: This research was supported by the "Development of assessment technology on the structure variations in marine ecosystem (R2022073)" funded by the National Institute of Fisheries Science (NIFS), Korea.

Data Availability Statement: Not applicable.

Acknowledgments: We appreciate the captains and crew of *R/V Tamgu 3* for their assistance with collecting our samples. We would also like to thank the researchers for their assistance with sample analysis.

Conflicts of Interest: The authors declare no conflict of interest.

References

1. Zhang, Q.L.; Weng, X.C. Analysis of water masses in the south Yellow Sea in spring. *Oceanol. Limnol. Sin.* **1996**, *27*, 421–428.
2. Rho, H.K. Studies on Marine Environment of Fishing Grounds in the Waters around Cheju Island. Ph.D. Thesis, Tokyo University, Tokyo, Japan, 1985.
3. Miyaji, K. Studies on the eddies associated with the meander of the Kuroshio in the waters off southwest coast of Kyushu and their effects on egg and larval transport. *Bull. Seikai Nat. Fish. Res. Inst.* **1991**, *69*, 1–77.
4. Guo, Y.J.; Zhang, Y.S. Characteristics of phytoplankton distribution in Yellow Sea. *J. Yellow Sea* **1996**, *2*, 90–103.
5. Simpson, J.H.; Pingree, R.D. Shallow sea fronts produced by tidal stirring. In *Oceanic Fronts in Coastal Processes*; Bowman, M.J., Esaias, W.E., Eds.; Springer: Berlin/Heidelberg, Germany, 1978; pp. 29–47.
6. Beardsley, R.; Limebumer, R.; Yu, H.; Cannon, G.A. Discharge of the Changjiang (Yangtze River) into the East China Sea. *Cont. Shelf Res.* **1985**, *4*, 57–76. [CrossRef]
7. Su, Y.S.; Weng, X.C. Water masses in China Seas. In *Oceanography of China Seas*; Zhou, D., Liang, Y.B., Zeng, C.K., Eds.; Springer: Dordrecht, The Netherlands, 1994; pp. 3–16.

8. Edmond, J.M.; Spivack, A.; Grant, B.C.; Chen, Z.; Chen, S.; Zeng, X. Chemical dynamics of the Changjiang Estuary. *Cont. Shelf Res.* **1985**, *4*, 17–36. [CrossRef]
9. Hama, T.; Shin, K.H.; Handa, N. Spatial variability in the primary productivity in the East China Sea and its adjacent water. *J. Oceanogr.* **1997**, *53*, 41–51. [CrossRef]
10. Gong, G.C.; Wen, Y.H.; Wang, B.W.; Liu, G.J. Seasonal variation of chlorophyll a concentration, primary production and environmental conditions in the subtropical East China Sea. *Deep Sea Res. Part II* **2003**, *50*, 1219–1236. [CrossRef]
11. Chen, Y.L.; Chen, H.Y.; Gong, G.C.; Lin, Y.H.; Jan, S.; Takahashi, M. Phytoplankton production during a summer coastal upwelling in the East China Sea. *Cont. Shelf Res.* **2004**, *24*, 1321–1338. [CrossRef]
12. Koroji, K.; Matsusaki, M. Present activity on marine biology in Japan Meteorological Agency. *Sokuko Jiho* **1980**, *47*, 19–21.
13. Chin, T.G.; Cheng, Z.M.; Lin, J.; Lin, S. Diatoms from the surface sediments of the east China sea. *Acta Oceanol. Sin.* **1980**, *2*, 97–110.
14. Huang, R. The influence of hydrography on the distribution of phytoplankton in the southern Taiwan strait. *Estuar. Coast. Shelf Sci.* **1988**, *26*, 643–656. [CrossRef]
15. Koizumi, I. Pliocene and Pleistocene diatom datum levels related with Paleoceanography in the northwest Pacific. *Mar. Micropaleontol.* **1986**, *10*, 309–325. [CrossRef]
16. Yu, J.L.; Li, R.X. Species composition and quantitative distribution of planktonic diatom in the Kuroshio current and its adjacent waters. In *Selected Papers on the Investigation of Kuroshio*; Su, J.L., Ed.; Ocean Press: Beijing, China, 1990; Volume 2, pp. 57–66.
17. Guo, Y.J. The Kuroshio, part 2. Primary productivity and phytoplankton. *Oceanogr. Mar.* **1991**, *29*, 155–189.
18. Huang, R. Phytoplankton distribution in the South China Sea and Kuroshio-flowing region of Taiwan. *Acta Oceanogr. Taiwan.* **1993**, *31*, 73–82.
19. Matsuda, O.; Nishi, Y.; Yoon, Y.H.; Endo, T. Observation of thermohaline structure and phytoplankton biomass the shelf front of East China Sea during early summer. *J. Fac. Appl. Biol. Sci.* **1989**, *28*, 27–35.
20. Kamiya, H. The correlation between appearance of phytoplankton and the sea condition. *Umi Sora* **1991**, *67*, 153–162.
21. Ueno, S. Phytoplankton communities and their relation to the water types in the vicinity of the Kuroshio front in the East China. *Sea J. Shimonoseki Univ. Fish.* **1993**, *41*, 251–256.
22. Meng, F.; Huang, F.; Zhaodang, M.; Qinliang, L. On the composition and distribution of zooplankton species in the Kuroshio region in the north of the East China Sea. In *Selected Papers on the Investigation of Kuroshio*; Su, J.L., Ed.; Ocean Press: Beijing, China, 1991; Volume 3, pp. 150–161.
23. Zhang, H.Q. Relationship between zooplankton distribution and hydrographic characteristics of the southern Yellow Sea. *Yellow Sea* **1995**, *1*, 50–67.
24. Choi, C.H.; Lee, C.R.; Kang, H.K.; Kang, K.A. Characteristics and variation of size-fractionated zooplankton biomass in the northern East China Sea. *Ocean Polar Res.* **2011**, *33*, 135–147. [CrossRef]
25. Oh, G.S. Characteristics and Variation of Zooplankton Community Structure in the Northern East China Sea. Master's Thesis, Chonnam National University, Yeosu, Korea, 2012.
26. Dong, H.; Ren, D.; Li, B. The distribution and influential factors of oxygen and phosphate in Kuroshio area of the East China Sea north to 30 degree N in summer and winter 1987. In *Selected Papers on the Investigation of Kuroshio*; Su, J.L., Ed.; Ocean Press: Beijing, China, 1990; Volume 2, pp. 208–217.
27. Noh, J.H. A Study on the Phytoplankton Distribution in the Yellow Sea and the East China Sea. Master's Thesis, Inha University, Incheon, Korea, 1995.
28. Yoon, Y.H.; Park, J.S.; Soh, H.Y.; Hwang, D.J. Spatial distribution of phytoplankton community and red tide of *Dinoflagellate, Prorocentrum donghaiense* in the East China Sea during early summer. *Korean J. Environ. Biol.* **2003**, *21*, 132–141.
29. Yoon, Y.H.; Park, J.S.; Park, Y.G.; Soh, H.Y.; Hwang, D.J. A characteristics of thermohaline structure and phytoplankton community from southwestern parts of the East China Sea during early summer, 2004. *Bull. Korean Soc. Fish.* **2005**, *41*, 129–139.
30. Noh, J.H.; Yoo, S.J.; Lee, J.A.; Kim, H.C.; Lee, J.H. Phytoplankton in the waters of the Ieodo Ocean Research Station determined by microscopy, flow cytometry, HPLC pigment data and remote sensing. *Ocean Polar Res.* **2005**, *27*, 397–417.
31. Kim, D.S.; Shim, J.H.; Lee, J.A.; Kang, Y.C. The distribution of nutrients and chlorophyll in the northern East China Sea during the spring and summer. *Ocean Polar Res.* **2005**, *27*, 251–263. [CrossRef]
32. Kim, D.S.; Kim, K.H.; Shim, J.H.; Yoo, S.J. The distribution and interannual variation in nutrients, chlorophyll-a, and suspended solids in the northern East China Sea during the summer. *Ocean Polar Res.* **2007**, *29*, 193–204. [CrossRef]
33. Son, Y.B.; Ryu, J.H.; Noh, J.H.; Ju, S.J.; Kim, S.H. Climatological variability of satellite-derived sea surface temperature and chlorophyll in the south sea of Korea and East China Sea. *Ocean Polar Res.* **2012**, *34*, 201–218. [CrossRef]
34. Gong, G.C.; Chang, J.; Chiang, K.P.; Hsiung, T.M.; Hung, C.C.; Duan, S.W.; Codispti, L.A. Reduction of primary production and changing of nutrient ratio in the East China Sea: Effect of the Three Gorges Dam? *Geophys. Res. Lett.* **2006**, *33*. [CrossRef]
35. Jiaoa, N.; Zhang, Y.; Zeng, Y.; Gardner, W.D.; Mishonov, A.V.; Richardson, M.J.; Hong, N.; Pan, D.; Yan, X.H.; Jo, Y.H.; et al. Ecological anomalies in the East China Sea: Impacts of the Three Gorges Dam? *Water Res.* **2007**, *41*, 1287–1293. [CrossRef]
36. Smayda, T.J. Biogeographical meaning indicators. In *Phytoplankton Manual*; Sournia, A., Ed.; United Nations Educational, Scientific, and Cultural Organization: Paris, France, 1978; pp. 225–229.

37. Rines, J.E.B.; Hargraves, P.E. *The Chaetoceros Ehrenberg (Bacillariophyceae) Flora of Narragansett Bay*; Schweizebart Science: Stuttgart, Germany, 1988.
38. Round, F.E.; Crawfor, R.M.; Mann, D.G. *The Diatoms*; Cambridge: New York, NY, USA, 1990.
39. Shim, J.H. Flora and fauna of Korea. In *Marine Phytoplankton*; Ministry of Education: Seoul, Korea, 1994; Volume 34.
40. Tomas, C.R. *Identifying Marine Phytoplankton*; Academic Press: Cambridge, MA, USA, 1997.
41. Parsons, T.R.; Maita, Y.; Lalli, C.M. *A Manual of Biological and Chemical Methods for Seawater Analysis*; Pergamon Press: Oxford, UK, 1984.
42. Oh, H.J. Ecological Characteristics of Phytoplankton in the Northern Part of East China Sea. Ph.D. Thesis, Pusan National University, Busan, Korea, 1998.
43. Lin, C.; Ning, X.; Su, J.; Lin, Y.; Xu, B. Environmental changes and the responses of the ecosystems of the Yellow Sea during 1976–2000. *J. Mar. Syst.* **2005**, *55*, 223–234. [CrossRef]
44. Li, M.K.; Xu, K.; Watanabe, M.; Chen, Z. Long-term variations in dissolved silicate, nitrogen, and phosphorus flux from the Changjiang River into the East China Sea and impacts on estuarine ecosystem. *Estuar. Coast. Shelf Sci.* **2007**, *71*, 3–12. [CrossRef]
45. Li, H.M.; Tang, H.J.; Shi, X.Y.; Zhang, C.S.; Wang, X.L. Increased nutrient loads from the Changjiang (Yangtze) River have led to increased Harmful Algal Blooms. *Harmful Algae* **2014**, *39*, 92–101. [CrossRef]
46. Ministry of Land, Transport and Maritime Affairs. *The Study of Oceanographic Environmental Impact in the South Sea (East China Sea) due to the Three Gorges Dam*; Ministry of Land, Transport and Maritime Affairs: Seoul, Korea, 2009; pp. 7–175.
47. Jiang, Z.; Liu, J.; Chen, J.; Chen, Q.; Yan, X.; Xuan, J.; Zeng, J. Responses of summer phytoplankton community to drastic environmental changes in the 664 Changjiang (Yangtze River) estuary during the past 50 years. *Water Res.* **2014**, *54*, 1–11. [CrossRef]
48. Youn, S.C.; Suk, H.Y.; Whang, J.D.; Suh, Y.S.; Yoon, Y.Y. Long-term variation in ocean environmental conditions of the northern East China Sea. *J. Korean Soc. Mar. Environ. Energy* **2015**, *18*, 189–206. [CrossRef]
49. Zhou, M.J.; Shen, Z.L.; Yu, R.C. Responses of a coastal phytoplankton community to increased nutrient input from the Changjiang (Yangtze) River. *Cont. Shelf Res.* **2008**, *28*, 1483–1489. [CrossRef]
50. Lee, Y.J. Phytoplankton Dynamics and Primary Production in the Yellow Sea during Winter and Summer. Ph.D. Thesis, Inha University, Incheon, Korea, 2012.
51. Dortch, Q.; Whitledge, T.E. Does nitrogen or silicon limit phytoplankton production in the Mississippi River plume and nearby regions? *Cont. Shelf Res.* **1992**, *12*, 1293–1309. [CrossRef]
52. Chen, Y.L.; Lu, H.; Shiah, F.; Gong, G.; Liu, K.; Kanda, J. New production and f-ratio on the continental shelf of the East China Sea: Comparisons between nitrate inputs from the subsurface Kuroshio Current and the Changjiang River. *Estuar. Coast. Shelf Sci.* **1999**, *48*, 59–75. [CrossRef]
53. Magazzu, G.; Decembrini, F. Primary production, biomass and abundance of phototrophic picoplankton in the Mediterranean Sea: A review. *Aquat. Microb. Ecol.* **1995**, *9*, 97–104. [CrossRef]
54. Uitz, J.; Stramski, D.; Gentili, B.; D'Ortenzio, F.; Claustre, H. Estimates of phytoplankton class-specific and total primary production in the Mediterranean Sea from satellite ocean color observations. *Glob. Biogeochem. Cycles* **2012**, *26*. [CrossRef]
55. Agawin, N.S.R.; Duarte, C.M.; Agustí, S. Nutrient and temperature control of the contribution of picoplankton to phytoplankton biomass and production. *Limnol. Oceanogr.* **2000**, *45*, 591–600. [CrossRef]
56. Zohary, T.; Brenner, S.; Krom, M.D.; Angel, D.L.; Kress, N.; Li, W.K.W.; Neori, A.; Yacobi, Y.Z. Buildup of microbial biomass during deep winter mixing in a Mediterranean warm-core eddy. *Mar. Ecol. Prog. Ser.* **1998**, *167*, 47–57. [CrossRef]
57. Decembrini, F.; Caroppo, C.; Azzaro, M. Size structure and production of phytoplankton community and carbon pathways channeling in the southern Tyrrhenian Sea (western Mediterranean). *Deep Sea Res. Part II* **2009**, *56*, 687–699. [CrossRef]
58. Kwak, J.H.; Lee, S.H.; Hwan, J.; Suh, Y.-S.; Park, H.J.; Chang, K.-I.; Kim, K.-R.; Kang, C.K. Summer primary productivity and phytoplankton community composition driven by different hydrographic structures in the East/Japan Sea and the Western Subarctic Pacific. *J. Geophys. Res. Oceans* **2014**, *119*, 4505–4519. [CrossRef]
59. Revelante, N.; Gilmartin, M. The relative increase of larger phytoplankton in a subsurface chlorophyll maximum of the northern Adriatic Sea. *J. Plankton Res.* **1995**, *17*, 1535–1562. [CrossRef]
60. Marañón, E.; Holligan, P.M.; Barciela, R.; González, N.; Mouriño, B.; Pazó, M.J.; Varela, M. Patterns of phytoplankton size structure and productivity in contrasting open-ocean environments. *Mar. Ecol. Prog. Ser.* **2001**, *216*, 43–56. [CrossRef]
61. Morán, X.A.G.; Taupier-Letage, I.; Vázquez-Domínguez, E.; Ruiz, S.; Arin, L.; Raimbault, P.; Estrada, M. Physical-biological coupling in the Algerian Basin (SW Mediterranean): Influence of mesoscale instabilities on the biomass and production of phytoplankton and bacterioplankton. *Deep Sea Res. Part I* **2001**, *48*, 405–437. [CrossRef]
62. Ning, X.; Chai, F.; Xue, H.; Cai, Y.; Liu, C.; Shi, J. Physical-biological oceanographic coupling influencing phytoplankton and primary production in the South China Sea. *J. Geophys. Res.* **2005**, *109*, C10005. [CrossRef]
63. Lee, S.H.; Joo, H.T.; Lee, J.H.; Lee, J.H.; Kang, J.J.; Lee, H.W.; Lee, D.B.; Kang, C.K. Carbon uptake rates of phytoplankton in the Northern East/Japan Sea. *Deep Sea Res. Part II* **2017**, *143*, 45–53. [CrossRef]
64. Furuya, K.; Hayashi, M.; Yabushita, Y.; Ishikawa, A. Phytoplankton dynamics in the East China Sea in spring and summer as revealed by HPLC derived pigment signatures. *Deep Sea Res. Part II* **2003**, *50*, 367–387. [CrossRef]
65. Raven, J.R. The twelfth tansley lecture. Small is beautiful: The picophytoplankton. *Funct. Ecol.* **1998**, *12*, 503–513. [CrossRef]

66. Litchman, E.; Klausmeier, C.A.; Schofield, O.M.; Falkowski, P.G. The role of functional traits and trade-offs in structuring phytoplankton communities: Scaling from cellular to ecosystem level. *Ecol. Lett.* **2007**, *10*, 1170–1181. [CrossRef]
67. Longhurst, A.R. *Ecological Geography of the Sea*; Academic Press: London, UK, 2010.
68. Joo, H.T.; Son, S.H.; Park, J.W.; Kang, J.J.; Jeong, J.Y.; Lee, C.I.; Kang, C.K.; Lee, S.H. Long-term pattern of primary productivity in the East/Japan Sea based on ocean color data derived from MODIS-aqua. *Remote Sens.* **2016**, *8*, 25. [CrossRef]
69. Richardson, A.J. In hot water: Zooplankton and climate change. *ICES J. Mar. Sci.* **2008**, *65*, 279–295. [CrossRef]
70. Lee, C.R.; Kang, H.K.; Choi, K.H. Latitudinal distribution of mesozooplankton community in the northwestern Pacific Ocean. *Ocean Polar Res.* **2011**, *33*, 337–347. [CrossRef]

Journal of
Marine Science and Engineering

Article

Seasonal Compositions of Size-Fractionated Surface Phytoplankton Communities in the Yellow Sea

Yejin Kim [1], Seok-Hyun Youn [2], Hyun-Ju Oh [2], Huitae Joo [2], Hyo-Keun Jang [1], Jae-Joong Kang [1], Dabin Lee [1], Naeun Jo [1], Kwanwoo Kim [1], Sanghoon Park [1], Jaehong Kim [1] and Sang-Heon Lee [1,*]

[1] Department of Oceanography, Pusan National University, Busan 46241, Korea
[2] Oceanic Climate & Ecology Research Division, National Institute of Fisheries Science, Busan 46083, Korea
* Correspondence: sanglee@pusan.ac.kr; Tel.: +82-51-510-2256

Abstract: Little information on the phytoplankton community in the Yellow Sea (YS)—especially size-fractionated phytoplankton—is currently available, in comparison to the various physicochemical studies in the literature. Using high-performance liquid chromatography (HPLC), size-fractionated phytoplankton communities were seasonally investigated in the YS in 2019. In the study period, diatoms (55.0 ± 10.2%) and cryptophytes (16.9 ± 9.3%) were the dominant groups. Due to the recent alteration in inorganic nutrient conditions reported in the YS, the contribution of diatoms was lower than in previous studies. The large-sized phytoplankton group (>20 μm) was dominated mostly by diatoms (89.0 ± 10.6%), while the small-sized phytoplankton group (<20 μm) was also dominated by diatoms (41.9 ± 9.1%), followed by cryptophytes (19.2 ± 9.8%). The contributions of small-sized diatoms (<20 μm) have been overlooked in the past, as they are difficult to detect, but this study confirms significant amounts of small-sized diatoms, accounting for 62.3% of the total diatoms in the YS. This study provides an important background for assessing the seasonal variations in different-sized diatom groups in the YS. Further detailed studies on their potential ecological roles should be conducted, in order to better understand marine ecosystems under future warming scenarios.

Keywords: Yellow Sea; phytoplankton; HPLC; diatoms; size fraction

Citation: Kim, Y.; Youn, S.-H.; Oh, H.-J.; Joo, H.; Jang, H.-K.; Kang, J.-J.; Lee, D.; Jo, N.; Kim, K.; Park, S.; et al. Seasonal Compositions of Size-Fractionated Surface Phytoplankton Communities in the Yellow Sea. *J. Mar. Sci. Eng.* **2022**, *10*, 1087. https://doi.org/10.3390/jmse10081087

Academic Editor: Carmela Caroppo

Received: 28 June 2022
Accepted: 5 August 2022
Published: 8 August 2022

Publisher's Note: MDPI stays neutral with regard to jurisdictional claims in published maps and institutional affiliations.

1. Introduction

Phytoplankton, as primary producers, contribute half of global primary production and play a central role in marine ecosystems [1,2]. In the ocean, the size and the community structure of phytoplankton are affected by light intensity, water temperature, nutrients, and other physicochemical properties [3,4]. Generally, the dominance of micro-sized phytoplankton (>20 μm) has been associated with nutrient-rich environments, while the dominance of pico-sized phytoplankton (<2 μm) has been associated with stratified and oligotrophic waters [5–7]. Recently increasing temperatures and stratification due to global warming have caused a shift in phytoplankton size structures towards smaller phytoplankton [8,9]. The increase in small phytoplankton could result in lower total primary production [10]. Moreover, the efficiency of the biological pump [4] and the food web structure [11] may be altered. Therefore, phytoplankton community size structure variation can serve as an indicator for the response of phytoplankton to environmental changes [12,13].

The Yellow Sea (YS) is a temperate semi-enclosed marginal sea in the West Pacific Ocean, surrounded by Korea to the east and China to the west. The YS is a highly productive ecosystem and is recognized as an important global fishery resource region [14]. The YS is significantly influenced by its unique topography, hydrological and biochemical characteristics, nonlinear tidal effect, and frequent human activity [15,16]. The spatial-temporal distribution of the phytoplankton community in the YS is affected by this complex environment [17]. Studies on the composition of phytoplankton communities have traditionally been carried out using optical microscopy, but this method has limited ability to detect pico- and nano-sized phytoplankton [18,19]. To address this issue, new approaches based on pigment analysis using High-Performance Liquid Chromatography (HPLC) have been developed. HPLC provides an opportunity to estimate small-sized phytoplankton, which can be difficult to identify when conducting microscopic analysis [20]. Moreover, HPLC pigment analysis can produce reliable and consistent data in a short period of time.

Previous studies on the phytoplankton community in the YS have mainly focused on the morphological classification of large-sized phytoplankton [17,21,22]. Several studies have investigated the relationship between phytoplankton communities and environmental factors, but they have been limited to fragmentary seasons [23,24]. In addition, some studies assessing the size-fractionated phytoplankton biomass in the YS have been undertaken [25,26], but no study has considered size-fractionated phytoplankton communities in this area. The main objectives of this study are to (1) identify major environmental factors controlling the seasonal variations in the phytoplankton communities and (2) classify the seasonal pattern of different-sized phytoplankton groups in the YS.

2. Materials and Methods

2.1. Water Sampling and Study Area

Seasonal sampling was conducted in the Yellow Sea on an R/V Tamgu 8 during 4 seasons in 2019 (22 February to 5 March, 3–12 April, 14–25 August, and 6–16 October, representing winter, spring, summer, and autumn, respectively; see Figure 1 and Table 1). At the surface, water samples were collected using Niskin bottles to examine dissolved inorganic nutrients ($NO_3 + NO_2$, NH_4, SiO_2, and PO_4 ($n = 29$)), chlorophyll-a (total ($n = 29$) and size-fractionated ($n = 87$)) concentrations, and pigment (total ($n = 29$) and size-fractionated ($n = 58$)) concentrations on a conductivity–temperature–depth (CTD)/rosette sampler (SBE 911 plus, Seabird Electronics Inc., Bellevue, WA, USA). Physical properties (temperature and salinity) were also measured using the CTD/rosette sampler. The stability index (SI) for the euphotic zone was calculated by dividing the difference in density (sigma-t) between the surface and the bottom of the euphotic zone by the depth of the euphotic zone [27].

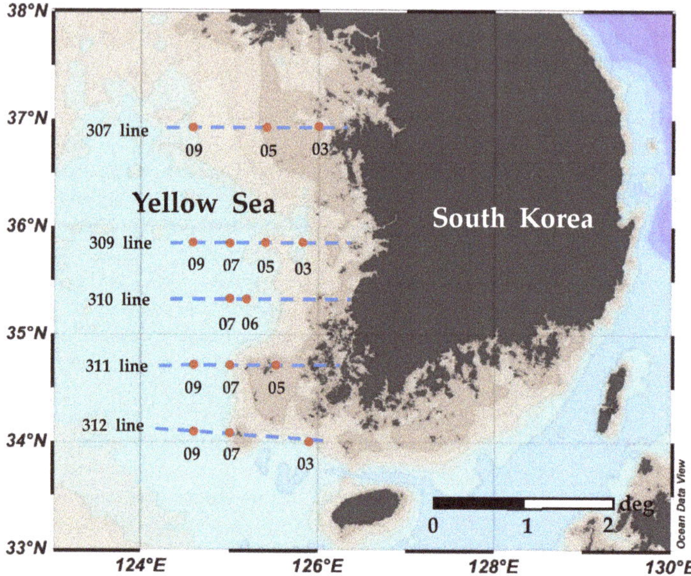

Figure 1. Sampling locations in the YS, 2019. The blue dotted lines are managed by the serial oceanographic observation project of the National Institute of Fisheries Science (NIFS) in Korea.

Table 1. Sampling locations and environmental parameters of the study sites in the YS, 2019.

Month	Station	Latitude (°N)	Longitude (°E)	Bottom Depth	Euphotic Depth	Stability Index	T (°C)	S (psu)	$NO_2 + NO_3$ (μM)	NH_4 (μM)	PO_4 (μM)	SiO_2 (μM)
Winter	307-05	36.92	125.42	54	11	0.00	4.88	31.34	10.04	0.33	0.68	8.27
	309-05	35.85	125.40	69	11	0.01	6.67	32.39	6.42	0.35	0.52	7.41
	309-09	35.85	124.59	82	11	0.00	7.60	32.36	5.52	0.38	0.40	8.23
	311-05	34.72	125.52	75	5	0.00	5.88	32.32	6.88	0.81	0.58	7.70
	311-09	34.72	124.59	89	14	0.01	8.27	32.35	6.44	0.41	0.41	7.67
	312-03	34.00	125.90	85	22	0.01	6.91	32.48	6.29	0.53	0.51	7.95
	312-07	34.10	125.00	92	5	0.00	8.14	32.50	4.65	0.39	0.41	5.60
Spring	307-03	36.92	126.00	37	8	0.01	5.76	31.60	11.95	0.59	0.63	8.32
	307-09	36.92	124.57	67	22	0.01	7.77	32.32	4.82	0.43	0.40	6.74
	309-03	35.85	125.82	54	16	0.00	7.22	32.12	5.37	1.06	0.73	3.74
	309-09	35.85	124.59	82	22	0.01	8.71	32.36	1.26	0.34	0.56	2.25
	310-07	35.34	125.00	82	22	0.00	8.66	32.37	3.25	0.32	0.34	7.56
	311-07	34.72	125.00	90	22	0.00	8.42	32.34	6.16	0.30	0.37	8.86
	312-09	34.09	124.60	89	22	0.00	8.81	32.50	5.51	0.32	0.33	6.51
Summer	307-09	36.92	124.57	67	24	0.13	26.81	31.93	1.35	0.17	0.05	2.82
	309-03	35.85	125.82	54	24	0.13	26.27	31.79	6.18	0.33	0.11	3.63
	309-09	35.85	124.59	82	46	0.11	27.65	32.07	1.10	0.30	0.06	3.29
	310-06	35.34	125.20	72	43	0.10	27.52	32.40	0.68	0.30	0.04	2.25
	311-05	34.72	125.52	75	27	0.03	21.99	32.17	0.92	0.30	0.09	4.29
	311-07	34.72	125.00	90	27	0.10	27.52	32.17	0.76	0.41	0.08	2.18
	312-03	34.00	125.90	85	19	0.09	24.07	31.72	0.51	0.55	0.11	3.58
	312-09	34.09	124.60	89	30	0.10	27.47	31.49	1.31	0.37	0.05	2.19
Autumn	307-03	36.92	126.00	37	11	0.00	21.43	31.88	3.57	0.59	0.57	6.22
	307-05	36.92	125.42	54	16	0.00	20.24	32.06	0.90	0.50	0.29	5.95
	307-09	36.92	124.57	67	16	0.00	20.23	31.83	1.07	0.47	0.08	5.56
	309-03	35.85	125.82	54	38	0.00	19.42	31.72	3.25	0.35	0.39	7.65
	309-07	35.86	125.00	66	22	0.01	20.54	31.51	1.16	0.38	0.04	6.58
	311-05	34.72	125.52	75	19	0.02	20.49	31.81	5.56	0.34	0.29	9.96
	312-07	34.10	125.00	92	38	0.03	21.28	32.01	0.92	0.29	0.04	3.26

2.2. Dissolved Inorganic Nutrients

To determine the dissolved inorganic nutrients ($NO_3 + NO_2$, NH_4, SiO_2, and PO_4), 100 mL of seawater was filtered through GF/F filters (07 μm; Whatman, Maidstone, UK) and the filtrate was immediately frozen at -20 °C. Concentrations of nutrients were analyzed using an automatic analyzer (Quattro, Seal Analytical, Norderstedt, Germany) at the National Institute of Fisheries Science (NIFS), Korea. Dissolved inorganic phosphate (DIP) was assessed on the basis of PO_4 and dissolved inorganic nitrogen (DIN) concentrations were calculated as the sum of NH_4, NO_2, and NO_3.

2.3. Chlorophyll-a Concentration

For total chlorophyll-a concentration, 300 mL of seawater was filtered through a 25 mm glass fiber filter (GF/F; Whatman). For the size-fractionated chlorophyll-a concentration, 500 mL of seawater was filtered through membrane filters with pore sizes of 20 and 2 μm and 47 mm GF/F, in sequence. After filtration, the filtered samples were wrapped with aluminum foil to prevent photolysis and stored at -80 °C in a freezer until further analysis. Chlorophyll-a was extracted with 90% acetone for 20–24 h in the dark at 4 °C, and the concentration was measured using a fluorometer (Turner Designs 10AU) [28].

2.4. High-Performance Liquid Chromatography Analysis

Water samples (1 L) were filtered through 47 mm GF/F for total pigment analysis, and 4–10 L of seawater was filtered through membrane filters with a pore size of 20 μm and 47 mm GF/F, in sequence, in order to perform size-fractionated pigment analysis. The samples for HPLC analysis were wrapped in aluminum foil to avoid degradation and stored at -80 °C in a freezer until analysis. In the laboratory, the frozen filters were extracted using 5 mL of 100% acetone at 4 °C over 20–24 h, and canthaxanthin was used as an internal standard [29]. After extraction, the extract was passed through a syringe filter (Polytetrafluoroethylene, PTFE; 0.2 μm, Hydrophobic, Advantec, Japan) and centrifuged (3500 rpm) for 10 min. Then, 1 mL of the extract was mixed with 300 μL of distilled water and analyzed using high-performance liquid chromatography (HPLC; Agilent Infinite 1260, Santa Clara, CA, USA) for qualitative and quantitative evaluation of pigments. Pigment concentrations were calculated using the equation suggested in [30]. The relative contributions of the total and size-fractionated phytoplankton community composition were estimated through the CHEMTAX program, as described in [31]. To separate the eight phytoplankton communities (i.e., diatoms, dinoflagellates, chrysophytes, cryptophytes, chlorophytes, prasinophytes, cyanobacteria) in the CHEMTAX program, we used the initial pigment ratios derived from [32].

To estimate the relative contributions of three pigment-based size classes (pico-, nano-, and micro-sized phytoplankton), diagnostic pigment analyses (DPAs) were carried out [33–36]. DPA was calculated based on the concentration of the seven diagnostic pigments—fucoxanthin (Fuco), peridinin (Peri), 19′-hexanoyloxyfucoxanthin (19Hexfuco), 19′-butanoyloxy-fucoxanthin (19Butfuco), alloxanthin (Allo), chlorophyll-b (Chlb), and zeaxanthin (Zea)—and the relative contributions of the three pigment-based size classes were calculated according to [37].

2.5. Statistical Analysis

Statistical analysis was performed using SPSS (version 24.0, SPSS Inc., Chicago, IL, USA) for Spearman's correlation, t-test, one-way ANOVA, and Kruskal–Wallis nonparametric one-way analysis of variance (ANOVA). Spearman rank-order correlation analysis was used to identify the environmental factors in relation to the concentrations of size-fractionated diatoms. The t-test was performed to identify the significant differences between the relative proportions of size-fractionated chlorophyll-a concentrations, as measured by fluorometer and pigment analysis. One-way ANOVA and Kruskal–Wallis (at an alpha value of <0.05) were applied to identify significant differences in physical properties for each season. Canonical correspondence analysis (CCA) was performed using CANOCO for Windows 4.5 (Biometris, Wageningen, The Netherlands), in order to investigate the relationships between the total phytoplankton community and environmental factors (i.e., temperature, salinity, stability index, and dissolved inorganic nutrients ($NO_3 + NO_2$, NH_4, SiO_2, and DIP)).

3. Results and Discussion

3.1. Physicochemical Characteristics in the Environment

The hydrographic conditions are presented in Table 1. The average surface temperature ranged from 6.91 ± 1.23 °C to 26.16 ± 2.06 °C, with a significant increase from spring to summer. On the other hand, the seasonal average salinity ranged from 31.83 ± 0.18 psu to 32.25 ± 0.41 psu, and significant seasonal changes were not observed (Kruskal–Wallis test, $p > 0.05$). The stratification intensity estimated from the stability index was significantly higher in summer than in other seasons (Kruskal–Wallis test, $p < 0.05$). The ranges of $NO_3 + NO_2$, NH_4, SiO_2, and DIP concentrations during the study period were 0.51–11.95, 0.17–1.06, 2.18–9.96, and 0.04–0.73 μM, respectively. The inorganic nutrient conditions in this study were within the range reported previously in the YS [25,38]. The average inorganic nutrient concentrations showed seasonal differences (Kruskal–Wallis test, $p < 0.05$), except for NH_4 (Kruskal–Wallis test, $p > 0.05$). The mean concentration of NH_4 showed no significant seasonal differences (0.46 ± 0.17 μM, 0.51 ± 0.29 μM, 0.38 ± 0.12 μM, and 0.42 ± 0.11 μM for winter, spring, summer, and autumn, respectively). The mean concentrations of inorganic nutrients, except for NH_4, were highest in winter (6.60 ± 1.69 μM, 0.5 ± 0.10 μM, and 7.55 ± 0.91 μM for $NO_3 + NO_2$, SiO_2, and DIP, respectively), whereas they were significantly lower in summer (1.60 ± 1.87 μM, 0.08 ± 0.03 μM, and 3.03 ± 0.79 μM for $NO_3 + NO_2$, SiO_2, and DIP, respectively) than in winter and spring (Kruskal–Wallis test, $p < 0.05$). This is because the strong water column stratification during summer prevents nutrient transfer from deep to surface water [25,39,40]. The depletion of surface water nutrients may be due to thermal stratification, as this study also showed a high summer SI (Kruskal–Wallis test, $p < 0.01$). In contrast, the water column was homogeneously mixed, supplying nutrients from the depths to the surface in the winter.

3.2. Total Chlorophyll-a Concentrations and Total Phytoplankton Community

The highest total HPLC chlorophyll-a concentration was detected in spring (2.76 ± 1.82 µg/L), followed by autumn (0.98 ± 0.40 µg/L), winter (0.55 ± 0.18 µg/L), and summer (0.29 ± 0.18 µg/L), respectively (Figure 2). Warmer temperatures, increased sunlight, and weak stratification triggers the spring phytoplankton bloom in the YS, resulting in higher chlorophyll-a concentrations in spring compared with winter and summer (Kruskal–Wallis test; $p < 0.05$) [25,41], whereas lower chlorophyll concentration was observed in summer compared with spring and autumn (Kruskal–Wallis test; $p < 0.05$) due to nutrient restriction caused by strong stratification [25].

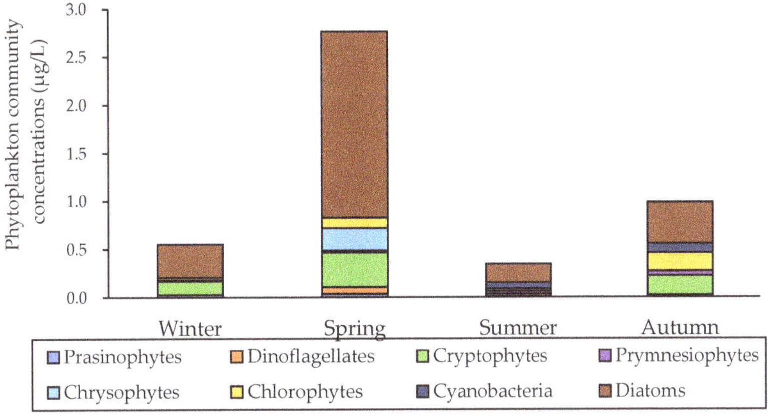

Figure 2. Phytoplankton community concentrations (µg/L) in the YS during the study period in 2019.

The total phytoplankton community results, calculated by the CHEXTAX program, are presented in Figure 2. During the study period, the major communities of total phytoplankton were diatoms (55.0 ± 10.2%), cryptophytes (16.9 ± 9.3%), cyanobacteria (9.5 ± 12.9%), and chlorophytes (8.4 ± 6.1%). Diatoms were predominant in all seasons (one-way ANOVA, $p > 0.05$), whereas the relative proportions of other phytoplankton communities changed seasonally except prasinophytes and chrysophytes (Kruskal–Wallis test; $p < 0.05$). The relative proportions of diatoms accounted for more than half of the total phytoplankton community in spring (65.3 ± 14.6%) and winter (61.9 ± 17.4%); meanwhile, in summer and autumn, the relative proportions of diatoms accounted for 48.9 ± 26.7%, and 43.9 ± 13.5%, respectively. The relative proportions of cryptophytes were high in winter (26.1 ± 12.0%), followed by autumn (20.4 ± 8.05%) and spring (16.8 ± 7.9%), whereas cryptophytes were rarely observed in summer (Kruskal–Wallis test; $p < 0.05$). In Bohai bay of the Bohai Sea, diatom dominates more than 80%, and the contribution of other communities are less than 6% in spring [42]. The relative proportions of cyanobacteria and prymnesiophytes were high in summer (27.3 ± 26.0% and 12.0 ± 7.6%, respectively) and autumn (10.5 ± 9.4% and 5.1 ± 3.2%, respectively), whereas they were rarely observed in winter and spring (Kruskal–Wallis test; $p < 0.05$). In Daya Bay, which is located in the west of the YS, dinoflagellates were dominant in spring and observed throughout the four seasons [43]. In contrast, dinoflagellates were hardly observed in this study except in spring (Kruskal–Wallis test; $p < 0.05$). The relationships between the total phytoplankton communities and environmental variables are presented in Figure 3. Taken together, axis 1 and axis 2 in the CCA explained 78.3% of the correlation between the total phytoplankton communities and environmental variables in the YS in 2019. Diatoms were situated adjacent to the center of the CCA ordinates and close to the arrows including salinity, DIP, and NH_4. Cyanobacteria and Prymnesiophytes were near the SI, N/P ratio, and temperature arrows, while Cryptophytes and Prasionophytes were closely related to the SiO_2 arrow.

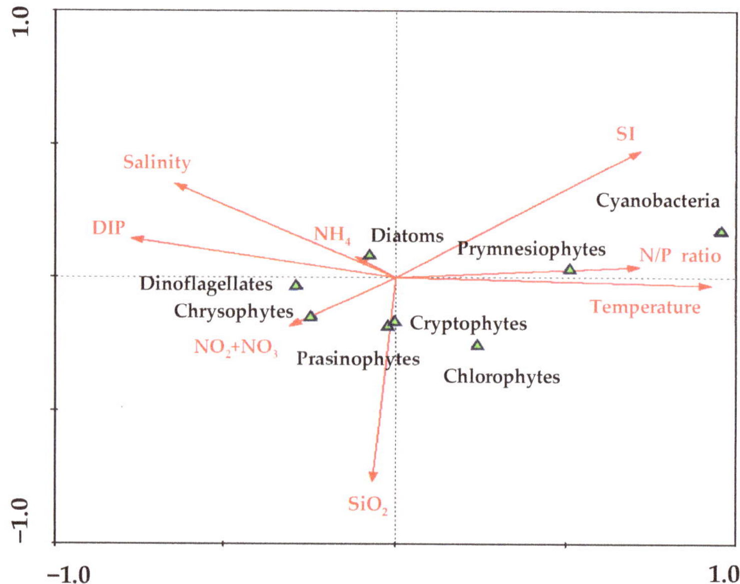

Figure 3. Canonical component analysis (CCA) to determine relationships among the environmental factors and phytoplankton communities.

The diatoms formed a dominant community in the YS, but the diatom composition in this study (55%) was distinctly lower than those reported previously in 1985 (80.3%) and 2000 (67.5%) [21]. According to [44,45], the positive trend of DIN concentration and the negative trend of DIP and SiO_2 concentrations have gradually increased the N/P ratio in the YS, gradually shifting to a P-limited environment. Although the P-limited environment was observed only in summer and autumn in this study, the gradually increasing N/P ratio and limited P might have caused the transition from a diatom-dominant environment to a small-sized flagellate-dominant environment [10,38,46], as DIP is a more important limiting factor for the growth of diatoms than other flagellates [47]. Indeed, the proportion of cryptophytes—a type of flagellate—was higher (16.9%) in this study than in previous studies in the YS [17,32]. Cryptophytes are known to be adaptable to a variety of environmental conditions, from stratified and well-lit environments to cool, well-mixed and light-limited environments [48]. Consequently, diatoms negatively respond to declining DIP and P-limitation, whereas cryptophytes are resistant to these conditions [49]. Therefore, the cryptophyte community could flourish, whereas the diatom community decreased under the current nutrient environment in the YS. Cyanobacteria are considered to be an important group in phytoplankton communities during summer in this study. Cyanobacteria have previously been reported to dominate at the surface layer in the YS during summer [24,50]. At the surface, low nutrient concentrations limit the growth of phytoplankton [51,52]. However, these conditions are favorable for the growth of small-sized cyanobacteria, in comparison to other phytoplankton communities [53,54]. In addition, many cyanobacteria have a competitive advantage under high temperatures [55,56]. Therefore, cyanobacteria were observed more under the high temperature and low nutrient conditions in summer, compared to other seasons. Unlike other phytoplankton communities, chlorophytes did not show a clear correlation with environmental factors such as water temperature, salinity, nutrients, and SI during this study (Figure 3). Therefore, the chlorophytes may have been more affected by other factors, such as competition with other phytoplankton groups and predation pressure, rather than physical and chemical environmental factors.

3.3. Size-Fractionated Chlorophyll-a Concentration and Phytoplankton Community Structure

Unlike the Southern Central YS dominated by pico-sized phytoplankton throughout the four seasons [25], the phytoplankton size structure in study area varied seasonally ($p < 0.05$; Kruskal–Wallis test) (see Figure 4). Based on the fluorometric size-fractionated chlorophyll-*a* concentrations in this study, pico-sized phytoplankton (0.7–2 µm) mainly dominated in summer (42.5 ± 14.1%) and autumn (51.0 ± 22.6%). The contribution of micro-sized phytoplankton (>20 µm) to the total chlorophyll-a was comparatively highest in spring (47.6 ± 27.1%), whereas the nano-sized phytoplankton (2–20 µm) contribution was comparatively dominant in winter (46.4 ± 9.9%). The relatively high contribution of micro-sized phytoplankton in spring compared to other seasons is similar to those of previous studies [38].

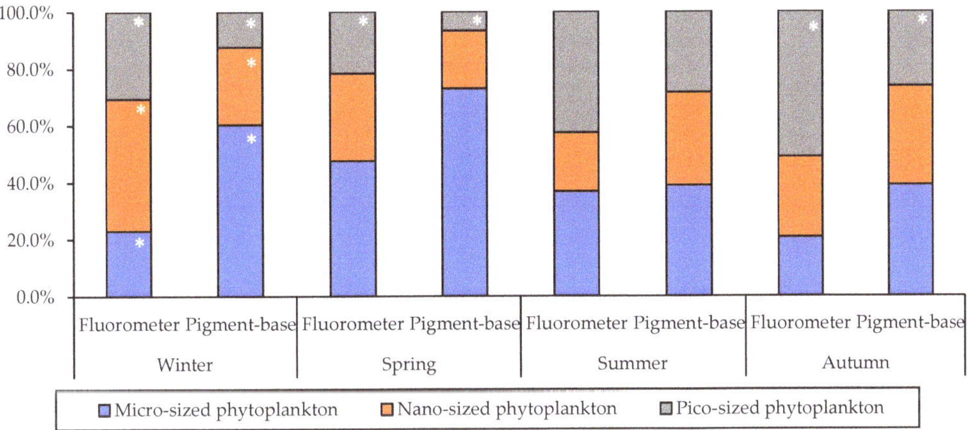

Figure 4. The relative contributions of size-fractionated chlorophyll-a concentrations, as determined by two methods. The white star (*) implies a significant difference at the level of $p < 0.05$.

Previously, Choi [57] found that micro-sized phytoplankton were dominant (approximately 50%) during the winter and spring seasons. In contrast, the nano-sized phytoplankton were dominant, instead of micro-sized phytoplankton, during winter in this study. Furthermore, the proportions of nano- and pico-sized phytoplankton during autumn were distinctly higher in this study (79.3%) than in 1992 (64.4%) [58]. Higher contributions of small-sized phytoplankton (<2 µm) to the chlorophyll-a concentration and primary production have consistently been observed in the YS recently [10]. However, the size-fractionated phytoplankton proportions vary largely seasonally, spatially, and inter-annually [30]. Further research should be carried out to verify the increasing contribution of small-sized phytoplankton in the YS.

In comparison to the fluorometric results for the relative contributions of size-fractionated chlorophyll-*a* concentrations, pigment-based size chlorophyll-*a* concentrations were obtained based on DPA, as presented in Figure 4. Micro-sized phytoplankton were dominant during the study period, especially winter and spring (60.3 ± 16.6% and 73.0 ± 13.8%, respectively). In summer and autumn, the contributions of micro- and nano-sized phytoplankton were collectively dominant, but the contribution of micro-sized phytoplankton (summer 38.9 ± 28.3%; autumn 39.0 ± 20.4%) was somewhat higher than that of nano-sized phytoplankton (summer

32.6 ± 20.5%; autumn 34.6 ± 15.2%). Comparing the two different methods, the relative contributions of phytoplankton size classes were significantly different in winter and autumn (*t*-test, $p < 0.05$). While micro-sized phytoplankton were the predominant community in the DPA results in winter, nano-sized phytoplankton were found to be dominant, based on the fluorometer analysis results. In addition, pico-sized phytoplankton were a predominant community in a fluorometer results, whereas their composition was lowest in the DPA results in autumn. These discrepancies may have been caused by the pigment-based size fractionation of DPA, which could lead to over-estimation of micro-sized phytoplankton and under-estimation of the contribution of nano-sized phytoplankton, as further discussed below.

The results for the size-fractionated phytoplankton communities calculated by the CHEXTAX program are presented in Figure 5. The large-sized phytoplankton group (>20 μm) was dominated by diatoms (89.0 ± 10.6%) during the study period. In particular, large-sized diatoms accounted for the majority (99.4 ± 1.0%) in winter, while the proportion decreased from winter to fall. Among the large-sized phytoplankton group, cyanobacteria were first detected in summer (14.4 ± 27.5%), but their relative proportion increased in autumn (20.9 ± 29.7%). In comparison, other phytoplankton communities were rarely observed in the large-sized phytoplankton group, with contribution rates of less than 1%. Compared to the large-sized phytoplankton group, various phytoplankton communities were observed in the small-sized phytoplankton group (<20 μm). Diatoms were also the dominant phytoplankton community (41.9 ± 9.1%) in the small-sized phytoplankton group. The relative proportion of small-sized diatoms was highest in winter (54.7 ± 16.0%), followed by spring (42.1 ± 10.8%), summer (36.2 ± 12.4%), and autumn (34.6 ± 16.7%). The relative proportion of small-sized cryptophytes was high in winter (28.6 ± 10.8%) and low in summer (5.8 ± 9.2%). The proportion of small-sized chlorophytes was low in winter (8.0 ± 5.1%) and increased along the seasons. The small-sized cyanobacteria were first observed in summer (25.2 ± 18.1%) and decreased to 9.4 ± 7.6% in autumn.

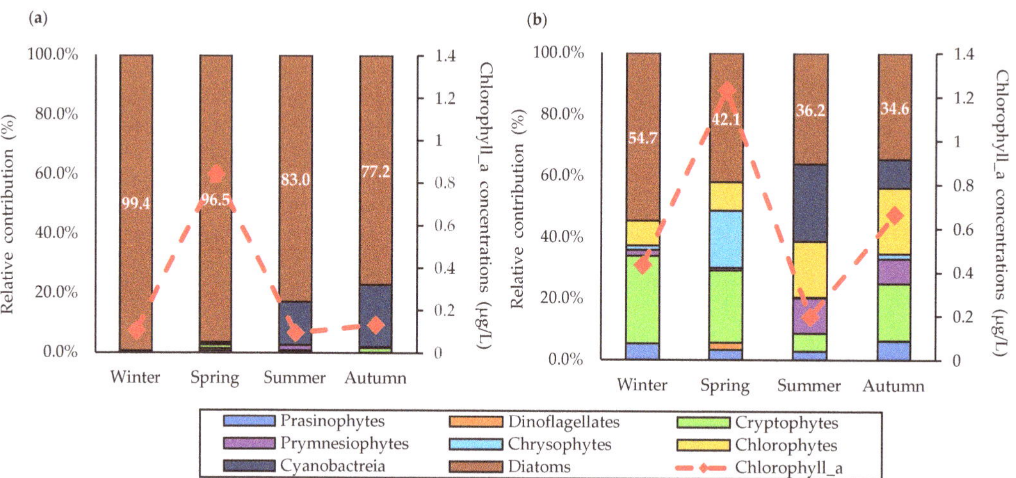

Figure 5. Relative contributions of the size-fractionated phytoplankton community in the YS during the study period in 2019: (**a**) large-sized group; and (**b**) small-sized group.

In contrast to the assumption for the DPA that diatoms mainly make up the micro-sized phytoplankton [35,43,59], a significant presence of diatoms (41.9 ± 9.1%) was found in the small-sized phytoplankton group based on the results from the size-fractionated pigment analysis in this study. In particular, the over-estimated contribution of micro-sized phytoplankton was conspicuous in winter, where the small-sized diatoms accounted for more than half of the total diatom community (72.4 ± 12.1%). In [60], it was reported that many diatoms belong to the nano-sized phytoplankton group, and a few species even overlapped with the pico-sized phytoplankton class. According to [21], the most dominant diatom species in the YS are *Chaetoceros* sp., *Cylindrocystis* sp., *Proboscia*, and *Skeletonema costatum*. In particular, *Proboscia* and *Skeletonema costatum* are nano-sized diatoms with a diameter of less than 20 μm [33]. Therefore, the existence of these nano-sized diatoms should be considered when conducting DPA, as DPA requires a careful interpretation of the size-fractionated phytoplankton biomass.

3.4. Seasonal Variation in the Size Structure of Diatoms

During the study period, the concentrations of large-sized diatoms (0.82 ± 0.65 μg/L) and small-sized diatoms (0.54 ± 0.35 μg/L) were significantly higher in spring than other seasons (*t*-test, $p < 0.05$). In contrast, similar concentrations of large- and small-sized diatoms were observed in other seasons (Figure 6).

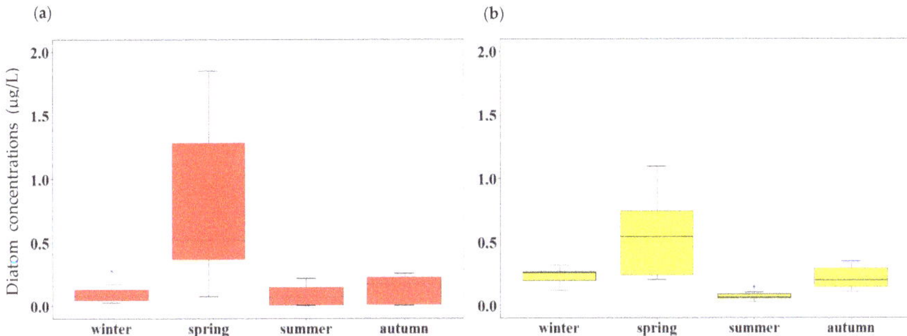

Figure 6. The concentrations of size-fractionated diatoms: (**a**) large-sized diatoms; and (**b**) small-sized diatoms. An asterisk (*) indicates an outlier.

In the YS, diatom blooming is observed during spring [32,41,61]. The proportion of small-sized diatoms in the total diatoms during the study period varied between 14.8% and 97.9% (Figure 7). The seasonally averaged proportion of small-sized diatoms was highest in winter (72.4 ± 12.1%), followed by autumn (70.9 ± 23.0%), summer (61.6 ± 27.4%), and spring (44.4 ± 23.7%). Overall, small-sized diatoms were dominant in the YS (62.3 ± 12.9%) during our study period.

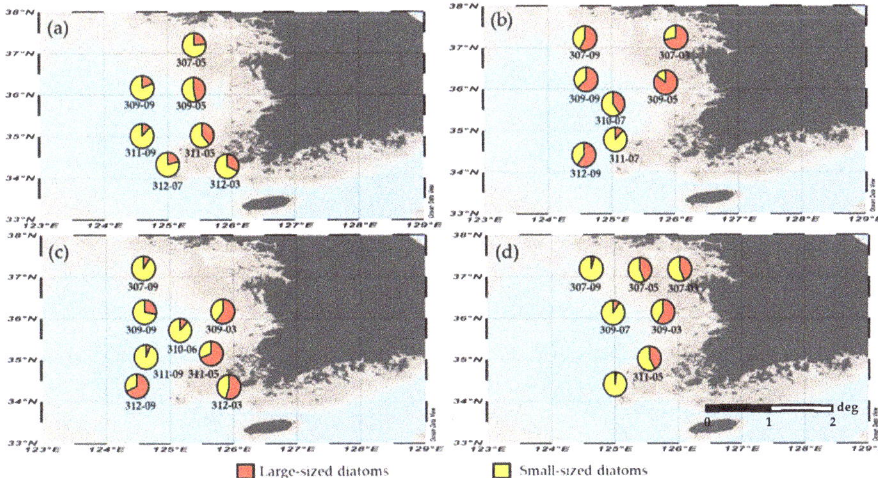

Figure 7. Relative contributions of size-fractionated diatoms during the study period in 2019: (**a**) winter; (**b**) spring; (**c**) summer; and (**d**) autumn.

In [62], it was reported that environmental changes (e.g., global warming and human impacts) decrease the cell size structure of phytoplankton, especially centric diatoms. In agreement with this, [63,64] reported that the size of the diatoms decreases as the water temperature increases. The annual SST in the YS has gradually increased over the past few decades (from 10.5 °C in 1976 to 13.5 °C in 2000) [44], and an increasing trend of 0.0135 °C/yr was observed from 1982 to 2017 [65]. Therefore, the increasing SST could affect the transition in the phytoplankton size structure. Furthermore, subsequent increasing stratification and nutrient limitations could affect cell size reductions in phytoplankton communities [66]. Therefore, if the water temperature continues to increase and stratification is strengthened in the YS, the proportion of small-sized diatoms may gradually increase. In small-sized phytoplankton-dominated ecosystems, the total primary production and energy transfer efficiency could be altered [38,67]. A further monitoring programs for phytoplankton communities, especially considering small-sized phytoplankton groups, should be conducted in the YS with respect to ongoing environmental changes.

4. Summary and Conclusions

This study is the first to report seasonal variations in the size-fractionated phytoplankton community structure in the YS. Overall, diatoms were dominant (55%) in the YS, with the community size being strongly associated with salinity and the concentrations of DIP and NH_4, based on CCA analysis. Compared to previous studies, the contribution of diatoms was lower and the contribution of cryptophytes was higher in this study, which may be due to the recent shift in inorganic nutrient conditions in the YS [44,45]. The size-fractionated phytoplankton community results demonstrated that the large-sized phytoplankton group was largely dominated by diatoms (89.0 ± 10.6%), whereas the small-sized phytoplankton group was dominated by small-sized diatoms (41.9 ± 9.1%) and cryptophytes (19.2 ± 9.8%). Until recently, the importance of small-sized diatoms has been overlooked, as it is difficult to detect and identify them. However, this study showed that small-sized diatoms contributed greatly to the total diatoms in the YS, accounting for 62.3%. Environmental changes due to continuous climate warming are expected to decrease the cell size of phytoplankton—especially centric diatoms—which could largely affect the quantity and quality of food sources in marine ecosystems. This study provides an important baseline for understanding the seasonal variations in different size groups of the phytoplankton community, especially different sizes of diatoms. However, this study was conducted only at the surface layer, and specific identification of diatom species was lacking. Therefore, further detailed studies should be conducted in order to better understand the potential ecological roles of small-sized diatoms.

Author Contributions: Conceptualization, Y.K. and S.-H.L.; methodology, Y.K. and J.-J.K.; formal analysis, Y.K.; investigation, Y.K., H.-K.J., J.-J.K., D.L., N.J., K.K., S.P. and J.K.; data curation, Y.K.; writing—original draft preparation, Y.K.; writing—review and editing, S.-H.L.; visualization, Y.K.; supervision, S.-H.L.; project administration, S.-H.Y. and S.-H.L.; funding acquisition, S.-H.Y., H.-J.O., H.J. and S.-H.L. All authors have read and agreed to the published version of the manuscript.

Funding: This research was supported by the "Development of assessment technology on the structure variations in marine ecosystem (R2022073)" funded by the National Institute of Fisheries Science (NIFS), Korea.

Institutional Review Board Statement: Not applicable.

Informed Consent Statement: Not applicable.

Data Availability Statement: Not applicable.

Acknowledgments: We appreciate the captains and crew of R/V Tamgu 8 for their assistance in collecting our samples. We would also like to thank the researchers in the NIFS for their assistance with sample analysis. We thank the anonymous reviewers who greatly improved an earlier version of manuscript.

Conflicts of Interest: The authors declare no conflict of interest.

References

1. Falkowski, P.G.; Barber, R.T.; Smetacek, V. Biogeochemical controls and feedbacks on ocean primary production. *Science* **1998**, *281*, 200–206. [CrossRef] [PubMed]
2. Behrenfeld, M.J.; Randerson, J.T.; McClain, C.R.; Feldman, G.C.; Los, S.O.; Tucker, C.J.; Falkowski, P.G.; Field, C.B.; Frouin, R.; Esaias, W.E. Biospheric primary production during an enso transition. *Science* **2001**, *291*, 2594–2597. [CrossRef] [PubMed]
3. Winder, M.; Sommer, U. Phytoplankton response to a changing climate. *Hydrobiologia* **2012**, *698*, 5–16. [CrossRef]
4. Froneman, P.; Pakhomov, E.; Balarin, M.G. Size-fractionated phytoplankton biomass, production and biogenic carbon flux in the eastern Atlantic sector of the Southern Ocean in late austral summer 1997–1998. *Deep-Sea Res. II: Top. Stud. Oceanogr.* **2004**, *51*, 2715–2729. [CrossRef]
5. Chisholm, S.W. Phytoplankton Size. In *Primary Productivity and Biogeochemical Cycles in the Sea*; Falkowski, P.G., Woodhead, A.D., Eds.; Springer: New York, NY, USA, 1992.
6. Agustí, S.; González-Gordillo, J.I.; Vaqué, D.; Estrada, M.; Cerezo, M.I.; Salazar, G.; Gasol, J.M.; Duarte, C.M. Ubiquitous healthy diatoms in the deep sea confirm deep carbon injection by the biological pump. *Nat. Commun.* **2015**, *6*, 1–8. [CrossRef]
7. Falkowski, P.G.; Laws, E.A.; Barber, R.T.; Murray, J.W. Phytoplankton and Their Role in Primary, New, and Export Production. In *Ocean Biogeochemistry*; Fasham, M.J.R., Ed.; Springer: Berlin/Heidelberg, Germany, 2003; pp. 99–121.
8. Peter, K.H.; Sommer, U.J.P.o. Phytoplankton cell size reduction in response to warming mediated by nutrient limitation. *PLoS ONE* **2013**, *8*, e71528K. [CrossRef]
9. Morán, X.A.G.; López-Urrutia, Á.; Calvo-Díaz, A.; LI, W.K.W. Increasing importance of small phytoplankton in a warmer ocean. *Glob. Chang. Biol.* **2010**, *16*, 1137–1144. [CrossRef]
10. Jang, H.K.; Kang, J.J.; Lee, J.H.; Kim, M.; Ahn, S.H.; Jeong, J.Y.; Yun, M.S.; Han, I.S.; Lee, S.H. Recent Primary Production and Small Phytoplankton Contribution in the Yellow Sea during the Summer in 2016. *Ocean Sci. J.* **2018**, *53*, 509–519. [CrossRef]
11. Finkel, Z.V. Does phytoplankton cell size matter? The evolution of modern marine food webs. In *Evolution of Primary Producers in the Sea*; Falkowski, P.G., Knoll, A.H., Eds.; Academic Press: New York, NY, USA, 2007; pp. 333–350.
12. F Finkel, Z.V.; Beardall, J.; Flynn, K.J.; Quigg, A.; Rees, T.A.V.; Raven, J.A. Phytoplankton in a changing world: Cell size and elemental stoichiometry. *J. Plankton Res.* **2010**, *32*, 119–137. [CrossRef]
13. G Gin, K.Y.-H.; Lin, X.; Zhang, S. Dynamics and size structure of phytoplankton in the coastal waters of Singapore. *J. Plankton Res.* **2000**, *22*, 1465–1484. [CrossRef]
14. Yoo, S.; An, Y.-R.; Bae, S.; Choi, S.; Ishizaka, J.; Kang, Y.-S.; Kim, Z.G.; Lee, C.; Lee, J.B.; Li, R.; et al. Status and trends in the Yellow Sea and East China Sea region. In *Marine Ecosystems of the North Pacific Ocean, 2003–2008*; McKinnell, S.M., Dagg, M.J., Eds.; PICES Special Publication: Sidney, BC, Canada, 2010; Volume 4, pp. 360–393.
15. B Belkin, I.M.; Cornillon, P.C.; Sherman, K. Fronts in Large Marine Ecosystems. *Prog. Oceanogr.* **2009**, *81*, 223–236. [CrossRef]
16. Bian, C.W.; Jiang, W.S.; Greatbatch, R.J. An exploratory model study of sediment transport sources and deposits in the Bohai Sea, Yellow Sea, and East China Sea. *J. Geophys. Res. Ocean.* **2013**, *118*, 5908–5923. [CrossRef]
17. Jiang, Z.; Gao, Y.; Zhai, H.; Jin, H.; Zhou, F.; Shou, L.; Yan, X.; Chen, Q. Regulation of spatial changes in phytoplankton community by water column stability and nutrients in the southern Yellow Sea. *J. Geophys. Res. Biogeosci.* **2019**, *124*, 2610–2627. [CrossRef]
18. Wright, S.; Jeffrey, S.; Mantoura, R. *Phytoplankton Pigments in Oceanography: Guidelines to Modern Methods*; Jeffrey, S., Mantoura, R., Wright, S., Eds.; Unesco Pub.: Paris, France, 2005.
19. Agirbas, E.; Feyzioglu, A.M.; Kopuz, U.; Llewellyn, C.A. Phytoplankton community composition in the south-eastern Black Sea determined with pigments measured by HPLC-CHEMTAX analyses and microscopy cell counts. *J. Mar. Biol. Assoc. UK* **2015**, *95*, 35–52. [CrossRef]
20. Schlüter, L.; Lauridsen, T.; Krogh, G.; Jørgensen, T. Identification and quantification of phytoplankton groups in lakes using new pigment ratios—A comparison between pigment analysis by HPLC and microscopy. *Freshw. Biol.* **2006**, *51*, 1474–1485. [CrossRef]
21. Luan, Q.; Kang, Y.; Wang, J. Long-term changes within the phytoplankton community in the Yellow Sea (1985–2015). *J. Fish. Sci. China* **2020**, *27*, 1–11.
22. Gao, Y.; Jiang, Z.; Liu, J.; Chen, Q.; Zeng, J.; Huang, W. Seasonal variations of net-phytoplankton community structure in the southern Yellow Sea. *J. Ocean Univ. China* **2013**, *12*, 557–567. [CrossRef]
23. Liu, H.J.; Huang, Y.J.; Zhai, W.D.; Jin, H.; Sun, J. Phytoplankton communities and its controlling factors in summer and autumn in the southern Yellow Sea, China. *Acta Oceanol. Sin.* **2015**, *34*, 114–123. (In Chinese) [CrossRef]
24. Kang, J.-J.; Min, J.-O.; Kim, Y.; Lee, C.-H.; Yoo, H.; Jang, H.-K.; Kim, M.-J.; Oh, H.-J.; Lee, S.-H. Vertical Distribution of Phytoplankton Community and Pigment Production in the Yellow Sea and the East China Sea during the Late Summer Season. *Water* **2021**, *13*, 3321. [CrossRef]
25. Fu, M.Z.; Wang, Z.L.; Li, Y.; Li, R.X.; Sun, P.; Wei, X.H.; Lin, X.Z.; Guo, J.S. Phytoplankton biomass size structure and its regulation in the Southern Yellow Sea (China): Seasonal variability. *Cont. Shelf Res.* **2009**, *29*, 2178–2194. [CrossRef]
26. Ming, F.; Ping, S.; Zong, W.; Yan, L.; Rui, L. Seasonal variations of phytoplankton community size structures in the Huanghai (Yellow) Sea Cold Water Mass area. *Haiyang Xuebao* **2010**, *32*, 120–129.
27. Cho, B.C.; Park, M.G.; Shim, J.H.; Choi, D.H. Sea-surface temperature and f-ratio explain large variability in the ratio of bacterial production to primary production in the Yellow Sea. *Mar. Ecol. Prog. Ser.* **2001**, *216*, 31–41. [CrossRef]
28. Parsons, T.R.; Maita, Y.; Lalli, C.M. *A Manual of Biological and Chemical Methods for Seawater Analysis*; Pergamon Press: Oxford, UK, 1984.

29. Kang, J.J.; Lee, J.H.; Kim, H.C.; Lee, W.C.; Lee, D.; Jo, N.; Min, J.-O.; Lee, S.H. Monthly Variations of Phytoplankton Community in Geoje-Hansan Bay of the Southern Part of Korea Based on HPLC Pigment Analysis. *J. Coast. Res.* **2018**, *85*, 356–360. [CrossRef]

30. Lee, Y.-W.; Park, M.-O.; Im, Y.-S.; Kim, S.-S.; Kang, C.-K. Application of photosynthetic pigment analysis using a HPLC and CHEMTAX program to studies of phytoplankton community composition. *Sea* **2011**, *16*, 117–124. [CrossRef]

31. Mackey, M.D.; Mackey, D.J.; Higgins, H.W.; Wright, S.W. CHEMTAX—A program for estimating class abundances from chemical markers: Application to HPLC measurements of phytoplankton. *Mar. Ecol. Prog. Ser.* **1996**, *144*, 265–283. [CrossRef]

32. Liu, X.; Huang, B.Q.; Huang, Q.; Wang, L.; Ni, X.B.; Tang, Q.S.; Sun, S.; Wei, H.; Liu, S.M.; Li, C.L.; et al. Seasonal phytoplankton response to physical processes in the southern Yellow Sea. *J. Sea Res.* **2015**, *95*, 45–55. [CrossRef]

33. Claustre, H. The trophic status of various oceanic provinces as revealed by phytoplankton pigment signatures. *Limnol. Oceanogr.* **1994**, *39*, 1206–1210. [CrossRef]

34. Vidussi, F.; Claustre, H.; Manca, B.B.; Luchetta, A.; Marty, J.-C. Phytoplankton Pigment Distribution in Relation to Upper Thermocline Circulation in the Eastern Mediterranean Sea during Winter. *J. Geophys. Res.* **2001**, *106*, 19939–19956. [CrossRef]

35. Uitz, J.; Claustre, H.; Morel, A.; Hooker, S.B. Vertical Distribution of Phytoplankton Communities in Open Ocean: An Assessment Based on Surface Chlorophyll. *J. Geophys. Res.* **2006**, *111*, C08005. [CrossRef]

36. Sun, X.; Shen, F.; Brewin, R.J.; Liu, D.; Tang, R. Twenty-year variations in satellite-derived chlorophyll-a and phytoplankton size in the Bohai Sea and Yellow Sea. *J. Geophys. Res. Ocean.* **2019**, *124*, 8887–8912. [CrossRef]

37. Sun, D.; Huan, Y.; Wang, S.; Qiu, Z.; Ling, Z.; Mao, Z.; He, Y. Remote sensing of spatial and temporal patterns of phytoplankton assemblages in the Bohai Sea, Yellow Sea, and east China sea. *Water Res.* **2019**, *157*, 119–133. [CrossRef] [PubMed]

38. Jang, H.K.; Youn, S.H.; Joo, H.; Kim, Y.; Kang, J.J.; Lee, D.; Jo, N.; Kim, K.; Kim, M.-J.; Kim, S.; et al. First Concurrent Measurement of Primary Production in the Yellow Sea, the South Sea of Korea, and the East/Japan Sea, 2018. *J. Mar. Sci. Eng.* **2021**, *9*, 1237. [CrossRef]

39. Tian, T.; Hao, W.; Jian, S.; Changsoo, C.; Wenxin, S. Study on cycle and budgets of nutrients in the Yellow Sea. *Adv. Mar. Biol.* **2003**, *21*, 1–11.

40. Goldman, C.R.; Elser, J.J.; Richards, R.C.; Reuter, J.E.; Priscu, J.C.; Levin, A.L. Thermal stratification, nutrient dynamics and phytoplankton productivity during the onset of spring phytoplankton growth in Lake Baikal, Russia. *Hydrobiologia* **1996**, *331*, 9–24. [CrossRef]

41. Tang, Q.; Su, J.; Zhang, J.; Tong, L. Spring blooms and the ecosystem processes: The case study of the Yellow Sea. *Deep-Sea Res. II Top. Stud. Oceanogr.* **2013**, *97*, 1–3. [CrossRef]

42. Lu, L.; Jiang, T.; Xu, Y.; Zheng, Y.; Chen, B.; Cui, Z.; Qu, K. Succession of phytoplankton functional groups from spring to early summer in the central Bohai Sea using HPLC–CHEMTAX approaches. *J. Oceanogr.* **2018**, *74*, 381–392. [CrossRef]

43. Wang, L.; Ou, L.; Huang, K.; Chai, C.; Wang, Z.; Wang, X.; Jiang, T. Determination of the spatial and temporal variability of phytoplankton community structure in Daya Bay via HPLC-CHEMTAX pigment analysis. *Chin. J. Oceanol. Limnol.* **2018**, *36*, 750–760. [CrossRef]

44. Lin, C.; Ning, X.; Su, J.; Lin, Y.; Xu, B. Environmental changes and the responses of the ecosystems of the Yellow Sea during 1976–2000. *J. Mar. Syst.* **2005**, *55*, 223–234. [CrossRef]

45. Wei, Q.; Yao, Q.; Wang, B.; Wang, H.; Yu, Z. Long-term variation of nutrients in the southern Yellow Sea. *Cont. Shelf Res.* **2015**, *111*, 184–196. [CrossRef]

46. Jin, J.; Liu, S.M.; Ren, J.L.; Liu, C.G.; Zhang, J.; Zhang, G.L.; Huang, D.J. Nutrient dynamics and coupling with phytoplankton species composition during the spring blooms in the Yellow Sea. *Deep. Res. Part II Top. Stud. Oceanogr.* **2013**, *97*, 16–32. [CrossRef]

47. Egge, J.K. Are diatoms poor competitors at low phosphate concentrations? *J. Mar. Syst.* **1998**, *16*, 191–198. [CrossRef]

48. Cerino, F.; Zingone, A. A survey of cryptomonad diversity and seasonality at a coastal Mediterranean site. *Eur. J. Phycol.* **2006**, *41*, 363–378. [CrossRef]

49. Kang, Y.; Moon, C.-H.; Kim, H.-J.; Yoon, Y.H.; Kang, C.-K. Water Quality Improvement Shifts the Dominant Phytoplankton Group From Cryptophytes to Diatoms in a Coastal Ecosystem. *Front. Mar. Sci.* **2021**, 1125. [CrossRef]

50. Liu, X.; Huang, B.; Liu, Z.; Wang, L.; Wei, H.; Li, C.; Huang, Q. High-resolution phytoplankton diel variations in the summer stratified central Yellow Sea. *J. Oceanogr.* **2012**, *68*, 913–927. [CrossRef]

51. Tilman, D. *Resource Competition and Community Structure*; Princeton University Press: Princeton, NJ, USA, 1982.

52. Tilman, D. Resource competition between plankton algae: An experimental and theoretical approach. *Ecology* **1977**, *58*, 338–348. [CrossRef]

53. Smith, R.E.; Kalff, J. Size dependence of growth rate, respiratory electron transport system activity and chemical composition in marine diatoms in the laboratory. *J. Phycol.* **1982**, *18*, 275–284. [CrossRef]

54. Raven, J. The twelfth Tansley Lecture. Small is beautiful: The picophytoplankton. *Funct. Ecol.* **1998**, *12*, 503–513. [CrossRef]

55. Mur, R.; Skulberg, O.; Utkilen, H. Cyanobacteria in the environment. In *Toxic Cyanobacteria in Water*; Chorus, I., Batram, J., Eds.; E&FN Spon: London, UK, 1999; pp. 15–40.

56. Peperzak, L. Climate change and harmful algal blooms in the North Sea. *Acta Oecol.* **2003**, *24*, S139–S144. [CrossRef]

57. Choi, J. *Phytoplankton Ecology in the Yellow Sea*; Bumwoo Publishing Company: Seoul, Korea, 2002; pp. 311–330.

58. Choi, J.K.; Noh, J.H.; Shin, K.S.; Hong, K.H. The early autumn distribution of chlorophyll-a and primary productivity in the Yellow Sea, 1992. *Yellow Sea* **1995**, *1*, 68–80.

59. Lee, M.; Kim, Y.-B.; Park, C.-H.; Baek, S.-H. Characterization of Seasonal Phytoplankton Pigments and Functional Types around Offshore Island in the East/Japan Sea, Based on HPLC Pigment Analysis. *Sustainability* **2022**, *14*, 5306. [CrossRef]
60. Leblanc, K.; Queguiner, B.; Diaz, F.; Cornet, V.; Michel-Rodriguez, M.; de Madron, X.; Bowler, C.; Malviya, S.; Thyssen, M.; Gregori, G.; et al. Nanoplanktonic diatoms are globally overlooked but play a role in spring blooms and carbon export. *Nat. Commun.* **2018**, *9*, 953. [CrossRef]
61. December, P. Uncoupling of bacteria and phytoplankton during a spring diatom bloom in the mouth of the Yellow Sea. *Mar. Ecol. Prog Ser.* **1994**, *115*, 181–190.
62. Abonyi, A.; Kiss, K.T.; Hidas, A.; Borics, G.; Várbíró, G.; Ács, É. Cell Size Decrease and Altered Size Structure of Phytoplankton Constrain Ecosystem Functioning in the Middle Danube River Over Multiple Decades. *Ecosystems* **2020**, *23*, 1254–1264. [CrossRef]
63. Montagnes, D.J.S.; Franklin, D.J. Effect of temperature on diatom volume, growth rate, and carbon and nitrogen content: Reconsidering some paradigms. *Limnol. Oceanogr.* **2001**, *46*, 2008–2018. [CrossRef]
64. Jørgensen, E.G. The adaptation of plankton algae:II. Aspects of the temperature adaptation of Skeletonema costatum. *Physiol. Plant* **1968**, *21*, 423–427. [CrossRef]
65. Sun, W.; Zhang, J.; Meng, J.; Liu, Y. Sea surface temperature characteristics and trends in China offshore seas from 1982 to 2017. *J. Coast. Res.* **2019**, *90*, 27–34. [CrossRef]
66. Finkel, Z.V.; Vaillancourt, C.J.; Irwin, A.J.; Reavie, E.D.; Smol, J.P. Environmental control of diatom community size structure varies across aquatic ecosystems. *Proc. R. Soc. B: Biol. Sci.* **2009**, *276*, 1627–16344. [CrossRef]
67. Lee, S.H.; Yun, M.S.; Kim, B.K.; Saitoh, S.I.; Kang, C.K.; Kang, S.H.; Whitledge, T. Latitudinal carbon productivity in the Bering and Chukchi Seas during the summer in 2007. *Cont. Shelf Res.* **2013**, *59*, 28–36.

Journal of
Marine Science and Engineering

Article

First Concurrent Measurement of Primary Production in the Yellow Sea, the South Sea of Korea, and the East/Japan Sea, 2018

Hyo-Keun Jang [1], Seok-Hyun Youn [2], Huitae Joo [2], Yejin Kim [1], Jae-Joong Kang [1], Dabin Lee [1], Naeun Jo [1], Kwanwoo Kim [1], Myung-Joon Kim [1], Soohyun Kim [1] and Sang-Heon Lee [1,*]

[1] Department of Oceanography, Pusan National University, Geumjeong-gu, Busan 46241, Korea; janghk@pusan.ac.kr (H.-K.J.); yejini@pusan.ac.kr (Y.K.); jaejung@pusan.ac.kr (J.-J.K.); ldb1370@pusan.ac.kr (D.L.); nadan213@pusan.ac.kr (N.J.); goanwoo7@pusan.ac.kr (K.K.); mjune@pusan.ac.kr (M.-J.K.); kksh1122@pusan.ac.kr (S.K.)

[2] Oceanic Climate & Ecology Research Division, National Institute of Fisheries Science, Busan 46083, Korea; younsh@korea.kr (S.-H.Y.); huitae@korea.kr (H.J.)

* Correspondence: sanglee@pusan.ac.kr; Tel.: +82-51-510-2256

Abstract: Dramatic environmental changes have been recently reported in the Yellow Sea (YS), the South Sea of Korea (SS), and the East/Japan Sea (EJS), but little information on the regional primary productions is currently available. Using the ^{13}C-^{15}N tracer method, we measured primary productions in the YS, the SS, and the EJS for the first time in 2018 to understand the current status of marine ecosystems in the three distinct seas. The mean daily primary productions during the observation period ranged from 25.8 to 607.5 mg C m^{-2} d^{-1} in the YS, 68.5 to 487.3 mg C m^{-2} d^{-1} in the SS, and 106.4 to 490.5 mg C m^{-2} d^{-1} in the EJS, respectively. In comparison with previous studies, significantly lower (*t*-test, $p < 0.05$) spring and summer productions and consequently lower annual primary productions were observed in this study. Based on PCA analysis, we found that small-sized (pico- and nano-) phytoplankton had strongly negative effects on the primary productions. Their ecological roles should be further investigated in the YS, the SS, and the EJS under warming ocean conditions within small phytoplankton-dominated ecosystems.

Keywords: primary production; phytoplankton; Yellow Sea; East/Japan Sea; South Sea of Korea

Citation: Jang, H.-K.; Youn, S.-H.; Joo, H.; Kim, Y.; Kang, J.-J.; Lee, D.; Jo, N.; Kim, K.; Kim, M.-J.; Kim, S.; et al. First Concurrent Measurement of Primary Production in the Yellow Sea, the South Sea of Korea, and the East/Japan Sea, 2018. *J. Mar. Sci. Eng.* **2021**, *9*, 1237. https://doi.org/10.3390/jmse9111237

Academic Editor: Antonello Sala

Received: 30 September 2021
Accepted: 5 November 2021
Published: 8 November 2021

Publisher's Note: MDPI stays neutral with regard to jurisdictional claims in published maps and institutional affiliations.

1. Introduction

Marine phytoplankton as primary producers play an important role as the base of the ecological pyramid in the ocean and are responsible for nearly a half of global primary production [1,2]. The primary production of phytoplankton is widely used as an important indicator to predict annual fishery yield in various oceanic regions [3–5], because it is one of key factors in determining amount of food source for upper-trophic-level consumers [6,7]. Lee et al. [8,9] also reported that an algorithm for estimation of the habitat suitability index for the mackerels and squids around the Korean peninsula was largely improved by including a primary production term. The physiological conditions and community structures of phytoplankton are closely related to physical and chemical factors (e.g., light regime, nutrients, and temperature) [10–12], which induce greatly different phytoplankton productions in various marine ecosystems [3,13,14]. Thus, the primary production measurements can provide fundamental backgrounds for better understanding marine ecosystems with different environmental conditions and detecting current potential ecosystem changes.

The Yellow Sea (hereafter YS), the South Sea of Korea (SS), and the East/Japan Sea (EJS), belonging to the East Asian marginal seas, have experienced 2–4 times faster increase (0.7–1.2 °C) in seawater temperature than that in global mean water temperature (0.4 °C) for 20 years (1983–2006) [15]. Moreover, some notable changes in physicochemical conditions were reported, such as increasing limitation of nutrients in the YS and rapid ocean acidification and shoaling of the mixed layer depth in the EJS [16–19]. These recent environmental changes could result in alterations in biological characteristics, including community structure and bloom pattern of phytoplankton and subsequently higher-trophic-level organisms [12,20–22]. Indeed, biological responses related to phytoplankton were observed in the YS [16,18,23,24] and the EJS [12,22,25,26]. For the YS, a remarkable decrease in chlorophyll-*a* (chl-*a*) concentration and phytoplankton diversity and abundance was observed between the periods 1983–1986 and 1996–1998, which affected the primary production [16]. The phytoplankton community assemblage was also dramatically changed in the YS, especially in the spring [18,23]. For the EJS, the patterns of timing, magnitude, and duration of the spring phytoplankton bloom were significantly different between 1998–2001 and 2008–2011 [25]. Joo et al. [26] found dramatic decreasing trends (1.3% each year) in annual primary productivities in various regions in the EJS for 1 decade (2003–2012) based on the satellite-based data. Nevertheless, we still have a lack of information on regional primary productions of phytoplankton for understanding the current status of the marine ecosystems in the YS, the SS, and the EJS.

In this study, one of our main objectives is to compare the seasonal and regional primary productions measured simultaneously in the YS, the SS, and the EJS for the first time in 2018 with those reported in previous studies in each region. The other one is to determine major controlling factors in the low primary production in the YS, the SS, and the EJS in 2018.

2. Materials and Methods

2.1. Water Sampling Collection

Seasonal cruise surveys were conducted onboard the R/V Tamgu 8 for the YS and the SS and R/V Tamgu 3 for the EJS from February to October 2018 (Figure 1). The data collected from the February, April, August, and October 2018 cruises were designated to represent winter, spring, summer, and autumn, respectively. At mid-morning, 9–10 different stations were determined in the YS and the EJS whereas 6–7 sampling stations were determined for the SS among the monitoring stations (Figure 1) managed by the National Institute of Fisheries Science (NIFS) in Korea (Table 1).

Table 1. Description of sampling sites in the YS, the SS, and the EJS for each cruise period, in 2018. (o) means investigation was conducted, while (-) means investigation was not conducted.

Region	Station	Latitude	Longitude	Bottom Depth (m)	Feb.	Apr.	Aug.	Oct.
	307-03	36.92	126.00	37	o	o	o	o
	307-05	36.92	125.42	54	o	o	o	o
	307-09	36.92	124.57	67	o	o	o	o
	308-06	36.33	125.21	58	-	-	o	o
	309-03	35.85	125.82	54	o	o	o	o
	309-05	35.85	125.40	69	o	o	o	-
	309-07	35.86	125.00	66	-	-	-	o
YS	309-09	35.85	124.59	82	o	o	o	-
	310-03	35.34	125.82	27	-	o	-	-
	310-06	35.34	125.20	72	-	-	-	o
	310-09	35.34	124.59	92	o	-	-	-
	311-05	34.72	125.52	75	o	-	-	-
	311-07	34.72	125.00	90	-	o	o	o
	311-09	34.72	124.59	89	o	-	o	o
	312-05	34.04	125.50	81	-	o	-	-
	312-09	34.09	124.60	89	o	o	o	o

Table 1. *Cont.*

Region	Station	Latitude	Longitude	Bottom Depth (m)	Feb.	Apr.	Aug.	Oct.
SS	203-03	33.64	126.36	133	-	-	-	o
	204-04	33.90	127.25	75	o	o	-	o
	205-03	34.08	127.94	82	o	o	o	o
	205-05	33.62	128.15	113	o	o	o	-
	206-03	34.37	128.82	92	o	o	o	o
	207-03	34.89	129.25	115	o	o	o	o
	400-14	34.21	128.40	75	o	-	o	o
	400-25	33.55	127.56	96	o	-	o	-
	400-27	33.51	127.08	124	-	o	-	-
EJS	102-06	36.08	129.80	700	-	o	-	-
	102-07	36.08	130.00	1390	-	-	-	o
	102-09	36.08	130.62	1880	-	-	o	-
	103-04	36.51	129.50	110	-	o	-	-
	103-05	36.51	129.59	205	-	-	-	o
	103-07	36.50	130.00	850	o	-	-	-
	103-09	36.51	130.62	2150	-	-	o	o
	103-10	36.51	130.93	1800	-	o	-	-
	103-11	36.50	131.24	2100	o	-	-	-
	104-04	37.06	129.48	110	-	-	o	-
	104-05	37.06	129.56	220	-	o	-	-
	104-08	37.06	130.31	720	-	-	-	o
	104-09	37.06	130.63	2340	-	o	-	-
	104-11	37.06	131.26	2325	-	-	o	-
	105-03	37.55	129.17	48	-	-	-	o
	105-05	37.55	129.37	280	o	-	-	-
	105-07	37.55	130.00	1480	o	-	o	-
	105-10	37.55	130.93	1503	-	o	-	-
	105-11	37.55	131.24	1140	o	-	o	o
	106-03	37.90	128.95	320	-	o	-	-
	106-05	37.90	129.37	1120	-	-	o	-
	106-07	37.90	130.00	1060	-	-	-	o
	106-10	37.90	130.94	1980	-	o	-	-
	107-03	38.21	128.84	1120	-	-	o	o
	107-05	38.20	129.37	1080	o	-	-	-
	107-07	38.20	130.00	846	o	o	o	o
	209-04	35.79	129.55	54	-	-	-	o
	209-05	35.75	129.64	150	o	-	-	-
	209-07	35.61	130.01	250	-	o	o	-
	209-08	35.60	130.00	200	o	-	-	-

The bottom depths at our sampling stations in the YS and the SS had relatively narrow range, whereas the EJS had a wide range of bottom depths (48–2340 m) in this study (Table 1). The six water depths were determined at each station by converting Secchi disc depth to 6 corresponding light depths (100, 50, 30, 12, 5, and 1% of surface photosynthetic active radiation; (PAR)). Then, each water sample was collected from 6 different depths using Niskin bottles (8 L) equipped with a conductivity, temperature, and depth (CTD)-rosette. The water temperature and salinity were obtained from SBE9/11 CTD (Sea-Bird Electronics, Bellevue, WA, USA). The mixed-layer depth (MLD) was defined as the depth at which the density is increased by 0.125 density units from the sea surface density [27,28]. Water samples for dissolved inorganic nutrients (NH_4, $NO_2 + NO_3$, PO_4, and SiO_2) and chl-*a* (total and size-fractionated) concentrations were collected at three light depths (100, 30, and 1% of PAR). Water samples for measuring the particle organic carbon (POC) and particle organic nitrogen (PON) concentrations and total carbon uptake rates (primary production) of phytoplankton were collected at six light depths (100, 50, 30, 12, 5, and 1% of PAR). The euphotic zone is defined as the depth from 100 to 1% of PAR.

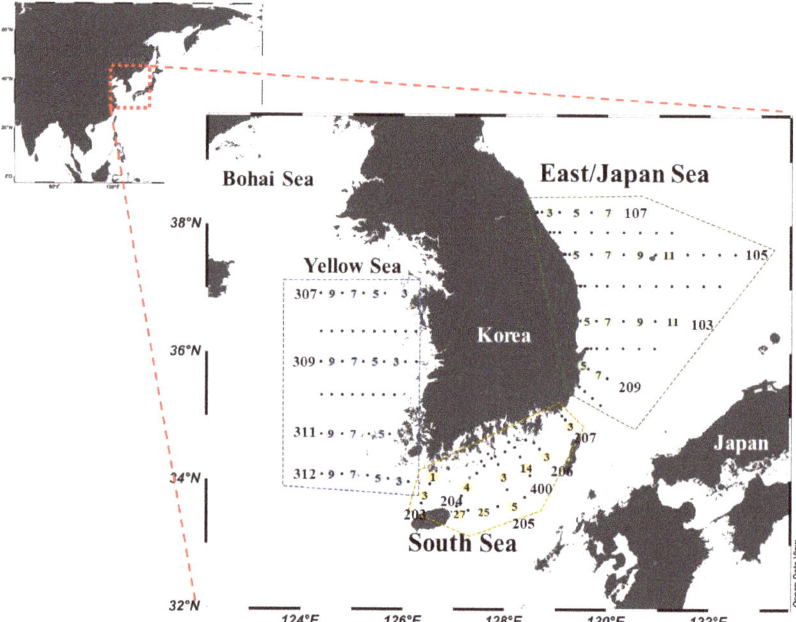

Figure 1. Locations of sampling regions in 2018. The station numbers are in consecutive order from coast to open sea as marked in each station line.

2.2. Inorganic Nutrients Concentrations

To measure concentrations of dissolved inorganic nutrients (NH_4, $NO_2 + NO_3$, PO_4, and SiO_2), 0.1 L water samples were filtered onto Whatman GF/F filters (ø = 47 mm) at a vacuum pressure lower than 150 mmHg. Filtered water samples were immediately frozen at $-20\ ^\circ$C for further analysis in our laboratory. An auto-analyzer (Quattro, Seal Analytical, Norderstedt, Germany) in the NIFS was used for the analysis of dissolved inorganic nutrients according to the manufacturer's instruction.

2.3. Chl-a Concentration

The primary method and calculation for determining the chl-*a* concentrations were conducted according to Parsons et al. [29]. Water samples (0.1–0.4 L) for total chl-*a* concentration were filtered through Whatman GF/F filters (ø = 25 mm), and samples (0.3–1 L) for three different size-fractionated chl-*a* concentrations were passed sequentially through 20 μm and 2 μm membrane filters (ø = 47 mm) and GF/F filters (ø = 47 mm) at low vacuum pressure. The filtered samples were then placed in a 15 mL conical tube, immediately stored in $-20\ ^\circ$C freezer until the analysis. In the laboratory, the frozen filters were extracted with 90% acetone at 4 $^\circ$C for 20–24 h, and chl-*a* concentrations were then measured using a fluorometer (Turner Designs, 10-AU, San Jose, CA, USA) calibrated based on commercially available reference material for chl-*a*.

2.4. Measurements of Phyoplankton Carbon and Nitrogen Uptake Rate

The ^{13}C-^{15}N dual stable isotope tracer technique was used for simultaneously measuring the carbon and nitrogen uptake rates of the phytoplankton as described by Dugdale and Goering [30] and Hama et al. [31]. In brief, water samples from each light depth (100%, 50%, 30%, 12%, 5%, and 1% of PAR) were immediately transferred to acid-rinsed polycarbonate incubation bottles (1 L) covered with neutral density screens (Lee Filters) [32] after passing through 333 μm sieves to eliminate the large zooplankton. The incubation bottles filled with seawater at each light depth were inoculated with the labeled carbon ($NaH^{13}CO_3$)

and nitrate ($K^{15}NO_3$) or ammonium ($^{15}NH_4Cl$), which correspond to 10–15% of the concentrations in the ambient water [30,31]. Then, the tracer-injected bottles were incubated in a large polycarbonate incubator at a constant temperature maintained by continuously circulating sea surface water under natural surface light for 4–5 h. The incubated water samples (0.1–0.4 L) were filtered onto Whatman GF/F filters (ø = 25 mm) precombusted at 450 °C, and the filters were then kept in a freezer (-20 °C) until mass spectrometer analysis. At the laboratory of Pusan National University, the filters were fumed with a strong hydro acid in a desiccator to remove the carbonate overnight and dried with a freeze drier for 2 h. Then, POC and PON concentrations and atom % of ^{13}C were analyzed by Finnigan Delta+XL mass spectrometer at the stable isotope laboratory of the University of Alaska (Fairbanks, AK, USA). The carbon uptake rates of the phytoplankton were estimated as described by Dugdale and Goering [30] and Hama et al. [31]. The final values of the carbon uptake rates of phytoplankton were then calculated by subtracting the carbon uptake rates of dark bottles to eliminate the heterotrophic bacterial production [33–35]. The daily primary productions of phytoplankton were calculated from the hourly primary productions observed in this study and 10-h photoperiod per day reported previously in the YS and EJS [22,24].

2.5. Statistical Analysis

The statistical analyses for Pearson's correlation, *t*-test, and one-way analysis of variance (one-way ANOVA) were performed using SPSS (version 12.0, SPSS Inc., Chicago, IL, USA). In the one-way ANOVA, a test to certify the homoscedasticity of variables was conducted by using Levene's test. To compare pairwise differences for the variables, Scheffe's (homogeneity) and Dunnett's (heteroscedasticity) post hoc tests were used, based on homogeneity of variances.

Principal component analysis (PCA) with the Varimax method with Kaiser normalization using the XLSTAT software (Addinsoft, Boston, MA, USA) was used to identify relatively significant factors affecting the total carbon uptake rates of phytoplankton in each sea during our observation time. Fourteen variables for PCA included physical (water temperature and salinity and euphotic and mixed-layer depths), chemical (NH_4, NO_2+NO_3, PO_4, and SiO_2 concentrations), and biological (total and size-fractionated chl-*a* and POC concentration) factors and carbon uptake rates of phytoplankton.

3. Results

3.1. Physicochemical Environmental Conditions

Seasonal vertical profiles of the mean temperatures and salinities at each light depth in the YS, the SS, and the EJS are presented in Figure 2. Seasonal water temperatures and salinities in the YS, the SS, and the EJS were evenly distributed within the euphotic zone except in August. The mean temperatures within the euphotic zone in the YS, the SS, and the EJS were lowest in February, with means of 5.9 (S.D. = $\pm$ 2.3), 13.6 ($\pm$ 1.3), and 9.9 ($\pm$ 1.7) °C, respectively, and gradually increased to their highest in August, with means of 23.2 ($\pm$ 1.4), 23.8 ($\pm$ 1.6), and 20.9 ($\pm$ 2.9) °C, respectively (Figure 2). The average water temperature in the YS was significantly lower than those in the SS and EJS during February and April (one-way ANOVA, $p < 0.05$). The highest mean salinities in the YS and EJS were observed in April (32.8 $\pm$ 0.8 and 34.4 $\pm$ 0.1 psu), whereas salinity in the SS was highest in February at 34.6 $\pm$ 0.0 psu (Figure 2). Overall, lower salinities were found in the YS than in the SS and the EJS throughout the observation period.

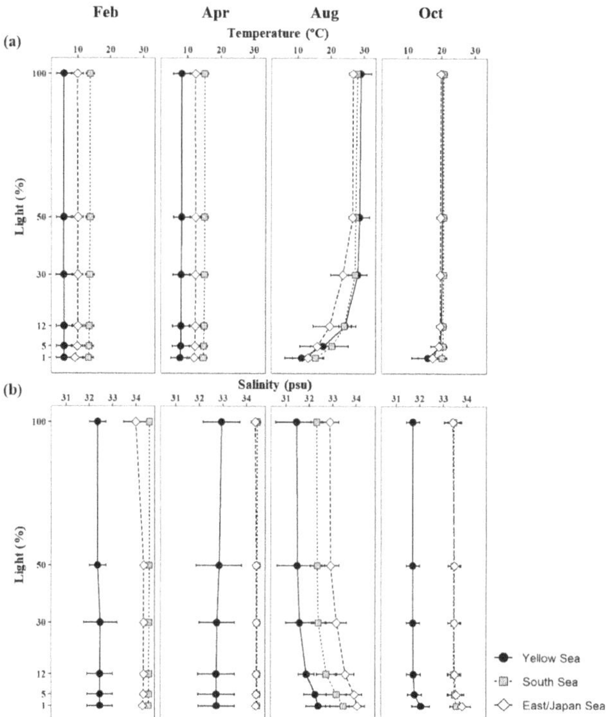

Figure 2. Vertical profiles of mean temperatures (**a**) and salinities (**b**) at six light depths (100, 50, 30, 12, 5, and 1%) in the YS, the SS, and the EJS, 2018.

The mean euphotic depths in the YS, the SS, and the EJS were deepest in August at 37.6 ± 15.6, 49.8 ± 11.3, and 54.4 ± 10.7 m, respectively (Figure 3). In particular, the euphotic depth in the EJS in February (51.0 ± 5.8 m) was significantly deeper (one-way ANOVA, $p < 0.01$) than those in the YS (12.8 ± 6.2 m) and the SS (28.1 ± 4.7 m). The deepest MLDs in the YS, the SS, and the EJS were observed in February, with means of 68.7 ± 15.7, 59.0 ± 40.5, and 80.6 ± 57.4 m, respectively (Figure 3). The MLDs in the YS, the SS, and the EJS became continuously shallow until August at 12.0 ± 14.2, 13.7 ± 6.6, and 13.2 ± 6.2 m, respectively, and then deepened in October to 26.3 ± 13.7, 30.2 ± 16.1, and 37.9 ± 14.2 m, respectively. In all regions, the differences between the MLDs and euphotic depths were greatest in February, decreased toward April, and then reversed in August when MLDs were significantly shallower than the euphotic depths (t-test, $p < 0.01$) (Figure 3). These results indicate that the euphotic zone was vertically well-mixed in all study regions during February and April, whereas strong stratifications were developed in the euphotic water columns during August.

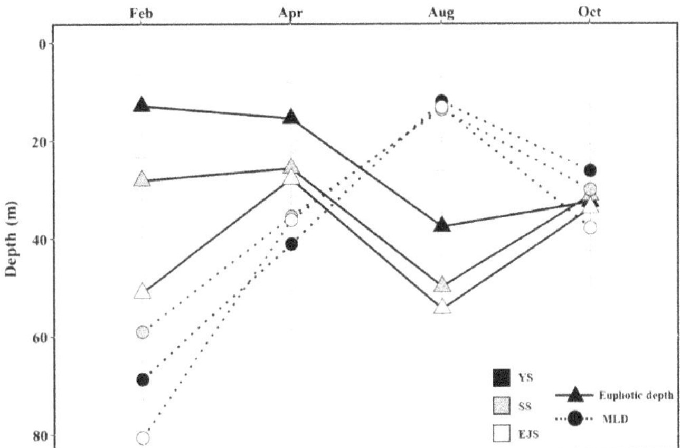

Figure 3. Variation in the mean euphotic and mixed-layer depths in the YS, the SS, and the EJS, 2018.

Major dissolved inorganic nutrient concentrations at each light depth (100%, 30%, and 1%) in the YS, the SS, and the EJS for each cruise are summarized in Table 2. The ranges of NO_2+NO_3, PO_4, and SiO_2 concentrations during the study period were 0.5–9.9, <0.1–0.6, and 2.4–10.0 μM in the YS; 0.9–8.1, 0.1–0.4, and 5.1–11.3 μM in the SS; and 0.2–8.7, 0.1–0.5, and 2.4–11.0 μM in the EJS, respectively. Ranges of nutrient concentrations except for NH_4 varied significantly in all regions during the study period, being generally high in February and low in other seasons except NO_2+NO_3 concentrations in the YS in April. The nutrient concentrations, except for NH_4 at 1% light depths in the YS and the EJS, were higher (one-way ANOVA, $p < 0.01$) than those at 100% and 30% light depths during August and October, whereas vertical differences in the SS were only detected in August. NH_4 concentrations ranged from 0.5 to 1.2 μM in the YS, 0.1 to 0.6 μM in the SS, and 0.4 to 0.9 μM in the EJS, respectively, during the observation period. Unlike other nutrients, NH_4 concentrations had no distinct seasonal and vertical characteristics in all study regions.

Table 2. The dissolved inorganic nutrient concentrations averaged from each light depth (100, 30, and 1%) in the YS, the SS, and the EJS, 2018.

Region	Month	Light Depth (%)	NH_4	NO_2+NO_3	PO_4	SiO_2
						μM
YS	Feb.	100	0.9 ± 0.6	8.8 ± 2.4	0.6 ± 0.1	8.9 ± 3.3
		30	0.9 ± 0.7	8.2 ± 2.1	0.5 ± 0.1	8.5 ± 3.4
		1	0.7 ± 0.3	9.9 ± 2.8	0.6 ± 0.1	10.0 ± 3.6
	Apr.	100	0.6 ± 0.4	6.6 ± 3.3	0.3 ± 0.1	6.8 ± 2.5
		30	0.5 ± 0.5	5.8 ± 3.0	0.2 ± 0.1	6.9 ± 2.9
		1	0.5 ± 0.5	7.1 ± 2.1	0.3 ± 0.1	7.8 ± 2.7
	Aug.	100	1.0 ± 1.1	1.2 ± 1.6	0.1 ± 0.1	2.8 ± 1.3
		30	1.2 ± 1.5	0.5 ± 0.5	0.0 ± 0.0	2.4 ± 0.5
		1	0.7 ± 0.5	3.4 ± 1.8	0.4 ± 0.2	8.7 ± 3.3
	Oct.	100	0.6 ± 0.1	1.7 ± 2.2	0.2 ± 0.2	4.6 ± 2.4
		30	0.5 ± 0.1	1.6 ± 2.1	0.2 ± 0.1	4.2 ± 1.6
		1	0.5 ± 0.1	5.0 ± 2.7	0.4 ± 0.2	7.0 ± 3.1

Table 2. *Cont.*

Region	Month	Light Depth (%)	NH$_4$	NO$_2$+NO$_3$	PO$_4$	SiO$_2$
					μM	
SS	Feb.	100	0.2 ± 0.0	5.3 ± 1.3	0.4 ± 0.0	9.2 ± 0.7
		30	0.2 ± 0.0	5.1 ± 0.8	0.4 ± 0.0	8.1 ± 1.7
		1	0.2 ± 0.1	5.1 ± 0.8	0.4 ± 0.0	8.9 ± 0.8
	Apr.	100	0.1 ± 0.1	0.9 ± 0.5	0.1 ± 0.0	5.6 ± 2.6
		30	0.1 ± 0.1	1.0 ± 0.5	0.1 ± 0.0	6.1 ± 2.4
		1	0.3 ± 0.2	2.1 ± 1.5	0.1 ± 0.1	6.7 ± 2.2
	Aug.	100	0.6 ± 0.3	1.7 ± 0.2	0.1 ± 0.1	6.8 ± 1.4
		30	0.5 ± 0.4	1.4 ± 0.3	0.1 ± 0.1	6.2 ± 1.0
		1	0.1 ± 0.0	8.1 ± 2.9	0.4 ± 0.2	11.3 ± 1.8
	Oct.	100	0.3 ± 0.2	1.6 ± 1.0	0.2 ± 0.0	5.1 ± 2.9
		30	0.3 ± 0.2	1.6 ± 1.0	0.2 ± 0.0	4.0 ± 2.1
		1	0.5 ± 0.3	3.0 ± 1.1	0.2 ± 0.1	5.2 ± 2.1
EJS	Feb.	100	0.4 ± 0.2	6.7 ± 0.9	0.4 ± 0.1	9.8 ± 1.4
		30	0.4 ± 0.2	6.4 ± 0.7	0.4 ± 0.0	8.6 ± 2.4
		1	0.3 ± 0.2	7.5 ± 1.1	0.5 ± 0.2	10.7 ± 2.8
	Apr.	100	0.9 ± 0.7	1.8 ± 1.0	0.1 ± 0.0	4.4 ± 1.4
		30	0.9 ± 0.6	1.9 ± 1.0	0.2 ± 0.1	4.2 ± 1.3
		1	0.9 ± 0.7	2.6 ± 1.1	0.2 ± 0.0	5.0 ± 1.4
	Aug.	100	0.4 ± 0.1	0.2 ± 0.1	0.1 ± 0.0	3.7 ± 1.7
		30	0.6 ± 0.6	0.4 ± 0.6	0.1 ± 0.0	3.8 ± 1.6
		1	0.4 ± 0.2	8.7 ± 3.3	0.5 ± 0.2	11.0 ± 4.1
	Oct.	100	0.6 ± 0.3	1.3 ± 1.1	0.1 ± 0.1	2.6 ± 1.3
		30	0.6 ± 0.2	1.3 ± 1.0	0.1 ± 0.1	2.4 ± 1.1
		1	0.5 ± 0.2	4.9 ± 4.3	0.4 ± 0.3	6.9 ± 6.0

3.2. Concentrations and Size-Fractionated Compositions of chl-a

The ranges of the total chl-*a* concentrations integrated throughout the euphotic water column in the YS, the SS, and the EJS were 1.3–96.6, 5.6–60.7, and 8.0–92.9 mg m^{-2}, respectively, during our observation period (Figure 4). The highest chl-*a* concentration in the YS was detected in April (mean ± S.D. = 31.1 ± 28.6 mg m^{-2}), followed by October (18.8 ± 8.0 mg m^{-2}), August (12.1 ± 5.0 mg m^{-2}), and February (3.3 ± 1.8 mg m^{-2}). The chl-*a* concentrations in the SS between April (47.6 ± 10.1 mg m^{-2}) and October (44.9 ± 15.3 mg m^{-2}) were similar, and the concentrations during these periods were higher (One-way ANOVA, $p < 0.05$) than those in February (11.7 ± 3.9 mg m^{-2}) and August (12.8 ± 3.9 mg m^{-2}). In the EJS, the highest chl-*a* concentration was observed in April (50.1 ± 12.5 mg m^{-2}), and the second-highest concentration was observed in February (38.0 ± 23.9 mg m^{-2}). The chl-*a* concentration was lowest in August (12.6 ± 4.6 mg m^{-2}).

Based on the size-fractionated chl-*a* concentrations, the compositions of micro- (>20 µm), nano- (2–20 µm), and pico-sized (0.7–2 µm) phytoplankton in the YS, the SS, and the EJS are shown in Figure 5a–c. Overall, the fraction of nano- and pico-sized phytoplankton was dominant in the YS (> 65%), the EJS (> 58%), and the SS (> 65%) during the study period except for October. In detail, the compositions of the nano- and pico-sized phytoplankton in the YS were 50.7 ± 6.1 and 27.0 ± 12.2% in February, 41.6 ± 14.1 and 23.6 ± 10.9% in April, 35.1 ± 7.6 and 45.9 ± 14.7% in August, and 23.4 ± 6.1 and 60.1 ± 21.4% in October, respectively. The compositions of micro-sized phytoplankton in the YS remained low (approximately 19%) during the study period except for April (34.8 ± 22.8%). The contributions of the nano- and pico-sized phytoplankton in the SS during February, April, August, and October were 27.7 ± 4.4 and 63.8 ± 3.5%, 43.3 ± 19.5 and 24.9 ± 13.2%, 36.7 ± 11.5 and 52.4 ± 8.8%, and 15.0 ± 2.6 and 35.6 ± 17.5%, respectively. The highest contribution of micro-sized phytoplankton in the SS was observed in October (49.4 ± 19.5%), followed by April (31.7 ± 30.5%), August (10.9 ± 3.8%), and February (8.4 ± 2.1%). The fractions of nano- and pico-sized phytoplankton in the total chl-*a* concentrations in the EJS in each season were 25.4 ± 9.2 and 42.0 ± 16.6% (February), 28.8 ± 8.3 and 43.8 ± 12.7% (April),

26.9 ± 13.3 and 55.9 ± 16.6% (August), and 19.0 ± 5.5 and 39.5 ± 18.0% (October), respectively. The fraction of micro-sized phytoplankton in the EJS was gradually decreased from February (32.7 ± 23.8%) to August (17.3 ± 11.4%) and then increased in October (41.5 ± 19.7%).

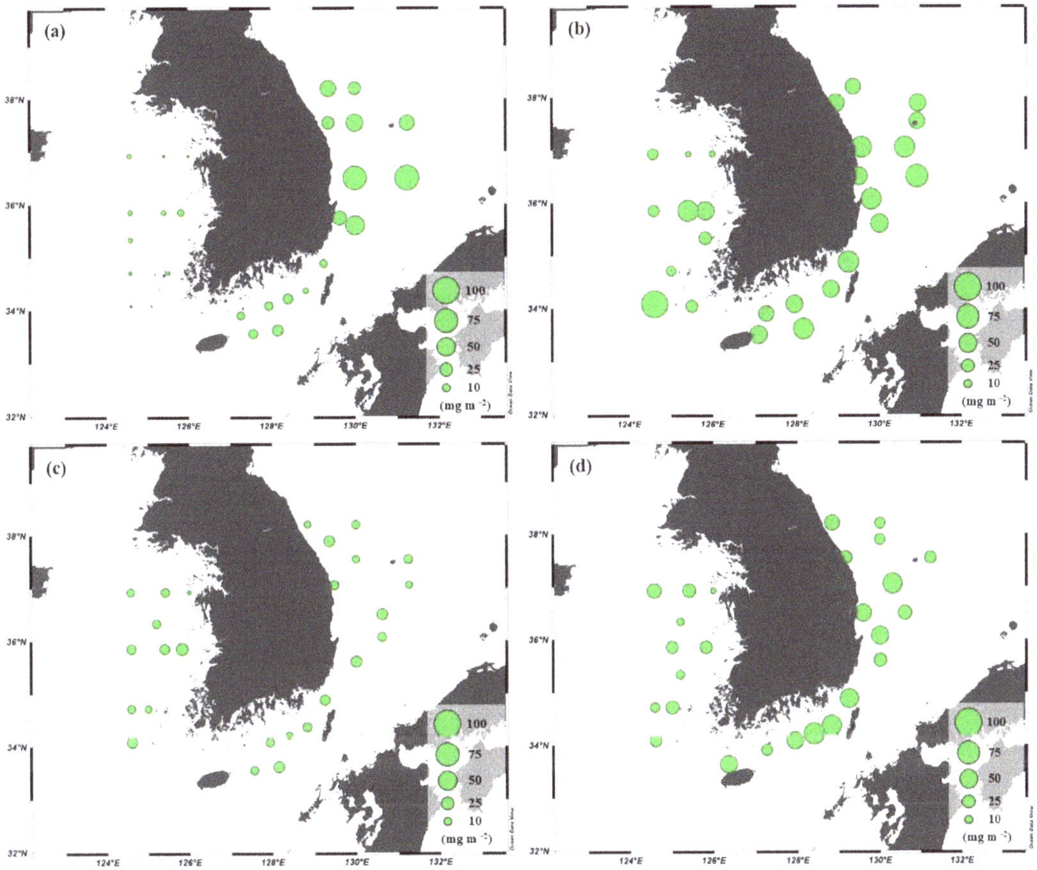

Figure 4. Spatial distributions of the chl-*a* concentrations integrated within the euphotic depth from 100 to 1% light depth in the YS, the SS, and the EJS during Feb. (**a**), Apr. (**b**), Aug. (**c**), and Oct. (**d**).

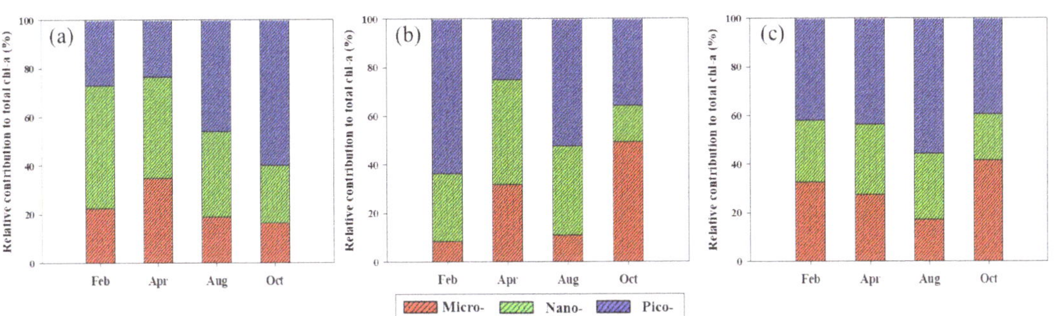

Figure 5. Seasonal variations in different size compositions of chl-*a* in the YS (**a**), the SS (**b**), and the EJS (**c**), 2018.

3.3. POC and PON Concentration

The mean POC concentrations integrated in the euphotic zone in the YS, the SS, and the EJS showed different seasonal patterns in comparison to the chl-*a* concentrations (Figure 6b). The POC concentrations in the YS and SS gradually increased from February, at 1.7 ± 0.5 and 2.7 ± 1.0 g C m^{-2}, to October, with 10.4 ± 3.7 and 7.5 ± 3.1 g C m^{-2}, respectively. In comparison, the POC concentrations in the EJS were the highest during August at 8.9 ± 1.5 g C m^{-2} but remained constant at an average of ~4 g C m^{-2} during other seasons.

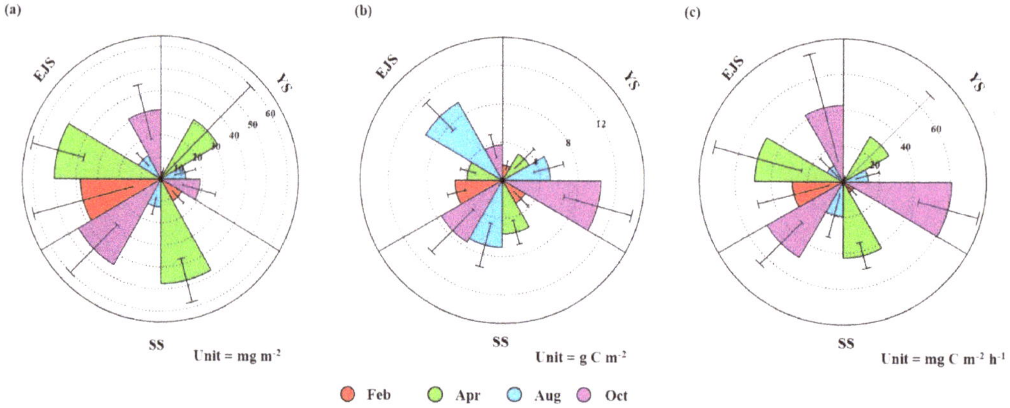

Figure 6. Nightingale rose diagrams of mean chl-*a* (**a**) and POC (**b**) concentrations and primary productions (**c**) in the YS, the SS, and the EJS, 2018.

The POC concentrations had significantly positive correlations ($R^2 = 0.7575$, $p < 0.01$ in the YS; $R^2 = 0.8105$, $p < 0.01$ in the SS; $R^2 = 0.5723$, $p < 0.01$ in the EJS) with the PON concentrations in this study. The average C/N ratios at each month (February, April, August, and October) were 9.2 ± 1.0 (mean $\pm$ S.D.), 7.7 ± 0.9, 9.3 ± 0.9, and 18.6 ± 2.3 in the YS; 10.8 ± 2.1, 8.1 ± 1.4, 9.2 ± 1.5, and 11.3 ± 3.0 in the SS; and 8.9 ± 1.2, 6.6 ± 0.8, 12.5 ± 0.9, and 6.3 ± 1.2 in the EJS, respectively.

3.4. Primary Production of Phytoplankton

The primary productions of phytoplankton integrated from different six-light depths (100, 50, 30, 12, 5, and 1%) ranged from 1.0 to 135.1 (YS), 1.8 to 63.7 (SS), and 2.3 to 119.3 (EJS) mg C m^{-2} h^{-1}, respectively (Figure 7). The ranges of the primary productions in the YS and the EJS were more variable than in the SS in this study. High mean primary productions in the YS, the SS, and the EJS were observed during April (29.3 ± 39.4, 42.6 ± 7.8, and 49.1 ± 25.2 mg C m^{-2} h^{-1}) and October (60.6 ± 17.8, 48.4 ± 15.4, and 43.3 ± 31.1 mg C m^{-2} h^{-1}) (Figure 6c). In comparison, the mean primary productions during February and August were low in the YS (2.6 ± 1.2, 9.3 ± 1.0 mg C m^{-2} h^{-1}), the SS (6.8 ± 3.5 and 19.5 ± 12.5 mg C m^{-2} h^{-1}), and the EJS (10.6 ± 7.7 and 28.4 ± 20.4 mg C m^{-2} h^{-1}) (Figure 6c). Overall, there were distinct seasonal variations in the primary productions, which were higher in spring and autumn than those in winter and summer in all waters of the littoral sea in Korea in 2018.

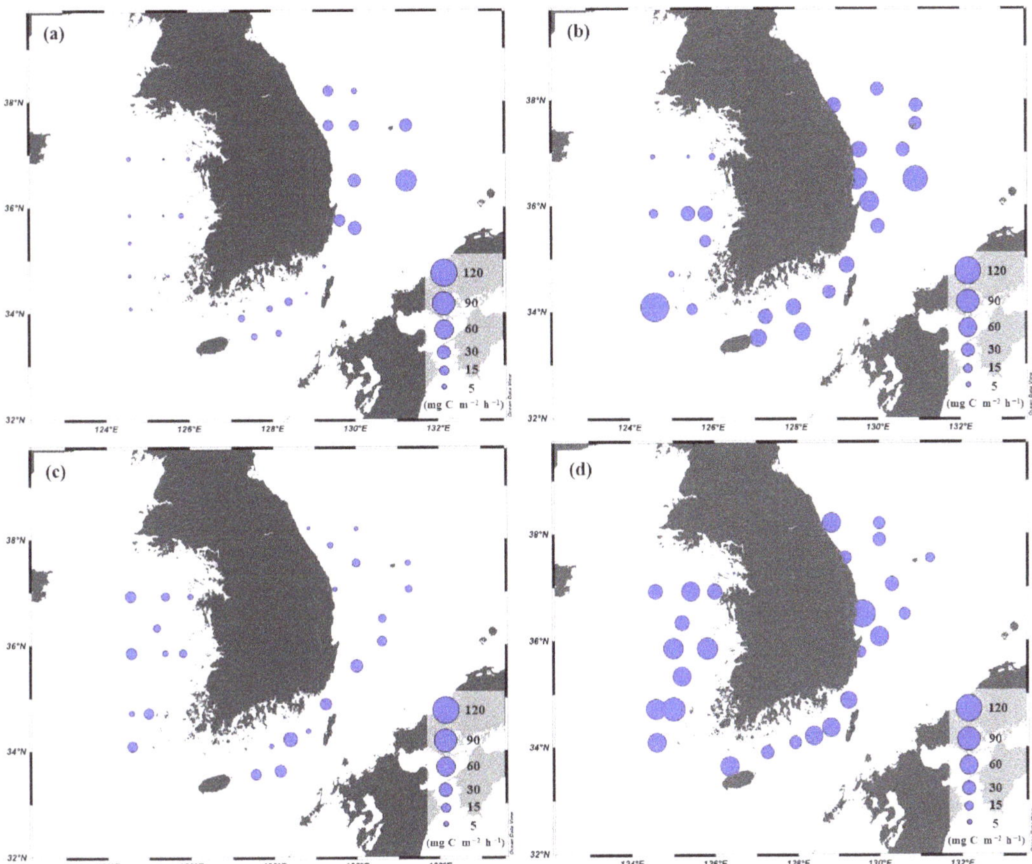

Figure 7. Spatial distributions of the primary production integrated within the euphotic depth from 100 to 1% light depth in the YS, the SS, and the EJS during Feb. (**a**), Apr. (**b**), Aug. (**c**), and Oct (**d**).

The results of PCA to determine major environmental and biological factors affecting the primary productions of phytoplankton in the YS, the SS, and the EJS throughout the observation period are shown in Figure 8a–c. The two ordination axes (PC1 and PC2) of principal components (PC) accounted for the cumulative variances of 61.6, 66.9, and 54.7% in the YS, the SS, and the EJS, respectively. Primary production in the YS was positively correlated with the chl-*a* and POC concentrations and temperature but negatively correlated with the MLD and compositions of nano-sized phytoplankton (Figure 8a). The positive relations between primary production and total chl-*a* concentrations and compositions of micro-sized phytoplankton were observed in the SS (Figure 8b). In contrast, pico-sized phytoplankton compositions and nutrients except for NH_4 were negatively related to primary production in the SS (Figure 8b). For the EJS, the total chl-*a* concentrations, compositions of the micro-sized plankton, and salinity had positive effects, whereas the pico-sized plankton and water temperature had negative effects on the primary production (Figure 8c).

No strong correlation ($R^2 = 0.1225$ $p > 0.05$) was found between the biomass contributions of pico-sized phytoplankton and the primary production of phytoplankton in the YS (Figure 9a). In contrast, significantly negative correlations between the biomass contributions of pico-sized phytoplankton and the primary production were observed in the SS ($R^2 = 0.791$, $p < 0.01$) and the EJS ($R^2 = 0.801$, $p < 0.01$) (Figure 9b,c).

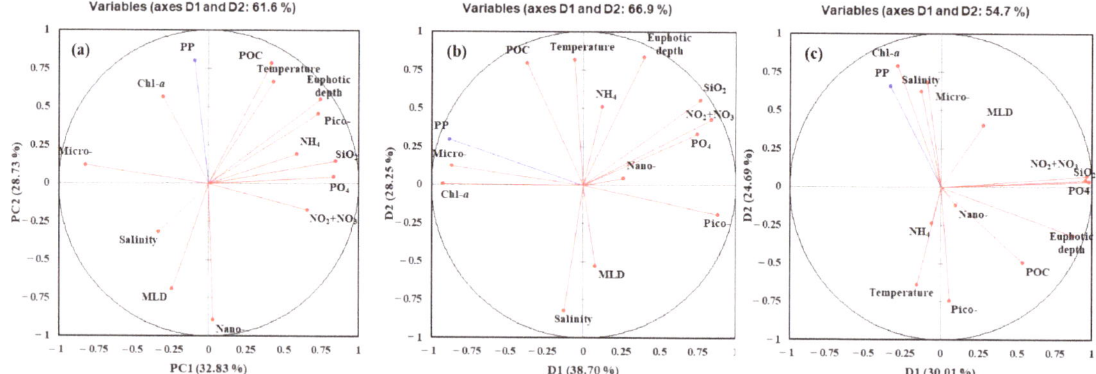

Figure 8. Principal components analysis (PCA) ordination plots showing primary production of phytoplankton in relation to environmental and biological conditions in the YS (**a**), the SS (**b**), and the EJS (**c**), 2018. Micro-, nano-, and pico-represent contributions of compositions of micro-, nano-, and pico-sized phytoplankton to total chl-*a*; PP represents primary production.

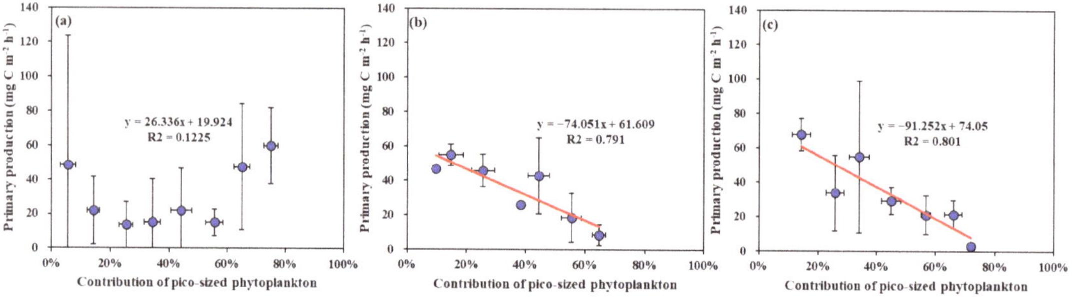

Figure 9. Relationships between primary production and contribution of pico-sized phytoplankton (<2 μm) to total chl-*a* concentrations in the YS (**a**), the SS (**b**), and the EJS (**c**), 2018.

4. Discussion

4.1. Comparisons of Primary Production between This and Previous Studies

Based on a 10-h photoperiod and the hourly primary productions obtained in this study (Figure 6), the mean daily primary productions in the YS were 25.8 ± 11.9, 292.7 ± 393.9, 139.0 ± 66.9, and 607.5 ± 172.6 mg C m^{-2} d^{-1} during winter, spring, summer, and autumn, respectively. Our values obtained in this study were slightly lower than the ranges (56–947 mg C m^{-2} d^{-1}) of the values reported previously in adjacent or nearly identical regions to our sites in the YS (Table 3). In particular, our spring (293 mg C m^{-2} d^{-1}) and summer (139 mg C m^{-2} d^{-1}) values in this study were significantly lower (*t*-test, $p < 0.05$) than the spring (851 ± 108 mg C m^{-2} d^{-1}) and summer (555 ± 231 mg C m^{-2} d^{-1}) values averaged from previous studies. These lower seasonal productions in 2018 might be explained by a recent change in the nutrient budgets in the YS. An increasing trend in dissolved inorganic nitrogen (DIN) concentration since the 1980s was reported, whereas a decreasing trend from the 1980s to 2000 followed by a slight increase in PO$_4$ concentration was observed in the YS [16,36]. These changes in DIN and PO$_4$ have induced a gradual increase in the N/P ratio and a shift from N-limitation to P-limitation in the YS [36]. The P-limited condition could convert dominant species of phytoplankton from diatoms to small-sized non-diatoms with higher growth rates in P-limited waters but lower photosynthetic efficiencies [18,24,37]. Lin et al. [16] reported that a dramatic decrease in primary production in the YS during all seasons between 1983–1986 and 1996–1998 periods could be one of the ecological responses caused by the increase in the N/P ratio. In this study, the N/P ratios (32 ± 14) during the spring period were significantly higher (one-sample t-test, $p < 0.01$) than the Redfield ratios (16) [38], which could have resulted in a limitation for diatom growth [39,40]. Indeed, the diatom compositions (approximately 50%) in the YS in spring based on the results from our parallel study (non-published data) were distinctly lower than those reported previously in 1986 (89%) and 1998 (70%) [23]. This shift in dominant species could have caused the low primary production in spring 2018. Jang et al. [24] reported that the high contribution of pico-sized (<2 μm) phytoplankton to the total primary production could induce a lower total primary production in the YS when the N/P ratio is higher than 30 during the summer period. We did not measure the production of pico-sized phytoplankton in this study, but the higher N/P ratio (54 ± 78) at upper euphotic depth (100 and 30%) accounted for about 75% of integrated primary production and could explain the lower primary production in the YS during summer 2018.

Since the primary production measurements have rarely been conducted in the SS section belonging to the northern part of the East China Sea, we compared our results with those measured previously in the entire East China Sea (Table 4). The average daily primary productions in the SS during this observation are within the range (102–1727 mg C m^{-2} d^{-1}) reported previously in the East China Sea (Table 4). However, the winter and summer values in this study were significantly lower (t-test, $p < 0.05$) than the mean winter (206 ± 93 mg C m^{-2} d^{-1}) and summer (621 ± 179 mg C m^{-2} d^{-1}) productions reported previously. In comparison, the autumn value (487 mg C m^{-2} d^{-1}) in this study was consistent with the previous findings (503 ± 186 mg C m^{-2} d^{-1}). For the springtime, our daily production (426 mg C m^{-2} d^{-1}) was not statistically different (t-test, $p > 0.05$) from the mean production (350 ± 161 mg C m^{-2} d^{-1}) in early spring (March), but our spring value was considerably lower than those reported previously in April (1727 mg C m^{-2} d^{-1}) and May (1375 mg C m^{-2} d^{-1}).

Table 3. Comparisons of daily primary production in the YS. PP represents daily primary production.

Resion	Method	Year	Month	Season	PP (mg C m^{-2} d^{-1})	Reference
Middle-eastern part	^{14}C method	1989	Aug.	summer	450	
			Oct.	autumn	130	
		1990	Feb.	winter	115	[11]
			Aug.	summer	486	
			Oct.	autumn	192	
Entire	^{14}C method	1992	Sep.	autumn	742	[42]
middle-eastern part	^{14}C method	1997	Feb.	winter	92	
			Apr.	spring	872	
			Aug.	summer	899	[43]
			Oct.	autumn	667	
			Dec.	winter	262	
Middle part	Satellite-based	1998–2003	May	spring	947	
			Sep.	autumn	723	
Middle-eastern part	Satellite-based	1998–2003	May	spring	734	[44]
			Sep.	autumn	558	
Middle part	^{14}C method	2008	Jan.	winter	56	[45]
			Aug.	summer	649	
Middle part	^{13}C-^{15}N method	2016	Aug.	summer	291	[24]
Middle-eastern part	^{13}C-^{15}N method	2018	Feb.	winter	26 ± 12	
			Apr.	spring	293 ± 394	This study
			Aug.	summer	139 ± 67	
			Oct.	autumn	606 ± 178	

Table 4. Comparisons of daily primary production in the East China Sea. PP represents daily primary production.

Region	Method	Year	Month	Season	PP (mg C m^{-2} d^{-1})	Reference
Northern part	^{14}C method	1989–1990	Mar.	spring	310	
			Apr.	spring	1727	[46]
			Nov.	autumn	517	
Changjjang river mouth to shelf edge (PN-line)	^{13}C method	1993–1994	Feb.	winter	282	[47]
			Aug.	summer	714	
			Oct.	autumn	573	
Entire shelf	^{14}C method	1997–1998	Dec.	winter	235	
			Mar.	spring	213	[4]
			Jul.	summer	734	
			Oct.–Nov.	autumn	355	
Entire shelf	^{13}C-^{15}N method	1998	Mar.	spring	528	[48]
			Oct.–Nov.	autumn	782	
Southern part	satellite-based	1999	Oct.	autumn	543	[49]
Entire	^{14}C method	2008–2009	May	spring	1375	
			Aug.	summer	414	[50]
			Nov.	autumn	245	
			Feb.	winter	102	
Northern part	^{13}C-^{15}N method	2018	Feb.	winter	68 ± 35	
			Apr.	spring	426 ± 74	This Study
			Aug.	summer	195 ± 125	
			Oct.	autumn	487 ± 161	

The daily primary productions measured in this study during four seasons are within the range (44–1505 mg C m^{-2} d^{-1}) obtained previously from the various regions in the EJS in different seasons (Table 5). However, our value (284 mg C m^{-2} d^{-1}) during the winter period was significantly higher (*t*-test, $p < 0.05$) than the winter mean value (75 ± 44 mg C m^{-2} d^{-1}) reported by Nagata [51] and Yoshie et al. [52], whereas our spring (491 mg C m^{-2} d^{-1}) and summer (106 mg C m^{-2} d^{-1}) rates were significantly lower (*t*-test, $p < 0.05$) than the spring (858 ± 376 mg C m^{-2} d^{-1}) and summer (519 ± 184 mg C m^{-2} d^{-1}) values averaged from various previous studies. A plausible mechanism for the difference might be related to the development of the MLD in the EJS during the wintertime. A vigorous vertical mixing driven by the Asian winter monsoon can limit the availability of light to phytoplankton in winter [53,54] but induces an increase in the nutrient availability in the upper euphotic layer from spring to summer [55,56]. However, the MLD has been gradually decreased by an increase in water temperature and weakened wind stress in the EJS [17,57,58], which could offer better light conditions for phytoplankton growth in winter but fewer nutrients for the spring phytoplankton bloom. In this way, the difference in seasonal primary production in the EJS mentioned above could be explained by the recent change in the MLD. However, because our surveys in the EJS were restricted to only 2018, this mechanism needs to be verified by a long-term observation. Another reason for the low primary production, especially in spring 2018, could be potentially having missed the bloom timing of the phytoplankton during our observation period. In general, the spring bloom in the EJS is mainly driven by the massive growth of diatoms, which account for the majority of large-sized (> 20 μm) phytoplankton [59–61]. Indeed, Kwak et al. [62] observed a significantly higher contribution (approximately 60%) of diatoms during the spring bloom period than in other seasons. In this study, the contribution of the large-sized phytoplankton was rather lower during the spring (Figure 5c). However, much lower diatom contributions were detected based on our parallel stud, showing that diatoms accounted for only 23.1% (± 9.9%) of total phytoplankton communities in the EJS in spring (non-published data). The other reason might be conspicuously low phytoplankton biomass in the EJS in April 2018. Based on MODIS (Moderate Resolution Imaging Spectrometer)-Aqua monthly level-3 datasets regarding chl-*a* (https://oceandata.sci.gsfc.nasa.gov/MODIS-Aqua/, accessed on 3 August 2021), the surface chl-*a* showed strong negative anomalies in the southwestern part of the EJS during April between 2003–2015 and 2018 (data not shown). As the chl-*a* concentrations in the EJS were one of the major factors controlling the primary production (Figure 8c), a noticeable low chl-*a* concentration could cause lower primary production in the EJS during the springtime in 2018. At the current stage, it is difficult to find out a solid reason for the low chl-*a* concentration in the EJS during April 2018, which should be further resolved for a better understanding of the EJS ecosystem.

Table 5. Comparisons of daily primary production in the EJS. PP represents daily primary production.

Region	Method	Year	Month	Season	PP (mg C m^{-2} d^{-1})	Reference
Southwestern part	^{14}C method	1986	Oct.	autumn	1505	[63]
Southwestern part	^{14}C method	1990	Oct.	autumn	1420	[64]
Yamato Basin	^{13}C method	1994–1996	Jan.	winter	44	
			Apr.	spring	1082	[51]
			Aug.	summer	353	
			Oct.	autumn	154	
West coast of Hokkaido		1997	Feb.	winter	106	
			Mar.–Apr.	spring	1419	[52]
			Jul.	summer	487	
			Oct.	autumn	254	
			Apr.	spring	1100	
Southwestern part	Satellite-based	1998–2002	Nov.	autumn	650	[65]
			Monthly		608	
Ulleng Basin	^{14}C method	2006	Apr.	spring	513	[66]
Ulleng Basin	^{13}C method	2010–2011	May	spring	1114	[67]
			Nov.	autumn	380	
Ulleng Basin	^{13}C method	2010	Jul.	summer	716	[68]
Northern part	^{13}C method	2012	Nov.	autumn	181	[22]
		2015	May	spring	442	
Ulleng Basin	^{13}C method	2016	Apr.	spring	790	[12]
Northwestern part			Apr.	spring	407	
Southwestern part	^{13}C-^{15}N method	2018	Feb.	winter	284 ± 203	
			Apr.	spring	491 ± 252	This study
			Aug.	summer	106 ± 77	
			Oct.	autumn	428 ± 307	

4.2. Main Factors Affecting the Primary Production in the YS, SS, and EJS in 2018

Based on the PCA results (Figure 8), the major factors controlling the phytoplankton productions were different among the three seas. Total chl-*a* concentrations (positively; +), temperature (+), MLD (negatively; −), and nano-sized phytoplankton contribution (−) are found to be major controlling factors in the YS. In comparison, total chl-*a* concentrations (+), pico- (−) and micro-sized (+) phytoplankton contributions, and nutrients (−) except for NH$_4$ can greatlyaffect the primary production in the SS. For the EJS, the primary production of phytoplankton can greatly vary due to total chl-*a* concentrations (+), micro-sized phytoplankton contribution (+), salinity (+), pico-sized phytoplankton (−), and water temperature (−). The effects of physical (temperature, salinity, and MLD) and chemical (nutrients) factors are different in the YS, the SS, and the EJS. Given the positive relationships between the primary productions and the total chl-*a* concentrations in this study, biomass-driven primary productions are characteristics in the YS, the SS, and the EJS ecosystems, at least in 2018. However, the effects of the three size groups of phytoplankton can be different among the three seas. The contribution of nano-sized phytoplankton in the YS and the contributions of pico-sized phytoplankton in the SS and the EJS are negatively correlated with the primary productions in this study. Choi et al. [42] reported nano-phytoplankton contributed greatly to the primary production in the YS, based on the large biomass contribution of nano-phytoplankton (approximately 60%). In this study, the negative relationship between the nano-sized phytoplankton contribution and the primary production indicates that increasing contributions of the nano-sized phytoplankton could decrease the primary production in the YS. In the EJS, several previous studies reported higher contributions of pico-sized phytoplankton could cause a decrease in the primary production [12,22,69]. Indeed, marked decreasing trends in the primary productions with increasing pico-sized phytoplankton biomass were observed in the SS and EJS during our observation period in 2018 (Figure 9b.c). This could be caused by the different primary productivities between pico- and large-sized (>2 μm) phytoplankton [22]. Generally, pico-sized phytoplankton have a lower primary productivity than large phytoplankton [14,22,70]. Therefore, the total primary production can be decreased by increasing contribution of pico-sized phytoplankton, with their lower productivity traits. Under ongoing warming ocean conditions, pico-sized phytoplankton are expected to be predominant in phytoplankton communities [71–74]. In this pico-sized-phytoplankton-dominated ecosystem, a lower total primary production could be expected in the SS and the EJS based on the negative relationships between the primary production and pico-sized phytoplankton observed in this study. The ecological roles of pico-sized phytoplankton in regional marine ecosystems should be further investigated in the YS, the SS, and the EJS with current environmental changes.

Author Contributions: Conceptualization, H.-K.J. and S.-H.L.; methodology, H.-K.J.; validation, H.-K.J. and S.-H.L.; formal analysis, H.-K.J.; investigation, H.-K.J., Y.K., H.J., J.-J.K., D.L., N.J., K.K., M.-J.K. and S.K.; data curation, H.-K.J.; writing—original draft preparation, H.-K.J.; writing—review and editing, S.-H.L.; visualization, H.-K.J.; supervision, S.-H.L.; project administration, S.-H.Y. and S.-H.L.; funding acquisition, S.-H.Y. and S.-H.L. All authors have read and agreed to the published version of the manuscript.

Funding: This research was fundedby the National Institute of Fisheries Science ('Development of marine ecological forecasting system for Korean waters'; R2021068).

Institutional Review Board Statement: Not applicable.

Informed Consent Statement: Not applicable.

Data Availability Statement: Not applicable.

Acknowledgments: We appreciate the captains and crew of R/V Tamgu 3 and 8 for their assistance in collecting our samples. We would also like to thank the researchers in the NIFS for their assistance with sample analysis. This work was supported by the National Institute of Fisheries Science (NIFS) grant ('Development of marine ecological forecasting system for Korean waters'; R2021068) funded by the Ministry of Oceans and Fisheries, Republic of Korea.

Conflicts of Interest: The authors declare no conflict of interest.

References

1. Falkowski, P.G.; Barber, R.T.; Smetacek, V. Biogeochemical controls and feedbacks on ocean primary production. *Science* **1998**, *281*, 200–206. [CrossRef]
2. Behrenfeld, M.J.; Randerson, J.T.; McClain, C.R.; Feldman, G.C.; Los, S.O.; Tucker, C.J.; Falkowski, P.G.; Field, C.B.; Frouin, R.; Esaias, W.E.; et al. Biospheric primary production during an ENSO transition. *Science* **2001**, *291*, 2594–2597. [CrossRef] [PubMed]
3. Nixon, S.; Thomas, A. On the size of the Peru upwelling ecosystem. *Deep. Res. Part I Oceanogr. Res. Pap.* **2001**, *48*, 2521–2528. [CrossRef]
4. Gong, G.C.; Wen, Y.H.; Wang, B.W.; Liu, G.J. Seasonal variation of chlorophyll a concentration, primary production and environmental conditions in the subtropical East China Sea. *Deep. Res. Part II Top. Stud. Oceanogr.* **2003**, *50*, 1219–1236. [CrossRef]
5. Tremblay, J.É.; Robert, D.; Varela, D.E.; Lovejoy, C.; Darnis, G.; Nelson, R.J.; Sastri, A.R. Current state and trends in Canadian Arctic marine ecosystems: I. Primary production. *Clim. Chang.* **2012**, *115*, 161–178. [CrossRef]
6. Kleppel, G.S.; Burkart, C.A. Egg production and the nutritional environment of Acartia tonsa: The role of food quality in copepod nutrition. *ICES J. Mar. Sci.* **1995**, *52*, 297–304. [CrossRef]
7. Kang, J.J.; Joo, H.T.; Lee, J.H.; Lee, J.H.; Lee, H.W.; Lee, D.; Kang, C.K.; Yun, M.S.; Lee, S.H. Comparison of biochemical compositions of phytoplankton during spring and fall seasons in the northern East/Japan Sea. *Deep. Res. Part II Top. Stud. Oceanogr.* **2017**, *143*, 73–81. [CrossRef]

8. Lee, D.; Son, S.H.; Kim, W.; Park, J.M.; Joo, H.; Lee, S.H. Spatio-temporal variability of the habitat suitability index for chub mackerel (Scomber Japonicus) in the East/Japan Sea and the South Sea of South Korea. *Remote Sens.* **2018**, *10*, 938. [CrossRef]
9. Lee, D.; Son, S.H.; Lee, C., Il; Kang, C.K.; Lee, S.H. Spatio-temporal variability of the habitat suitability index for the Todarodes pacificus (Japanese Common Squid) around South Korea. *Remote Sens.* **2019**, *11*, 2720. [CrossRef]
10. Lee, S.H.; Yun, M.S.; Kim, B.K.; Saitoh, S.I.; Kang, C.K.; Kang, S.H.; Whitledge, T. Latitudinal carbon productivity in the Bering and Chukchi Seas during the summer in 2007. *Cont. Shelf Res.* **2013**, *59*, 28–36. [CrossRef]
11. Kim, B.K.; Joo, H.T.; Song, H.J.; Yang, E.J.; Lee, S.H.; Hahm, D.; Rhee, T.S.; Lee, S.H. Large seasonal variation in phytoplankton production in the Amundsen Sea. *Polar Biol.* **2015**, *38*, 319–331. [CrossRef]
12. Kang, J.J.; Jang, H.K.; Lim, J.H.; Lee, D.; Lee, J.H.; Bae, H.; Lee, C.H.; Kang, C.K.; Lee, S.H. Characteristics of Different Size Phytoplankton for Primary Production and Biochemical Compositions in the Western East/Japan Sea. *Front. Microbiol.* **2020**, *11*, 1–16. [CrossRef]
13. Taniguchi, A. Geographical variation of primary production in the Western Pacific Ocean and adjacent seas with reference to the inter-relations between various parameters of primary production. *Mem. Fac. Fish. Hokkaido Univ.* **1972**, *19*, 1–33.
14. Lee, S.H.; Sun Yun, M.; Kyung Kim, B.; Joo, H.T.; Kang, S.H.; Keun Kang, C.; Whitledge, T.E. Contribution of small phytoplankton to total primary production in the Chukchi Sea. *Cont. Shelf Res.* **2013**, *68*, 43–50. [CrossRef]
15. Belkin, I.M. Rapid warming of Large Marine Ecosystems. *Prog. Oceanogr.* **2009**, *81*, 207–213. [CrossRef]
16. Lin, C.; Ning, X.; Su, J.; Lin, Y.; Xu, B. Environmental changes and the responses of the ecosystems of the Yellow Sea during 1976–2000. *J. Mar. Syst.* **2005**, *55*, 223–234. [CrossRef]
17. Chang, P.H.; Cho, C.H.; Ryoo, S.B. Recent changes of mixed layer depth in the East/Japan Sea: 1994-2007. *Asia-Pac. J. Atmos. Sci.* **2011**, *47*, 497–501. [CrossRef]
18. Jin, J.; Liu, S.M.; Ren, J.L.; Liu, C.G.; Zhang, J.; Zhang, G.L.; Huang, D.J. Nutrient dynamics and coupling with phytoplankton species composition during the spring blooms in the Yellow Sea. *Deep. Res. Part II Top. Stud. Oceanogr.* **2013**, *97*, 16–32. [CrossRef]
19. Kim, J.Y.; Kang, D.J.; Lee, T.; Kim, K.R. Long-term trend of CO2 and ocean acidification in the surface water of the Ulleung Basin, the East/Japan Sea inferred from the underway observational data. *Biogeosciences* **2014**, *11*, 2443–2454. [CrossRef]
20. Chiba, S.; Batten, S.; Sasaoka, K.; Sasai, Y.; Sugisaki, H. Influence of the Pacific Decadal Oscillation on phytoplankton phenology and community structure in the western North Pacific. *Geophys. Res. Lett.* **2012**, *39*, 2–7. [CrossRef]
21. Doney, S.C.; Ruckelshaus, M.; Emmett Duffy, J.; Barry, J.P.; Chan, F.; English, C.A.; Galindo, H.M.; Grebmeier, J.M.; Hollowed, A.B.; Knowlton, N.; et al. Climate change impacts on marine ecosystems. *Ann. Rev. Mar. Sci.* **2012**, *4*, 11–37. [CrossRef]
22. Lee, S.H.; Joo, H.T.; Lee, J.H.; Lee, J.H.; Kang, J.J.; Lee, H.W.; Lee, D.; Kang, C.K. Seasonal carbon uptake rates of phytoplankton in the northern East/Japan Sea. *Deep. Res. Part II Top. Stud. Oceanogr.* **2017**, *143*, 45–53. [CrossRef]
23. Wang, J. Study on phytoplankton in the Yellow Sea in spring. *Mar. Fisher. Res.* **2001**, *22*, 56–61.
24. Jang, H.K.; Kang, J.J.; Lee, J.H.; Kim, M.; Ahn, S.H.; Jeong, J.Y.; Yun, M.S.; Han, I.S.; Lee, S.H. Recent Primary Production and Small Phytoplankton Contribution in the Yellow Sea during the Summer in 2016. *Ocean Sci. J.* **2018**, *53*, 509–519. [CrossRef]
25. Lee, S.H.; Son, S.; Dahms, H.U.; Park, J.W.; Lim, J.H.; Noh, J.H.; Kwon, J.I.; Joo, H.T.; Jeong, J.Y.; Kang, C.K. Decadal changes of phytoplankton chlorophyll-a in the East Sea/Sea of Japan. *Oceanology* **2014**, *54*, 771–779. [CrossRef]
26. Joo, H.T.; Son, S.H.; Park, J.W.; Kang, J.J.; Jeong, J.Y.; Lee, C., Il; Kang, C.K.; Lee, S.H. Long-term pattern of primary productivity in the East/Japan sea based on ocean color data derived from MODIS-Aqua. *Remote Sens.* **2016**, *8*, 25. [CrossRef]
27. Levitus, S. *Climatological Atlas of the World Ocean*; NOAA Prof. Paper 13; U.S. Government Printing Office: Washington, DC, USA, 1982; 173p.
28. Gardner, W.D.; Chung, S.P.; Richardson, M.J.; Walsh, I.D. The oceanic mixed-layer pump. *Deep. Res. Part II* **1995**, *42*, 757–775. [CrossRef]
29. Parsons, T.R.; Maita, Y.; Lalli, C.M. *A Manual of Biological and Chemical Methods for Seawater Analysis*; Pergamon Press: Oxford, UK, 1984.
30. Dugdale, R.C.; Goering, J.J. Uptake of New and Regenerated Forms of Nitrogen in Primary Productivity. *Limnol. Oceanogr.* **1967**, *12*, 196–206. [CrossRef]
31. Hama, T.; Miyazaki, T.; Ogawa, Y.; Iwakuma, T.; Takahashi, M.; Otsuki, A.; Ichimura, S. Measurement of photosynthetic production of a marine phytoplankton population using as Table 13C isotope. *Mar. Biol.* **1983**, *73*, 31–36. [CrossRef]
32. Garneau, M.È.; Gosselin, M.; Klein, B.; Tremblay, J.É.; Fouilland, E. New and regenerated production during a late summer bloom in an Arctic polynya. *Mar. Ecol. Prog. Ser.* **2007**, *345*, 13–26. [CrossRef]
33. Li, W.K.W.; Irwin, B.D.; Dickie, P.M. Dark fixation of 14C: Variations related to biomass and productivity of phytoplankton and bacteria. *Limnol. Oceanogr.* **1993**, *38*, 483–494. [CrossRef]
34. Gosselin, M.; Levasseur, M.; Wheeler, P.A. Deep Sea Research Part II: Topical Studies in Oceanography—New measurements of phytoplankton and ice algal production in the Arctic Ocean. *Deep Sea Res. Part II Top. Stud. Oceanogr.* **1997**, *44*, 1623–1644. [CrossRef]
35. Lee, H.W.; Noh, J.H.; Choi, D.H.; Yun, M.; Bhavya, P.S.; Kang, J.J.; Lee, J.H.; Kim, K.W.; Jang, H.K.; Lee, S.H. Picocyanobacterial contribution to the total primary production in the northwestern pacific ocean. *Water* **2021**, *13*, 1610. [CrossRef]
36. Wei, Q.; Yao, Q.; Wang, B.; Wang, H.; Yu, Z. Long-term variation of nutrients in the southern Yellow Sea. *Cont. Shelf Res.* **2015**, *111*, 184–196. [CrossRef]

37. Tao, F.; Daoji, L.; Lihua, Y.; Lei, G.; Lihua, Z. Effects of irradiance and phosphate on growth of nanophytoplankton and picophytoplankton. *Acta Ecol. Sin.* **2006**, *26*, 2783–2789. [CrossRef]
38. Redfeild, A. The influence of organisms on the composition of sea water. *Sea* **1963**, *2*, 26–77.
39. Egge, J.K. Are diatoms poor competitors at low phosphate concentrations? *J. Mar. Syst.* **1998**, *16*, 191–198. [CrossRef]
40. Zhou, M.-j.; Shen, Z.-l.; Yu, R.-c. Responses of a coastal phytoplankton community to increased nutrient input from the Changjiang (Yangtze) River. *Cont. Shelf Res.* **2008**, *28*, 1483–1489. [CrossRef]
41. Kang, Y.S.; Choi, J.K.; Chung, K.H.; Park, Y.C. Primary productivity and assimilation umber in the Kyonggi bay and the mid-eastern coast of Yellow Sea. *J. Oceanogr. Soc. Korea* **1992**, *27*, 237–246.
42. Choi, J.K.; Noh, J.H.; Shin, K.S.; Hong, K.H. The early autumn distribution of chlorophyll-a and primary productivity in the Yellow Sea, 1992. *The Yellow Sea* **1995**, *1*, 68–80.
43. Choi, J.K.; Noh, J.H.; Cho, S.H. Temporal and spatial variation of primary production in the Yellow Sea. The present and the future of yellow sea environments. In Proceedings of the Yellow Sea International Symposium, Ansan, Korea, 6–7 November 2003; Korea Ocean R&D Institute: Seoul, Korea, 2003; pp. 103–115.
44. Son, S.H.; Campbell, J.; Dowell, M.; Yoo, S.; Noh, J. Primary production in the Yellow Sea determined by ocean color remote sensing. *Mar. Ecol. Prog. Ser.* **2005**, *303*, 91–103. [CrossRef]
45. Lee, Y. Phytoplankton Dynamics and Primary Production in the Yellow Sea during Winter and Summer. Unpublished Ph.D. Thsesis, Inha University, Incheon, Korea, August 2012.
46. Chung, C.S.; Yang, D.B. On the primary productivity in the southern sea of Korea. *J. Oceanogr. Soc. Korea* **1991**, *26*, 242–254.
47. Hama, T.; Shin, K.H.; Handa, N. Spatial variability in the primary productivity in the East China Sea and its adjacent waters. *J. Oceanogr.* **1997**, *53*, 41–51. [CrossRef]
48. Lee Chen, Y.L.; Chen, H.Y. Nitrate-based new production and its relationship to primary production and chemical hydrography in spring and fall in the East China Sea. *Deep. Res. Part II Top. Stud. Oceanogr.* **2003**, *50*, 1249–1264. [CrossRef]
49. Siswanto, E.; Ishizaka, J.; Yokouchi, K. Estimating chlorophyll-a vertical profiles from satellite data and the implication for primary production in the Kuroshio front of the East China Sea. *J. Oceanogr.* **2005**, *61*, 575–589. [CrossRef]
50. Zhang, Y.R.; Ding, Y.P.; Li, T.J.; Xue, B.; Euo, Y.M. Annual variations of chlorophyll a and primary productivity in the East China Sea. *Oceanol. Limno. Sin.* **2016**, *47*, 261–268. [CrossRef]
51. Nagata, H. the Yamato Rise, central Japan Sea. *Plankton Bio. Ecol.* **1998**, *45*, 159–170.
52. Yoshie, N.; Shin, K.H.; Noriki, S. Seasonal variations of primary productivity and assimilation numbers in the western North Pacific. *Spec. Rep. Reg. Stud. North-East Eurasia North Pac. Hokkaido Univ.* **1999**, *1*, 49–62.
53. Sverdrup, H.U. On conditions for the vernal blooming of phytoplankton. *ICES J. Mar. Sci.* **1953**, *18*, 287–295. [CrossRef]
54. Yentsch, C.S. Estimates of "new production" in the mid-North Atlantic. *J. Plankton Res.* **1990**, *12*, 717–734. [CrossRef]
55. Nishikawa, H.; Yasuda, I.; Komatsu, K.; Sasaki, H.; Sasai, Y.; Setou, T.; Shimizu, M. Winter mixed layer depth and spring bloom along the Kuroshio front: Implications for the Japanese sardine stock. *Mar. Ecol. Prog. Ser.* **2013**, *487*, 217–229. [CrossRef]
56. Mayot, N.; D'Ortenzio, F.; Uitz, J.; Gentili, B.; Ras, J.; Vellucci, V.; Golbol, M.; Antoine, D.; Claustre, H. Influence of the Phytoplankton Community Structure on the Spring and Annual Primary Production in the Northwestern Mediterranean Sea. *J. Geophys. Res. Ocean.* **2017**, *122*, 9918–9936. [CrossRef]
57. Lim, S.H.; Jang, C.J.; Oh, I.S.; Park, J.J. Climatology of the mixed layer depth in the East/Japan Sea. *J. Mar. Syst.* **2012**, *96–97*, 1–14. [CrossRef]
58. Bae, H.; Lee, D.; Kang, J.J.; Lee, J.H.; Jo, N.; Kim, K.; Jang, H.K.; Kim, M.J.; Kim, Y.; Kwon, J., Il; et al. Satellite-derived protein concentration of phytoplankton in the Southwestern East/Japan Sea. *J. Mar. Sci. Eng.* **2021**, *9*, 189. [CrossRef]
59. Kang, Y.-S.; Choi, H.-C.; Lim, J.-H.; Jeon, I.-S.; Seo, J.-H. Dynamics of the Phytoplankton Community in the Coastal Waters of Chuksan Harbor, East Sea. *Algae* **2005**, *20*, 345–352. [CrossRef]
60. Zuenko, Y.; Selina, M.; Stonik, I. On conditions of phytoplankton blooms in the coastal waters of the north-western East/Japan Sea. *Ocean Sci. J.* **2006**, *41*, 31–41. [CrossRef]
61. Chang, K.-I.; Zhang, C.-I.; Park, C.; Kang, D.-J.; Ju, S.-J.; Lee, S.-H.; Wimbush, M. (Eds.) *Oceanography of the East Sea (Japan Sea)*, 1st ed.; Springer International Publishing: Cham, Switzerland, 2016.
62. Kwak, J.H.; Han, E.; Lee, S.H.; Park, H.J.; Kim, K.R.; Kang, C.K. A consistent structure of phytoplankton communities across the warm–cold regions of the water mass on a meridional transect in the East/Japan Sea. *Deep. Res. Part II Top. Stud. Oceanogr.* **2017**, *143*, 36–44. [CrossRef]
63. Chung, C.S.; Shim, J.H.; Park, Y.H.; Park, S.G. Primary productivity and nitrogenous nutrient dynamics in the East Sea of Korea. *J. Oceanogr Soc. Korea* **1989**, *24*, 52–61.
64. Park, J.S.; Kang, C.K.; An, K.H. Community structure and spatial distribution of phytoplankton in the polar front region off the east coast of Korea in summer. *Bull. Korean Fish. Soc.* **1991**, *24*, 237–247.
65. Yamada, K.; Ishizaka, J.; Nagata, H. Spatial and temporal variability of satellite primary production in the Japan Sea from 1998 to 2002. *J. Oceanogr.* **2005**, *61*, 857–869. [CrossRef]
66. Hyun, J.H.; Kim, D.; Shin, C.W.; Noh, J.H.; Yang, E.J.; Mok, J.S.; Kim, S.H.; Kim, H.C.; Yoo, S. Enhanced phytoplankton and bacterioplankton production coupled to coastal upwelling and an anticyclonic eddy in the Ulleung basin. *East Sea Aquat. Microb. Ecol.* **2009**, *54*, 45–54. [CrossRef]

67. Kwak, J.H.; Lee, S.H.; Park, H.J.; Choy, E.J.; Jeong, H.D.; Kim, K.R.; Kang, C.K. Monthly measured primary and new productivities in the Ulleung Basin as a biological "hot spot" in the East/Japan Sea. *Biogeosciences* **2013**, *10*, 4405–4417. [CrossRef]
68. Kwak, J.H.; Hwang, J.; Choy, E.J.; Park, H.J.; Kang, D.J.; Lee, T.; Chang, K., Il; Kim, K.R.; Kang, C.K. High primary productivity and f-ratio in summer in the Ulleung basin of the East/Japan Sea. *Deep. Res. Part I Oceanogr. Res. Pap.* **2013**, *79*, 74–785. [CrossRef]
69. Joo, H.T.; Son, S.H.; Park, J.W.; Kang, J.J.; Jeong, J.Y.; Kwon, J., Il; Kang, C.K.; Lee, S.H. Small phytoplankton contribution to the total primary production in the highly productive Ulleung Basin in the East/Japan Sea. *Deep. Res. Part. II Top. Stud. Oceanogr.* **2017**, *143*, 54–61. [CrossRef]
70. Lim, Y.J.; Kim, T.W.; Lee, S.H.; Lee, D.; Park, J.; Kim, B.K.; Kim, K.; Jang, H.K.; Bhavya, P.S.; Lee, S.H. Seasonal Variations in the Small Phytoplankton Contribution to the Total Primary Production in the Amundsen Sea, Antarctica. *J. Geophys. Res. Ocean.* **2019**, *124*, 8324–8341. [CrossRef]
71. Agawin, N.S.R.; Duarte, C.M.; Agusti, S. Nutrient and temperature control of the contribution of picoplankton to phytoplankton biomass and production. *Limnol. Oceanogr.* **2000**, *45*, 591–600. [CrossRef]
72. Hilligsøe, K.M.; Richardson, K.; Bendtsen, J.; Sørensen, L.L.; Nielsen, T.G.; Lyngsgaard, M.M. Linking phytoplankton community size composition with temperature, plankton food web structure and sea-air CO2 flux. *Deep. Res. Part I Oceanogr. Res. Pap.* **2011**, *58*, 826–838. [CrossRef]
73. Morán, X.A.G.; López-Urrutia, Á.; Calvo-Díaz, A.; LI, W.K.W. Increasing importance of small phytoplankton in a warmer ocean. *Glob. Chang. Biol.* **2010**, *16*, 1137–1144. [CrossRef]
74. Mousing, E.A.; Ellegaard, M.; Richardson, K. Global patterns in phytoplankton community size Structure-evidence for a direct temperature effect. *Mar. Ecol. Prog. Ser.* **2014**, *497*, 25–38. [CrossRef]

Journal of
*Marine Science
and Engineering*

MDPI

Article

Estimation of Phytoplankton Size Classes in the Littoral Sea of Korea Using a New Algorithm Based on Deep Learning

Jae Joong Kang [1], Hyun Ju Oh [1], Seok-Hyun Youn [1], Youngmin Park [2], Euihyun Kim [2], Hui Tae Joo [1,*]
and Jae Dong Hwang [1,*]

[1]   National Institute of Fisheries and Sciences, Gijan Haean-ro, Gijang Gun, Busan 15807, Korea
[2]   Geosystem Research Corporation, Department of Marine Forecast, 306, 172 LS-ro, Gunpo-si 15807, Korea
*   Correspondence: huitae@korea.kr (H.T.J.); jdhwang@korea.kr (J.D.H.); Tel.: +82-51-720-2221(H.T.J.);
    +82-51-720-2240 (J.D.H.)

Abstract: The size of phytoplankton (a key primary producer in marine ecosystems) is known to influence the contribution of primary productivity and the upper trophic level of the food web. Therefore, it is essential to identify the dominant sizes of phytoplankton while inferring the responses of marine ecosystems to change in the marine environment. However, there are few studies on the spatio-temporal variations in the dominant sizes of phytoplankton in the littoral sea of Korea. This study utilized a deep learning model as a classification algorithm to identify the dominance of different phytoplankton sizes. To train the deep learning model, we used field measurements of turbidity, water temperature, and phytoplankton size composition (chlorophyll-a) in the littoral sea of Korea, from 2018 to 2020. The new classification algorithm from the deep learning model yielded an accuracy of 70%, indicating an improvement compared with the existing classification algorithms. The developed classification algorithm could be substituted in satellite ocean color data. This enabled us to identify spatio-temporal variation in phytoplankton size composition in the littoral sea of Korea. We consider this to be highly effective as fundamental data for identifying the spatio-temporal variation in marine ecosystems in the littoral sea of Korea.

Keywords: phytoplankton; phytoplankton size classes (PSCs); ocean color; deep neural network (DNN)

Citation: Kang, J.J.; Oh, H.J.; Youn, S.-H.; Park, Y.; Kim, E.; Joo, H.T.; Hwang, J.D. Estimation of Phytoplankton Size Classes in the Littoral Sea of Korea Using a New Algorithm Based on Deep Learning. *J. Mar. Sci. Eng.* **2022**, *10*, 1450. https://doi.org/10.3390/jmse10101450

Academic Editor: Carmela Caroppo

Received: 30 August 2022
Accepted: 5 October 2022
Published: 7 October 2022

Publisher's Note: MDPI stays neutral with regard to jurisdictional claims in published maps and institutional affiliations.

1. Introduction

Environmental condition (temperature, salinity, nutrients, and light availability) change impacts on the structure and functions of marine ecosystems, especially phytoplankton, which play a critical role as primary producers in the biogeochemical cycles of marine ecosystems [1–3]. Physical and chemical variations in the ocean are known to influence the primary production, photosynthetic properties, and size of phytoplankton [2]. The West, South, and East Seas comprise the littoral sea of Korea. Their surface water temperatures are increasing two to four times faster than global temperatures [4]. Previous studies have continuously reported variations in biological properties, such as the blooming cycle of phytoplankton, community fluctuations, and effects on the upper trophic level [5–8]. In particular, several recent papers have reported that the proportion of small phytoplankton is increasing, owing to warming water temperatures [9–12]. The dominant size distribution of phytoplankton plays a critical role in variations in energy efficiency and primary productivity for upper trophic marine organisms [13–17]. The physiological characteristics of phytoplankton are also highly related to its size; therefore, it plays a crucial role in the identification of the impacts of environmental change on marine ecosystems [17]. Nevertheless, there is inadequate long-term information on the community fluctuations and spatio-temporal distributions of phytoplankton according to size. A method to identify these long-term spatio-temporal distributions involves chlorophyll-a, using satellite ocean color [18]. However, previous studies have focused on investigations related to the total amount of phytoplankton. Recently, it has been considered more important to identify the impacts of change on variations in the environment and phytoplankton structure [19]. Although studies related to the contributions of the dominant phytoplankton size in the littoral sea of Korea have been reported, research on the long-term spatio-temporal distributions in the littoral sea is inadequate. Therefore, a method to identify the long-term spatio-temporal distributions of dominant phytoplankton size is essential.

The dominant size of phytoplankton is generally divided into three sizes (referred to as phytoplankton size classes (PSCs)): micro-size phytoplankton (>20 μM), nano-size phytoplankton (2–20 μM), and pico-size phytoplankton (<2 μM). Researchers are attempting to perform studies on PSCs using various methods (microscopy, flow-cytometry, filters, pigment analysis). However, it is difficult to identify the long-term spatio-temporal characteristics of variations in PSCs owing to the limitations of data observed in the field [17]. Researchers recently developed a method to identify the spatio-temporal distributions of phytoplankton size compositions using ocean color data [20–22]. There are two commonly used types of PSCs algorithms: spectral-based and abundance-based algorithms [21,23]. Spectral-based algorithms perform estimation using correlations with the properties of PSCs that vary with optical characteristics [24–27]. This method is highly sensitive to optical characteristics and is difficult use in ocean waters with high turbidity [28,29]. Abundance-based algorithms perform estimation using the statistical correlation between total chlorophyll-a concentration and composition according to phytoplankton size [30,31]. This method has difficulty reflecting both regional characteristics and characteristics in ocean waters with a complex environment [17]. To enhance the estimation accuracy of PSCs, an algorithm needs to be developed that can reflect the variations in phytoplankton communities according to time and environmental transitions by using an abundance-based algorithm that considers physical factors [18,32,33]. Deep learning techniques, distinct from existing methods, are drawing attention as tools to develop the algorithm. Furthermore, the distribution of PSCs in oceans worldwide can be determined by applying physical characteristics and spatiotemporal variations in PSCs and PFTs (Phytoplankton Functional Types) by using actual machine learning techniques [34].

The objective of this study is to develop a PSCs algorithm, which includes environmental factors, using deep learning techniques and lays a foundation for identifying the spatio-temporal distributions of PSCs. We develop a new algorithm to estimate PSCs from satellite ocean color data, which enables spatio-temporal analysis that is missing in observations and may contribute to understanding the variations in marine ecosystems caused by environmental changes.

2. Data and Methods

2.1. Field Observation Data

This study used observation data (2018–2020) from the National Institute of Fisheries Science (NIFS) to develop a new PSCs algorithm applicable to the East, West, South Seas (which comprise the littoral sea of Korea), and the East China Sea. Using the ocean observation data from the main institute and regional research centers of the NIFS, we developed the new algorithm based on data observed at the sea surface at 60 stations (Figure 1) in spring (April–May), summer (August), fall (October–November), and winter (February). Details of the survey stations are shown in Supplementary Table S1.

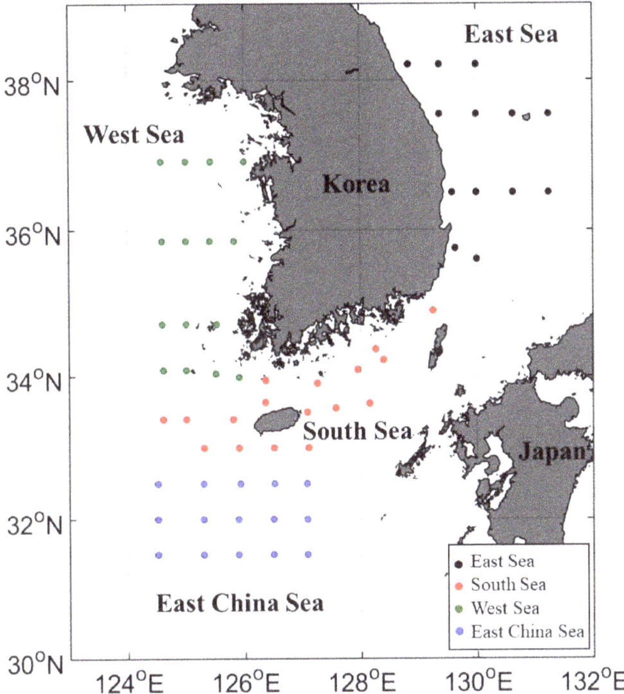

Figure 1. Locations of field observation stations.

The parameters used to develop the new algorithm were total chlorophyll-a, size-fractionated chlorophyll-a (micro, nano, and pico size), total suspended solids (TSS), and sea surface water temperature (SST). The data were collected and observed using a rosette sampler and CTD (SBE-911, Seabird Electronics Inc., Bellevue, WA, United States) installed on the vessel. To measure the total chlorophyll-a in the collected seawater, it was filtered with a 25 mm GF/F (Whatman) and stored in a freezer (-70 °C). To measure size-fractionated chlorophyll-a (micro, nano, and pico size), the seawater was sequentially filtered on 20 and 2 μm Nuclepore membrane filters (47 mm) and then 47 mm GF/F (pore size: 0.7 μm), and stored in a freezer (-70 °C). Then, chlorophyll-a was extracted in a laboratory with 90% acetone over 24 h using the method specified by Parson et al. [35] and was analyzed using the 10-AU device. The weight before and after filtering was measured using a precombusted 47 mm GF/F to analyze TSS.

2.2. Training and Model Structure

In this study, a deep neural network (DNN) based model was used to construct the new algorithm for phytoplankton size classification, which was station-based. As training data, the above-mentioned parameters (Section 2.1) were used to train the model, which were obtained by field observation. The total chlorophyll-a (total phytoplankton biomass; unit: μg L^{-1}, $n = 531$), TSS (determination factor of turbidity in study areas; unit: μg L^{-1}, $n = 531$), and SST (controlling factor of phytoplankton size; unit: °C, $n = 531$) were used as input variables, whereas the size fractionated chlorophyll-a, obtained in a laboratory using the Parson method [35], was used as ground truth data classifying the three categorical types (micro-size phytoplankton, $n = 126$; nano-size phytoplankton, $n = 99$; pico-size phytoplankton, $n = 306$; Table 1) corresponding with the dominant size. In other words, the ground truth data (dominant size of phytoplankton in each area; unit: character types) were used to train and validate the output layer of the DNN-based model for each input variable to develop the new algorithm for phytoplankton size classification (Figure 2). Of the total 531 pieces of station dataset input variables: total chlorophyll-a (n = 531), TSS (n = 531), and SST (n = 531); ground truth data: size fractionated chlorophyll-a (n = 531)], 210 pieces were used for the training dataset, and 30 pieces were used for validation dataset in the model training process. The training and validation datasets were composed of 70 and 10 pieces of each class, respectively, in order to prevent over-fitting to a specific class in the training process. The remaining 291 pieces were used as the test dataset, validating the final trained model.

Table 1. The number of dominant-sized phytoplankton (ground truth data for training DNN-based model) which appeared at field observation stations in each sea area (East, West, South, and East China Seas) during the study period (2018–2020).

	East Sea	West Sea	South Sea	East China Sea	Sum
Microsize phytoplankton (>20 µM)	39	13	21	53	126
Nanosize phytoplankton (2–20 µM)	42	39	8	10	99
Picosize phytoplankton (0.7–2 µM)	54	115	86	51	306
Total	135	167	115	114	531

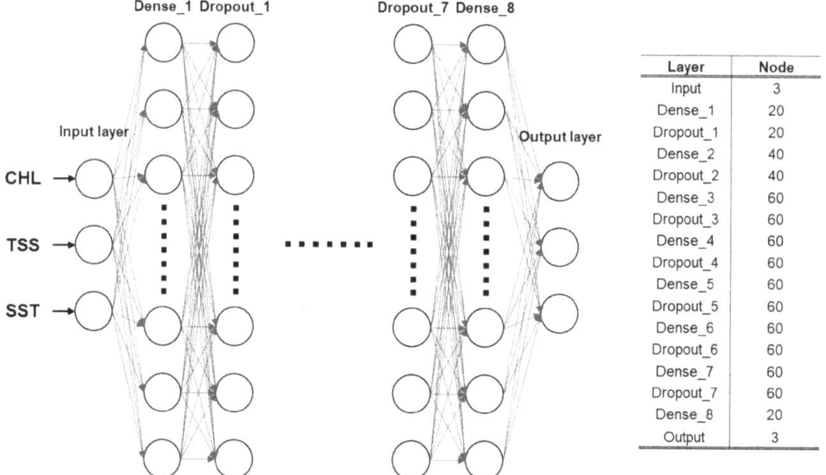

Figure 2. Deep learning framework of this study model. As training data, the total chlorophyll-a (CHL), total suspended solids (TSS), and sea surface temperature (SST), obtained by field observation, were used as input variables in this model.

Figure 2 shows the DNN-based model structure of this study. In an input layer, the three input variables (total chlorophyll-a, TSS, and SST) were used and a sigmoid function was used as an activation function (Equation (1)). The sigmoid function has a characteristic of a curve shape, so it prevents a divergence of each value. The hidden layers were comprised as eight dense layers, and a hyperbolic tangent function was used for the activation function of each layer (Equation (2)). In terms of shape, the hyperbolic tangent function is similar to the sigmoid function as used in the input layer. However, it is faster than the sigmoid function in terms of optimization. Therefore, it is possible to train the model by densely stacking each layer and solving a non-linearity problem efficiently. Each hidden layer consisted of 20–60 nodes per dense layer. Between the dense layers, the dropout layers were combined to prevent the model over-fitting to a specific class during training. The last layer, the output layer, used the soft-max function to classify the dominant phytoplankton size classes (Equation (3)). In the processing of training, an Adam and a categorical cross-entropy were applied for the model training optimizer and the loss function, respectively. The training epochs and batch size were set to 2000 and 10, respectively, so the weights were configured to update 21 times per epoch. In addition, training was configured to stop early if the loss function value for the validation dataset did not improve within 80 epochs, in order to reduce the over-fitting and the model execution time.

$$\sigma(x) = \frac{1}{1 + exp^{-x}} \tag{1}$$

$$\tanh(x) = \frac{exp^x - exp^{-x}}{exp^x + exp^{-x}} \tag{2}$$

$$y_k = \frac{exp(a_k)}{\sum_{i=1}^{n} exp(a_i)} \tag{3}$$

2.3. Satellite Data

Using the new algorithm based on the final trained model, this study used satellite ocean color data to identify the spatio-temporal distribution of dominant phytoplankton size classes. The collected satellite data was a VIIRS-SNPP (Visible and Infrared Imager/Radiometer Suite-Suomi National Polar-orbiting Partnership) by the OBPG (Ocean Biology Processing Group) at NASA Goddard Space Flight Center accessed on 9 May 2022 (https://oceandata.sci.gsfc.nasa.gov/VIIRS-SNPP/), and consisted of monthly level-3 data on reflectance of remote sensing at 551 nm (Rrs551), total chlorophyll-a, and SST from 2018 to 2020. TSS, which needs a trained model as an input variable, was estimated using a previously developed algorithm [36] with Rrs551. By assumption, and due to no significant difference between the satellite ocean color data and in-situ data, this study used the satellite data to identify the spatio-temporal distribution of dominant phytoplankton size classes.

3. Results and Discussion

3.1. Results of DNN-Based Model for PSCs

We verified the accuracy of the trained DNN-based model results using classification performance evaluation indicators. Then, the model was analyzed using Equations (4)–(7) with the classification of the confusion matrix in Table 2. According to the confusion matrix, a result is considered true positive (TP) if both prediction and measurement are correct, false positive (FP) if the actual incorrect answer is predicted as correct, false negative (FN) if the actual correct answer is predicted as incorrect, and true negative (TN) if the actual incorrect answer is predicted as incorrect.

Table 2. Confusion matrix.

		In Situ Results	
		True	**False**
Model results	True	True Positive (TP)	False Positive (FP)
	False	False Negative (FN)	True Negative (TN)

We used four parameters to verify the model performance: precision (Equation (4)), recall (Equation (5)), accuracy (Equation (6)), and F1-score (Equation (7)). Table 2 lists the results obtained using the equations.

$$Precision = \frac{TP}{TP + FP} \tag{4}$$

$$\text{Recall} = \frac{TP}{TP + FN} \tag{5}$$

$$\text{Accuracy} = \frac{TP + TN}{TP + FN + FP + TN} \tag{6}$$

$$\text{F1} - \text{score} = 2 \times \frac{Precision \times Recall}{Precision + recall} \tag{7}$$

We used 291 pieces of data excluding the training and validation data (46 micro-size phytoplankton (16%), 19 nano-size phytoplankton (6%), and 226 pico-size phytoplankton (78%)) to verify the accuracy of the developed classification model. The precision, recall, and F1-score for micro-size phytoplankton, nano-size phytoplankton, and pico-size phytoplankton are 34.8%, 45.7%, and 39.5; 42.1%, 17.8%, and 25%; and 80.9%, 85.8%, and 82.8%, respectively. The model yielded an overall accuracy of 70.5% (Table 3). An examination of the test dataset results of the trained DNN-based model revealed that the accuracy of 70% exceeds that of the existing methods for the waters surrounding the Korean Peninsula. However, the accuracy of the training data, validation data, and test data of the DNN-based model differ moderately. With regard to the accuracy of the training set, the precision for pico-size phytoplankton is high. The high accuracy is attributed partially to the large proportion of pico-size phytoplankton in the test dataset at 78%. This is because the model mainly estimates pico-size phytoplankton, which is the dominant phytoplankton size observed in the littoral sea of Korea. Regular acquisition of observations to increase the data would help improve the accuracy of the DNN-based model.

Table 3. Precision, recall, F1-score, and accuracy of training, validation, and test datasets for classification model.

Data		Precision (%)	Recall (%)	F1_Score	Accuracy (%)
Training data (210)	Micro-size phytoplankton (70)	37.1	65.0	47.3	
	Nano-size phytoplankton (70)	44.3	67.4	53.5	55.7
	Pico-size phytoplankton (70)	85.7	48.4	61.9	
Validation data (30)	Micro-size phytoplankton (10)	20.0	66.7	30.8	
	Nano-size phytoplankton (10)	30.0	60.0	40.0	46.7
	Pico-size phytoplankton (10)	90.0	40.9	56.3	
Test data (291)	Micro-size phytoplankton (46)	34.8	45.7	39.5	
	Nano-size phytoplankton (19)	42.1	17.8	25.0	70.5
	Pico-size phytoplankton (226)	80.1	85.8	82.8	

3.2. Estimation of Phytoplankton Size Classes in the Littoral Sea of South Korea Using Satellite

To identify the spatio-temporal distributions of the dominant phytoplankton size in the littoral sea (the purpose of this study), we applied the developed algorithm using a DNN-based model to the satellite ocean color data, and obtained monthly distributions of dominant phytoplankton size from 2018 to 2020. These are shown in Figures 3–5. Although there are marginal spatial differences in the dominant sea areas from 2018 to 2020, the distribution trends of the dominant size are similar. An examination of the distributions of dominant phytoplankton size by sea area revealed that micro-size phytoplankton are dominant in the West Sea from January to March, whereas nano-size phytoplankton are dominant in April and May. In contrast, according to a seasonal survey of the West Sea (2018) by Jang et al. [37], nano-size phytoplankton are dominant in both February (approximately 50%; micro-size: approximately 22%) and April (approximately 42%; micro-size: approximately 37%) during the research period. This was because the contributions of chlorophyll-a according to size, reported by Jang et al. [37], reflected the chlorophyll-a concentrations in the depth of the photic zone, which includes the surface layer. With regard to the phytoplankton community structure in the East Sea, identified using the algorithm developed in this study, nano-size phytoplankton are dominant from January to April. From May, the dominant sea area of pico-size phytoplankton expands gradually. According to prior research, the key dominant size of phytoplankton communities in spring (March–April) in the East Sea are micro-size diatoms [38–40]. However, previous studies on phytoplankton communities in the East Sea reported differences in the dominant communities depending on the study period and survey station [41–45]. An observation shared among most of these studies is that the size of the key dominant size of the phytoplankton communities tends to decrease as the water temperature increases after spring [41,43–45]. According to the observations of phytoplankton communities in the South Sea and East China Sea from January to May, micro-size phytoplankton are dominant in the west, whereas pico-size phytoplankton are dominant in the central region. Although this spatio-temporal distribution pattern in the East China Sea is similar to those observed in previous studies [46,47], there is insufficient comparable prior research data on variations in phytoplankton communities in the South Sea during this period. From June to September, pico-size phytoplankton showed high distributions in all the sea areas (Figures 3–5). In summer (June–August) when strong stratification develops, pico-size phytoplankton are the dominant size in the littoral sea at surface depths (East, West, South, and East China Seas) [37,44,45,47–49]. However, certain studies conducted in the South Sea observed that the contribution of nano-size phytoplankton or micro-size phytoplankton were high [19,50]. This is likely because the sea area where the studies were conducted is a bay with a high inflow of nutrients from the outside through precipitation [50,51]. From November, sea areas dominant with nano-size phytoplankton gradually begin to appear in the central West Sea and East Sea littoral seas. The dominant sea areas increased until December (Figures 3–5). For the West Sea, comparable prior research results on the spatio-temporal variations in phytoplankton communities during this period are insufficient. In contrast, as mentioned above, the East Sea shows different distribution characteristics depending on the study period and station. However, according to Jo et al. [44], nano-size was the dominant size during September and October after summer, whereas micro-size was dominant in November, followed by nano-size. By substituting the developed PSCs algorithm for all the sea areas of South Korea, the most dominant size in the East, West, South, and East China Seas was observed to be pico-size phytoplankton. Meanwhile, nano-size phytoplankton was dominant in the northern waters of the West Sea, and micro-size phytoplankton in the littoral seas of the West Sea.

To determine the accuracy and usability of the developed algorithm applied to ocean color, we compared the dominant size estimated via satellite using new PSCs algorithm and the dominant size by sea area (East, West, South, and East China Seas) in the field data, as well as previously developed dominant size estimation algorithms (Table 4). The accuracy was approximately 69.5% according to comparisons of the most dominant phytoplankton size by sea area observed in the field and the most dominant size analyzed through ocean color (the algorithm developed in this study) (Table 4). The dominant distribution of pico-size phytoplankton showed high accuracy in the developed algorithm. One of the main causes of this result is because pico-size phytoplankton are the most dominant size in Korean waters, and were the most prevalent in the sample. With regard to accuracy by sea area, the accuracies for the East, West, South, and East China Seas were 50%, 66.6%, 90%, and 67%, respectively. The lowest was for the East Sea, and the highest was for the East China Sea. To compare the accuracy of the results applied to ocean color using the new algorithm with existing models, we performed a comparison using a dominant size estimation method applying an absorption model (A_{ph}) [25] and dominant size estimation results using the three components model [52] (Table 4). According to the dominant size results, the A_{ph} model yielded an accuracy of approximately 54% in the littoral sea. Meanwhile, the three-component model showed an accuracy of 13%. The previous field observations indicated that pico-size phytoplankton were dominant in most of the sea areas, whereas the three-component model estimated that micro- and nano-size phytoplankton were dominant in most of the sea areas. In particular, all the cases were incorrect in the South Sea. The algorithm developed in this study showed a higher accuracy than the existing algorithms in identifying variations in the dominant size in the littoral sea of South Korea. We consider that real AI learning-based algorithms could attain even higher performance if they were continuously trained with new data.

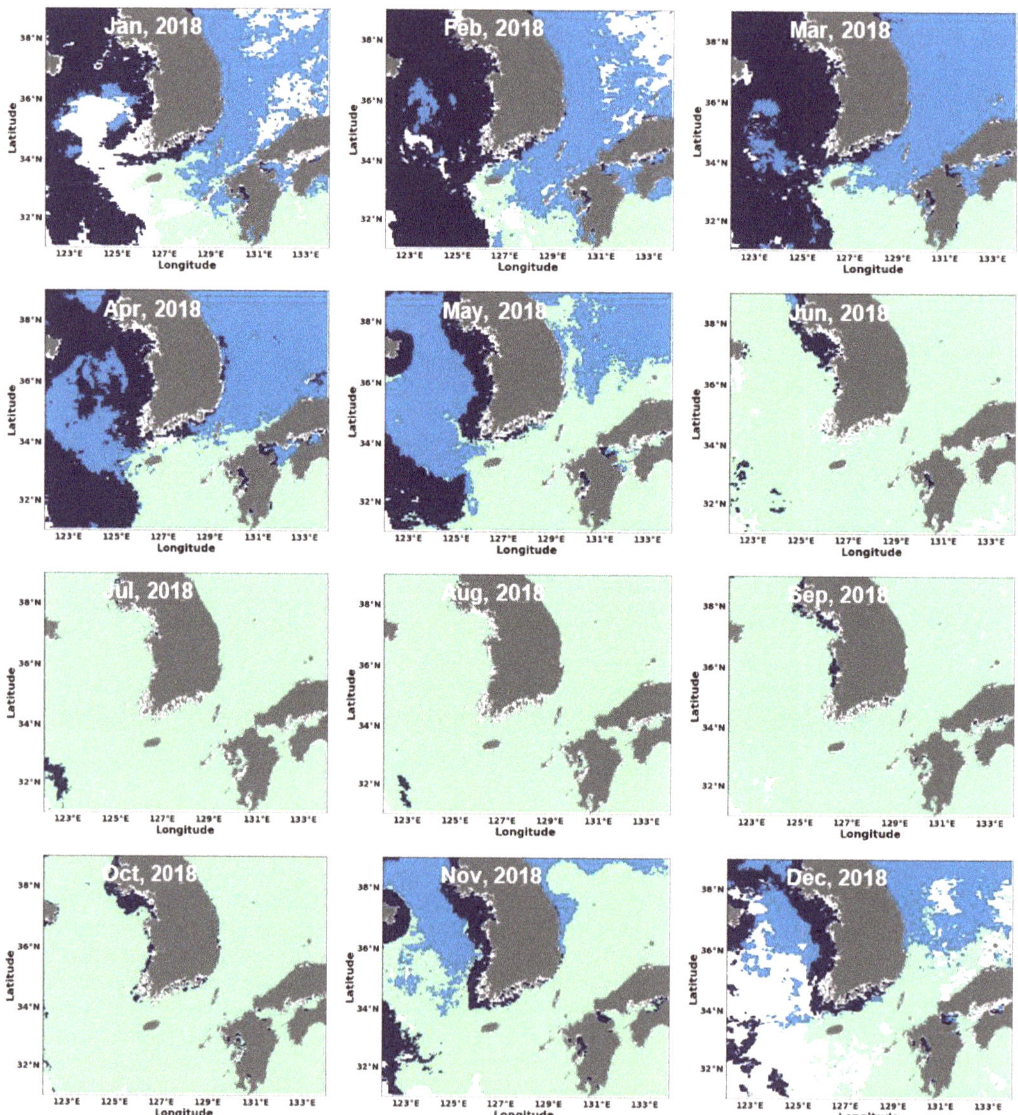

Figure 3. Classification based monthly PSCs distribution using L3 mapped VIIRS data in 2018. White color indicates missing area by cloud. Indigo, blue, and green colors indicate micro-size phytoplankton, nano-size phytoplankton, and pico-size phytoplankton, respectively.

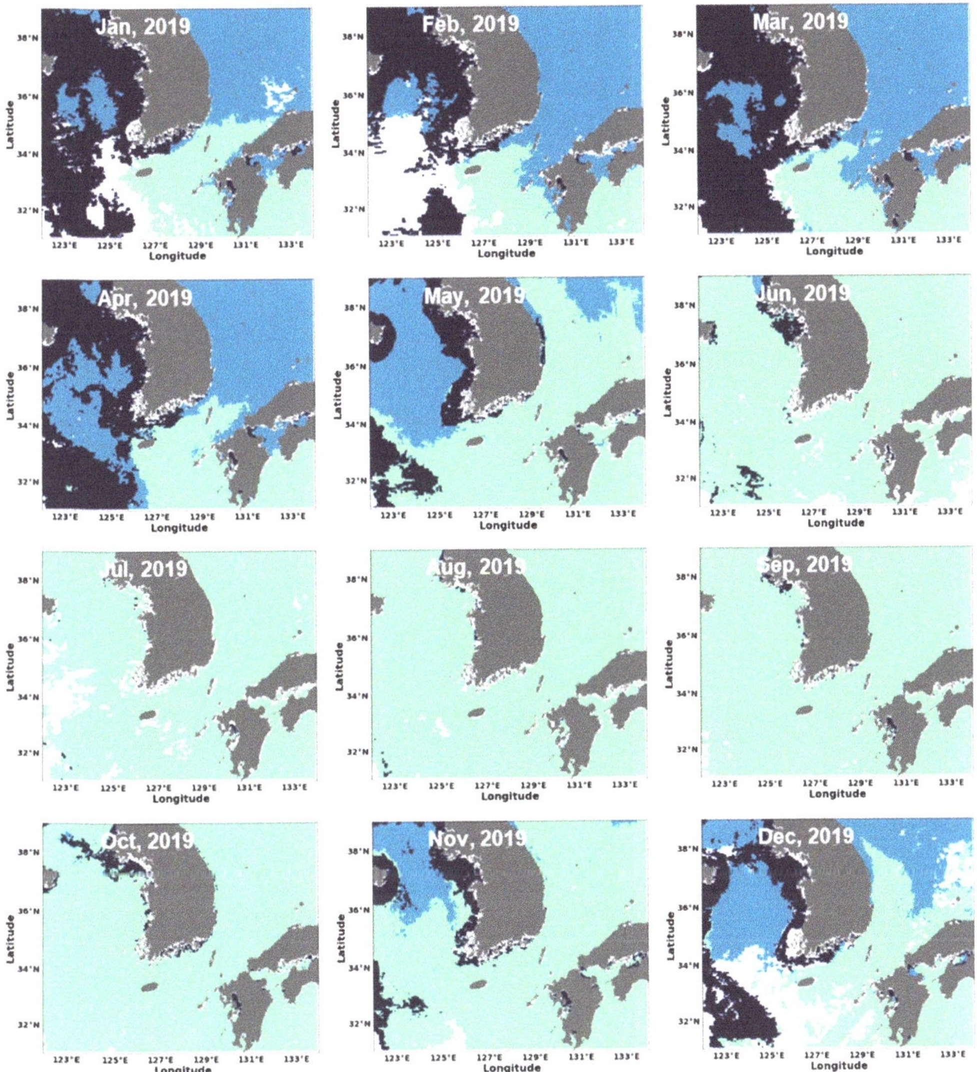

Figure 4. Classification based monthly PSCs distribution using L3 mapped VIIRS data in 2019. White color indicates missing area by cloud. Indigo, blue, and green colors indicate micro-size phytoplankton, nano-size phytoplankton, and pico-size phytoplankton, respectively.

Table 4. Matched percent between field measurement Chl-a size fractionated with New AI algorithm (this study), A_{ph} algorithm, and three-component model derived from satellite data.

Field Measurments	New AI Algorithm (This Study)	A_{ph} Algorithm	Three-Component Model
Study area	67.3%	54.3%	13.4%
East Sea	50.0%	41.7%	16.7%
West Sea	66.7%	41.7%	33.3%
South Sea	90%	90%	0%
East China sea	66.7%	50%	0%

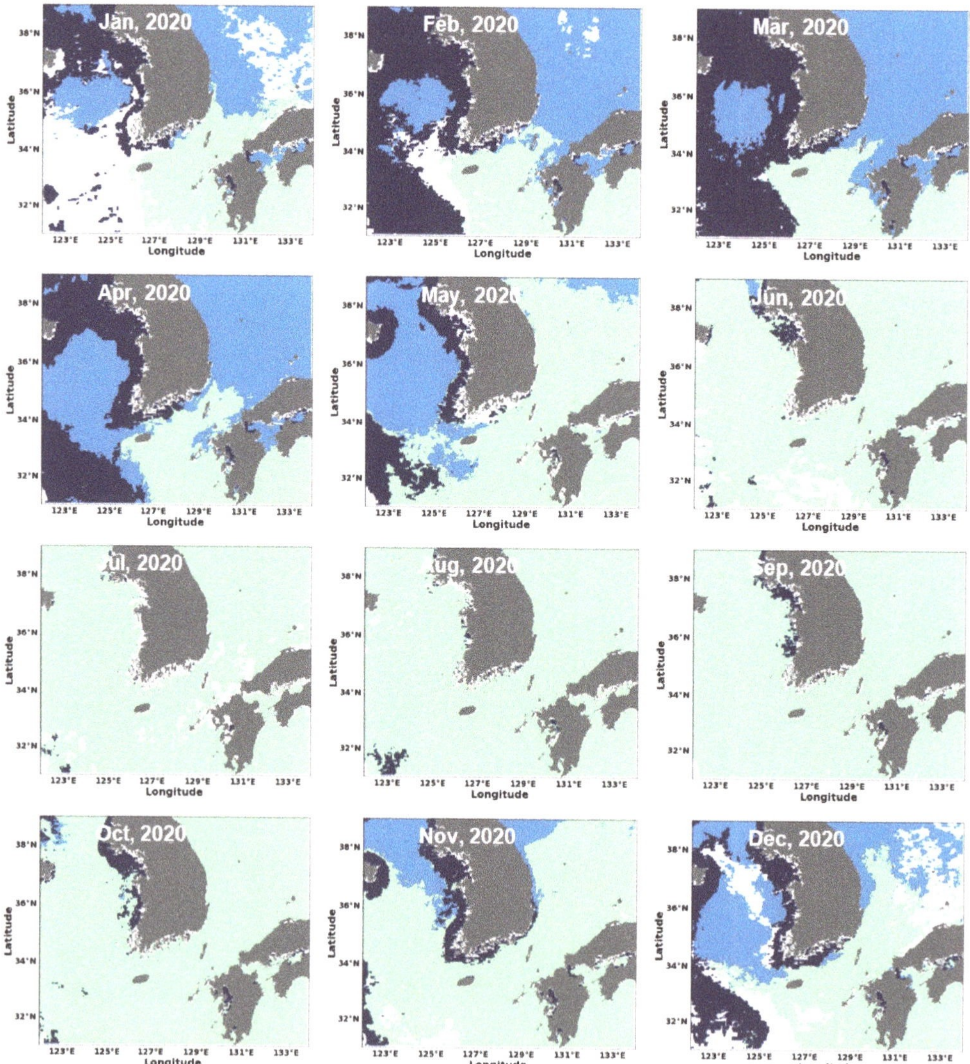

Figure 5. Classification based monthly PSCs distribution using L3 mapped VIIRS data in 2020. White color indicates a missing area by the cloud. Indigo, blue, and green colors indicate micro-size phytoplankton, nano-size phytoplankton, and pico-size phytoplankton, respectively.

4. Summary and Conclusions

It is crucial to identify the spatio-temporal distributions of dominant phytoplankton size to understand the variations in coastal marine ecosystems occurring alongside environmental change. The PSCs classification models in prior studies have limited usability owing to their low accuracy in sea areas surrounding the Korean Peninsula. Accordingly, this study attempted to develop a DNN-based PSCs classification algorithm suitable for sea areas around the Korean Peninsula by utilizing data collected from continuous observations as training data for the DNN model.

For this purpose, we collected data observed at the sea surface at 60 stations in spring (April–May), summer (August), fall (October–November), and winter (February) over three years (2018–2020), from the ocean observation data obtained from the NIFS main institute and regional research centers. Data on total chlorophyll-a, size-fractionated chlorophyll-a, TSS, and SST were collected and observed using a rosette sampler and CTD (SBE-911) installed on the vessel. The collected data were used as training data for the DNN-based model.

The available data was collected at different times and locations, and is not continuous. Therefore, the DNN-based model was used for development of new PSCs algorithm. To prevent biased learning owing to imbalanced data, we acquired 70 samples of each class from the data by plankton size for the training process. In addition, we selected four sea areas (the West, South, East, and East China Seas) according to the characteristics of waters around the Korean Peninsula, and configured the ratios of data identically for each sea area. The developed DNN-based PSCs algorithm achieved an accuracy of 70%, which exceeds that of the existing algorithms. However, the high accuracy is partially attributed to the large proportion of pico-size phytoplankton in the test dataset, at 78%. This aspect of the model must be improved by securing additional data in the future.

To examine the distribution characteristics of PSCs in the sea areas surrounding the Korean Peninsula through the developed DNN-based PSCs algorithm, we input satellite data to express spatio-temporal distribution characteristics. Monthly averages and three-year averages of the satellite water temperature, turbidity (i.e., TSS), and total chlorophyll were applied as input values to the new algorithm. The results verified that micro-size phytoplankton were dominant on the West Sea coast, nano-size phytoplankton in the northern East Sea, and pico-size phytoplankton in the South and East China Seas (Figure 6). By season, micro-size phytoplankton and nano-size phytoplankton are dominant in winter and spring in the East, West, and South Seas. From summer, pico-size phytoplankton are dominant throughout the littoral sea of Korea (Figures 3–5).

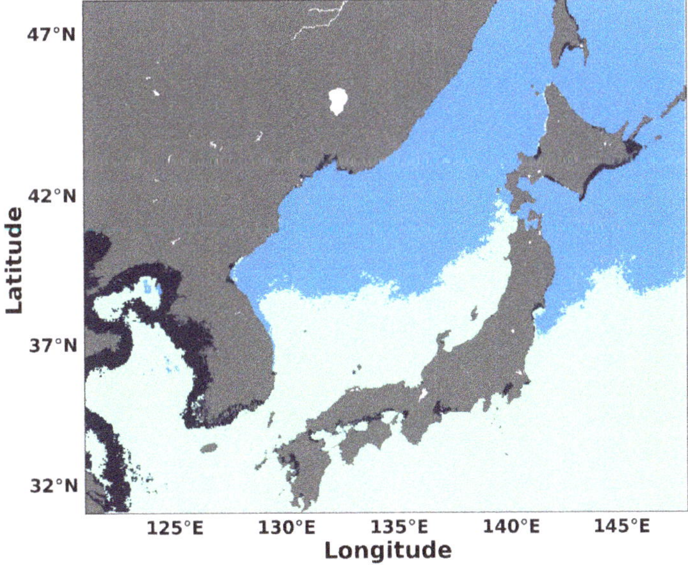

Figure 6. Dominant area of each class. Indigo, blue, and green colors indicate micro-size phytoplankton, nano-size phytoplankton, and pico-size phytoplankton, respectively.

This study developed a DNN-based PSCs algorithm that classifies the phytoplankton in the littoral sea of Korea into three size levels for the first time. In addition, it presented results on the spatio-temporal distribution of the dominant size based on satellite data. Based on these results, we consider that it can be made capable of producing important data for understanding the variations in coastal marine ecosystems occurring alongside environmental change. However, continuous improvement of the DNN-based PSCs algorithm accuracy through comparison with in situ data is necessary for more precise satellite observations.

Supplementary Materials: The following supporting information can be downloaded at: https://www.mdpi.com/article/10.3390/jmse10101450/s1, Table S1: Marine ecosystem survey stations of NIFS; Table S2: Dominant phytoplankton size (micro: M, nano: N, and pico: P) in the littoral water of Korea from field measurements, new algorithm (this study), Aph algorithm, and three-component model.

Author Contributions: Conceptualization, H.T.J. and Y.P.; methodology, H.T.J. and E.K.; validation, Y.P. and J.D.H. formal analysis, J.J.K. and H.J.O.; investigation, H.T.J., H.J.O., J.J.K., E.K. and J.D.H.; data curation, H.T.J.; writing-original draft preparation, H.T.J. and J.J.K.; writing-review and editing, J.J.K.; S.-H.Y.; E.K.; visualization, H.T.J. and E.K.; supervision, J.D.H.; project administration, J.D.H. and H.J.O.; funding acquisition, S.-H.Y. and H.J.O. All authors have read and agreed to the published version of the manuscript.

Funding: This research was funded by the National Institute of Fisheries and Science ('Development of marine ecological forecasting system for Korean waters'; R2022075).

Institutional Review Board Statement: Not applicable.

Informed Consent Statement: Not applicable.

Data Availability Statement: Not applicable.

Acknowledgments: This work was supported by the National Institute of Fisheries Science (NIFS) grant ('Development of marine ecological forecasting system for Korean waters'; R2022075) funded by the ministry of oceans and fisheries, republic of Korea.

Conflicts of Interest: The authors declare no conflict of interest.

References

1. D'Alelio, D.; Libralato, S.; d'Alcalà, M.R. Ecological-network models link diversity, structure and function in the plankton food-web. *Sci. Rep.* **2016**, *6*, 21806. [CrossRef]
2. D'Alelio, D.; Rampone, S.; Cusano, L.M.; Morfino, V.; Russo, L.; Saneverino, N.; Cloern, J.E.; Lomas, M.W. Machine learning identifies a strong association between warming and reduced primary productivity in an oligotrophic ocean gyre. *Sci. Rep.* **2020**, *10*, 3287. [CrossRef]
3. Harris, G. *Phytoplankton Ecology: Structure, Function and Fluctuation*; Chapman and Hall: London, UK, 1986.
4. Belkin, I.M. Rapid warming of Large Marine Ecosystems. *Prog. Oceanogr.* **2009**, *81*, 207–213. [CrossRef]
5. Kang, J.J.; Jang, H.K.; Lim, J.-H.; Lee, D.; Lee, J.-H.; Bae, H.; Lee, C.H.; Kang, C.-K.; Lee, S.H. Characteristics of different size phytoplankton for primary production and biochemical compositions in the western East/Japan Sea. *Front. Microbiol.* **2020**, *11*, 560102. [CrossRef] [PubMed]
6. Chiba, S.; Batten, S.; Sasaoka, K.; Sasai, Y.; Sugisaki, H. Influence of the Pacific Decadal Oscillation on phytoplankton phenology and community structure in the western North Pacific. *Geophys. Res. Lett.* **2012**, *39*, 2–7. [CrossRef]
7. Doney, S.C.; Ruckelshaus, M.; Duffy, J.E.; Barry, J.P.; Chan, F.; English, C.A.; Galindo, H.M.; Grebmeier, J.M.; Hollowed, A.B.; Knowlton, N.; et al. Climate change impacts on marine ecosystems. *Ann. Rev. Mar. Sci.* **2012**, *4*, 11–37. [CrossRef]
8. Lee, S.H.; Joo, H.T.; Lee, J.H.; Lee, J.H.; Kang, J.J.; Lee, H.W.; Lee, D.; Kang, C.K. Seasonal carbon uptake rates of phytoplankton 494 in the northern East/Japan Sea. *Deep. Res. Part II Top. Stud. Oceanogr.* **2017**, *143*, 45–53. [CrossRef]
9. Agawin, N.; Duarte, C.; Agustí, S. Nutrient and temperature control of the contribution of picoplankton to phytoplankton biomass and production. *Limnol. Oceanogr.* **2000**, *45*, 591–600. [CrossRef]
10. Morán, X.A.G.; López-urrutia, Á.; Calvo-díaz, A.; Li, W.K. Increasing importance of small phytoplankton in a warmer ocean. *Glob. Change Biol.* **2010**, *16*, 1137–1144. [CrossRef]
11. Hilligsøe, K.M.; Richardson, K.; Bendtsen, J.; Sørensen, L.L.; Nielsen, T.G.; Lyngsgaard, M.M. Linking phytoplankton community size composition with temperature, plankton food web structure and sea–air CO_2 flux. *Deep Sea Res. Part I Oceanogr. Res. Pap.* **2011**, *58*, 826–838. [CrossRef]
12. Mousing, E.A.; Ellegaard, M.; Richardson, K. Global patterns in phytoplankton community size structure—Evidence for a direct temperature effect. *Mar. Ecol. Prog. Ser.* **2014**, *497*, 25–38. [CrossRef]
13. Legendre, L.; Rassoulzadegan, F. Food-web mediated export of biogenic carbon in oceans: Hydrodynamic control. *Mar. Ecol. Prog. Ser.* **1996**, *145*, 179–193. [CrossRef]

14. Falkowski, P.G.; Oliver, M.J. Mix and match: How climate selects phytoplankton. *Nat. Rev. Microbiol.* **2007**, *5*, 813–819. [CrossRef] [PubMed]
15. Finkel, Z.V.; Beardall, J.; Flynn, K.J.; Quigg, A.; Rees, T.A.V.; Raven, J.A. Phytoplankton in a changing world: Cell size and elemental stoichiometry. *J. Plankton Res.* **2010**, *32*, 119–137. [CrossRef]
16. Marañón, E.; Cermeño, P.; Latasa, M.; Tadonléké, R.D. Temperature, resources, and phytoplankton community size structure in the ocean. *Limnol. Oceanogr.* **2012**, *57*, 1266–1278. [CrossRef]
17. Liu, H.; Liu, X.; Xiao, W.; Laws, E.A.; Huang, B. Spatial and temporal variations of satellite-derived phytoplankton size classes using a three-component model bridged with temperature in marginal seas of the western pacific ocean. *Prog. Oceanogr.* **2021**, *191*, 102511. [CrossRef]
18. Brewin, R.J.W.; Sathyendranath, S.; Jackson, T.; Barlow, R.; Brotas, V.; Airs, R.; Lamont, T. Influence of light in the mixed-layer on the parameters of a threecomponent model of phytoplankton size class. *Remote Sens. Environ.* **2015**, *168*, 437–450. [CrossRef]
19. Lee, J.H.; Lee, W.C.; Kim, H.C.; Jo, N.; Kim, K.; Lee, D.; Kang, J.J.; Sim, B.-R.; Kwon, J.-I.; Lee, S.H. Temporal and Spatial Variations of the Biochemical Composition of Phytoplankton and Potential Food Material (FM) in Jaran Bay, South Korea. *Water* **2020**, *12*, 3093. [CrossRef]
20. Alvain, S.; Moulin, C.; Dandonneau, Y.; Breon, F.M. Remote sensing of phytoplankton groups in case 1 waters from global SeaWiFS imagery. *Deep-Sea Res. I* **2005**, *52*, 1989–2004. [CrossRef]
21. IOCCG. *Phytoplankton Functional Types from Space*; Sathyendranath, S., Ed.; Reports of the International Ocean-Colour Coordinating Group, No. 15; IOCCG: Dartmouth, Canada, 2014.
22. Zhang, H.; Wang, S.; Qiu, Z.; Sun, D.; Ishizaka, J.; Sun, S.; He, Y. Phytoplankton size class in the East China Sea derived from MODIS satellite data. *Biogeosciences* **2018**, *15*, 4271–4289. [CrossRef]
23. Mouw, C.; Hardman-Mountford, N.J.; Alvain, S.; Bracher, A.; Brewin, R.; Bricaud, A.; Ciotti, A.M.; Devred, E.; Fujiwara, A.; Hirata, T.; et al. A consumer's guide to satellite remote sensing of multiple phytoplankton groups in the global ocean. *Front. Mar. Sci.* **2017**, *4*, 41. [CrossRef]
24. Ciotti, A.M.; Lewis, M.R.; Cullen, J.J. Assessment of the relationships between dominant cell size in natural phytoplankton communities and the spectral shape of the absorption coefficient. *Limnol. Oceanogr.* **2002**, *47*, 404–417. [CrossRef]
25. Hirata, T.; Aiken, J.; Hardman-Mountford, N.; Smyth, T.; Barlow, R. An absorption model to determine phytoplankton size classes from satellite ocean colour. *Remote Sens. Environ.* **2008**, *112*, 3153–3159. [CrossRef]
26. Kostadinov, T.S.; Siegel, D.A.; Maritorena, S. Retrieval of the particle size distribution from satellite ocean color observations. *J. Geophys. Res. Ocean.* **2009**, *114*, C09015. [CrossRef]
27. Roy, S.; Sathyendranath, S.; Bouman, H.; Platt, T. The global distribution of phytoplankton size spectrum and size classes from their light-absorption spectra derived from satellite data. *Remote Sens. Environ.* **2013**, *139*, 185–197. [CrossRef]
28. Garver, S.A.; Siegel, D.A.; Greg, M.B. Variability in near-surface particulate absorption spectra: What can a satellite ocean color imager see? *Limnol. Oceanogr.* **1994**, *39*, 1349–1367. [CrossRef]
29. Sun, D.; Huan, Y.; Wang, S.; Qiu, Z.; Ling, Z.; Mao, Z.; He, Y. Remote sensing of spatial and temporal patterns of phytoplankton assemblages in the Bohai Sea, Yellow Sea, and east China sea. *Water Res.* **2019**, *157*, 119–133. [CrossRef]
30. Brewin, R.J.W.; Sathyendranath, S.; Hirata, T.; Lavender, S.J.; Barciela, R.M.; Hardman-Mountford, N.J. A three-component model of phytoplankton size class for the Atlantic Ocean. *Ecol. Model* **2010**, *221*, 1472–1483. [CrossRef]
31. Hirata, T.; Hardman-Mountford, N.J.; Brewin, R.J.W.; Aiken, J.; Barlow, R.; Suzuki, K.; Isada, T.; Howell, E.; Hashioka, T.; Noguchi-Aita, M.; et al. Synoptic relationships between surface Chlorophyll-a and diagnostic pigments specific to phytoplankton functional types. *Biogeosciences* **2011**, *8*, 311–327. [CrossRef]
32. Brewin, R.J.W.; Ciavatta, S.; Sathyendranath, S.; Jackson, T.; Tilstone, G.; Curran, K.; Airs, R.L.; Cummings, D.; Brotas, V.; Organelli, E.; et al. Uncertainty in ocean-color estimates of chlorophyll for phytoplankton groups. *Front. Mar. Sci.* **2017**, *4*, 104. [CrossRef]
33. Ward, B.A. Temperature-correlated changes in phytoplankton community structure are restricted to polar waters. *PLoS ONE* **2015**, *10*, e0135581. [CrossRef] [PubMed]
34. Hu, S.; Liu, H.; Zhao, W.; Shi, T.; Hu, Z.; Li, Q.; Wu, G. Comparison of machine learning techniques in inferring phytoplankton size classes. *Remote Sens.* **2018**, *10*, 191. [CrossRef]
35. Parsons, T.R.; Maita, Y.; Lalli, C.M. *A Manual of Biological and Chemical Methods for Seawater Analysis*; Pergamon Press: Oxford, UK, 1984.
36. Moon, J.-E.; Ahn, Y.-H.; Ryu, J.-H.; Palanisamy, S. Development of Ocean environmental algorithms for Geostationary ocean color imager. *Korea J. Remote Sens.* **2010**, *26*, 198–207.
37. Jang, H.K.; Youn, S.H.; Joo, H.; Kim, Y.; Kang, J.J.; Lee, D.; Jo, N.; Kim, K.; Kim, M.-J.; Kim, S.; et al. First Concurrent Measurement of Primary Production in the Yellow Sea, the South Sea of Korea, and the East/Japan Sea, 2018. *J. Mar. Sci. Eng.* **2021**, *9*, 1237. [CrossRef]
38. Yamada, K.; Ishizaka, J.; Yoo, S.; Kim, H.-C.; Chiba, S. Seasonal and interannual variability of sea surface chlorophyll a concentration in the Japan/East Sea (JES). *Prog. Oceanogr.* **2004**, *61*, 193–211. [CrossRef]
39. Kim, T.-H.; Lee, Y.-W.; Kim, G. Hydrographically mediated patterns of photosynthetic pigments in the East/Japan Sea: Low N:P ratios and cyanobacterial dominance. *J. Mar. Syst.* **2010**, *82*, 72–79. [CrossRef]

40. Kwak, J.H.; Hwang, J.; Choy, E.J.; Park, H.J.; Kang, D.-J.; Lee, T.; Chang, K.-I.; Kim, K.-R.; Kang, C.-K. Monthly measured primary and new productivities in the Ulleung Basin as a biological "hot spot" in the East/Japan Sea. *Biogeosciences* **2013**, *10*, 4405–4417. [CrossRef]

41. Kwak, J.H.; Lee, S.H.; Hwang, J.; Suh, Y.S.; Park, H.J.; Chang, K.I.; Kim, K.-R.; Kang, C.K. Summer primary productivity and phytoplankton community composition driven by different hydrographic structures in the East/Japan Sea and the Western Subarctic Pacific. *J. Geophys. Res. Ocean.* **2014**, *119*, 4505–4519. [CrossRef]

42. Kang, J.J.; Joo, H.; Lee, J.H.; Lee, J.H.; Lee, H.W.; Lee, D.; Kang, C.K.; Yun, M.S.; Lee, S.H. Comparison of biochemical compositions of phytoplankton during spring and fall seasons in the northern East/Japan Sea. *Deep Sea Res. Part II Top. Stud. Oceanogr.* **2017**, *143*, 73–81. [CrossRef]

43. Kwak, J.H.; Han, E.; Lee, S.H.; Park, H.J.; Kim, K.R.; Kang, C.K. A consistent structure of phytoplankton communities across the warm–cold regions of the water mass on a meridional transect in the East/Japan Sea. *Deep Sea Res. Part II Top. Stud. Oceanogr.* **2017**, *143*, 36–44. [CrossRef]

44. Jo, N.; Kang, J.J.; Park, W.G.; Lee, B.R.; Yun, M.S.; Lee, J.H.; Kim, S.M.; Lee, D.; Joo, H.; Lee, J.H.; et al. Seasonal variation in the biochemical compositions of phytoplankton and zooplankton communities in the southwestern East/Japan Sea. *Deep Sea Res. Part II Top. Stud. Oceanogr.* **2017**, *143*, 82–90. [CrossRef]

45. Lee, M.; Kim, Y.B.; Park, C.H.; Baek, S.H. Characterization of Seasonal Phytoplankton Pigments and Functional Types around Offshore Island in the East/Japan Sea, Based on HPLC Pigment Analysis. *Sustainability* **2022**, *14*, 5306. [CrossRef]

46. Sun, X.; Shen, F.; Liu, D.; Bellerby, R.G.; Liu, Y.; Tang, R. In situ and satellite observations of phytoplankton size classes in the entire continental shelf sea, China. *J. Geophys. Res. Ocean.* **2018**, *123*, 3523–3544. [CrossRef]

47. Kim, Y.; Youn, S.H.; Oh, H.J.; Kang, J.J.; Lee, J.H.; Lee, D.; Kim, K.; Jang, H.K.; Lee, J.; Lee, S.H. Spatiotemporal variation in phytoplankton community driven by environmental factors in the northern East China Sea. *Water* **2020**, *12*, 2695. [CrossRef]

48. Fu, M.; Wang, Z.; Li, Y.; Li, R.; Sun, P.; Wei, X.; Lin, X.; Guo, J. Phytoplankton biomass size structure and its regulation in the Southern Yellow Sea (China): Seasonal variability. *Cont. Shelf Res.* **2009**, *29*, 2178–2194. [CrossRef]

49. Kang, J.J.; Min, J.O.; Kim, Y.; Lee, C.H.; Yoo, H.; Jang, H.K.; Kim, M.-J.; Oh, H.-J. Lee, S.H. Vertical Distribution of Phytoplankton Community and Pigment Production in the Yellow Sea and the East China Sea during the Late Summer Season. *Water* **2021**, *13*, 3321. [CrossRef]

50. Kim, Y.; Lee, J.H.; Kang, J.J.; Lee, J.H.; Lee, H.W.; Kang, C.K.; Lee, S.H. River discharge effects on the contribution of small-sized phytoplankton to the total biochemical composition of POM in the Gwangyang Bay, Korea. *Estuar. Coast. Shelf Sci.* **2019**, *226*, 106293. [CrossRef]

51. Shaha, D.C.; Cho, Y.-K. Comparison of empirical models with intensively observed data for prediction of salt intrusion in the Sumjin River estuary, Korea. Hydrol. *Earth Syst. Sci.* **2009**, *13*, 923–933. [CrossRef]

52. Ye, H.; Tang, D. A three component model of phytoplankton size classes for the south china sea. *Malays. J. Sc.* **2013**, *32*, 325–332.

Journal of *Marine Science and Engineering*

MDPI

Article

Spatiotemporal Distribution Characteristics of Copepods in the Water Masses of the Northeastern East China Sea

Sang Su Shin [1,†], Seo Yeol Choi [2,†] , Min Ho Seo [1], Seok Ju Lee [3], Ho Young Soh [2,*] and Seok Hyun Youn [4,*]

[1] Marine Ecology Research Center, Yeosu 59697, Korea; shinsangsu3@gmail.com (S.S.S.); copepod79@gmail.com (M.H.S.)
[2] Department of Marine Integrated Science, Chonnam National University, Yeosu 59626, Korea; seoyeol84@gmail.com
[3] Marine Biological Resource Center, Yeosu 59697, Korea; sjlee840425@gmail.com
[4] Oceanic Climate and Ecology Research Division, National Institute of Fisheries Science, Busan 46083, Korea
* Correspondence: hysoh@chonnam.ac.kr (H.Y.S.); younsh@korea.kr (S.H.Y.)
† These authors contributed equally to this work.

Abstract: To understand the effects of variable water masses in the northeastern East China Sea (Korea South Sea), planktonic copepods were seasonally sampled. Out of a total of 106 copepod species, 85 were oceanic warm-water species, and the number of species varied in summer, autumn, spring, and winter. The study area was divided into two or three regions according to the degree of influence of the water masses. *Canthocalanus pauper*, *Clausocalanus furcatus*, *Oithona plumifera*, *Oncaea venella*, *Oncaea venusta*, and *Paracalanus aculeatus* showed a positive correlation with water temperature and salinity and were indicator species of warm currents. *Calanus sinicus*, known as an indicator species of the Yellow Sea Bottom Cold Water, showed a high abundance and occurrence ratio in the western sea of the study area from spring to autumn. Moreover, *Acartia pacifica* indicated the extension of coastal waters to offshore areas. Several oceanic warm-water species (*A. danae*, *Centropages gracilis*, *Labidocera acuta*, *Rhincalanus nasutus*, and *Temoropia mayumbaensis*) were considered indicator species of the Taiwan Warm Current. Our results suggest that the spatiotemporal distribution patterns of indicator species are partly explained by different water masses.

Keywords: East China Sea; warm currents; copepods; indicators; spatiotemporal distribution

Citation: Shin, S.S.; Choi, S.Y.; Seo, M.H.; Lee, S.J.; Soh, H.Y.; Youn, S.H. Spatiotemporal Distribution Characteristics of Copepods in the Water Masses of the Northeastern East China Sea. *J. Mar. Sci. Eng.* **2022**, *10*, 754. https://doi.org/10.3390/jmse10060754

Academic Editor: Marco Uttieri

Received: 29 April 2022
Accepted: 27 May 2022
Published: 30 May 2022

Publisher's Note: MDPI stays neutral with regard to jurisdictional claims in published maps and institutional affiliations.

1. Introduction

In the marine ecosystem, zooplankton, particularly copepods, occupy an important intermediate position, transmitting energy to upper trophic levels as they feed on phytoplankton, grow, and are consumed by predators [1]. Moreover, copepods have limited locomotion ability to move against the flow of water masses and are sensitive to changes within them. They have been used as indicators of changes in water masses and ocean currents [2–6]. In particular, *Clausocalanus furcatus*, *Oithona plumifera*, *Paracalanus aculeatus*, and *Oncaea venusta* have been used as indicator species for the inflow of the Tsushima Warm Current [5,7–9], and *Calanus sinicus* has been used as an indicator species for the inflow of the Yellow Sea Bottom Cold Water (YSBCW) to the East China Sea [10,11]. Furthermore, the coastal species *Acartia omorii* and *Paracalanus parvus* s. l. were used as indicator species for the expansion of coastal waters into the open sea [12,13].

Changes in water masses generally cause a seasonal succession of zooplankton, fluctuations in abundance, and the distribution of diverse community structures or species [2,14]. Therefore, zooplankton have generally been used as indicators of water masses and ocean currents [4,5]. Changes in physical environmental factors, such as the temperature and salinity of marine ecosystems, can lead to rapid changes in the species richness and quantitative characteristics of zooplankton [4,5]. Water masses or currents vary in their extent of expansion according to the season, and when different water masses meet each other, they mix or form a front. The seasonal influence of water masses acts as a major factor in changing the distribution pattern and community structure of zooplankton [15,16]. In the northeastern East China Sea (nECS), zooplankton are expected to distinctly define their habitat characteristics in response to a variety of physical environmental variables. Various studies on the characteristics of water masses that affect the distribution of copepods and chaetognaths [16,17], the seasonal vertical distribution characteristics of copepods [18], and the environmental factors affecting neustonic zooplankton in summer [19] have been conducted in a part of the nECS. However, as these studies were conducted in a limited area or only in one season, there is a limited understanding of the effects of various water masses or currents in the nECS. Therefore, the purpose of this study was to understand the effects of these seasonal water masses on changes in the planktonic copepod community structure and to clarify the relationship between environmental factors and indicator species that could explain the effect of water masses or currents in the entire nECS.

2. Materials and Methods

2.1. Study Sites and Environmental Factors

The study area is a site with a widely developed continental shelf at a depth of 200 m or less, and is connected to the Yellow Sea, East China Sea, and East Sea (Sea of Japan). There are also various water masses or warm currents. The Tsushima Warm Current (TWC), characterized by high temperature and salinity, originates from the Kuroshio Water (KW) and flows into the East Sea from the East China Sea. The YSBCW, with a low water temperature, originates from the Yellow Sea in winter, moves south in summer, and affects the bottom layer. Moreover, in the summer, the Changjiang diluted water (CDW), which comprises a mixture of surface water and fresh water from the Changjiang River, has a low salinity effect on the study area. In addition, the Taiwan Warm Current (TAWC) and Chinese coastal waters affect the study area [20–22].

Field surveys were conducted seasonally in the nECS (Figure 1, Table 1). Water temperature and salinity were vertically measured using a calibrated SBE 9/11 plus CTD instrument (Sea-Bird Electronics, Bellevue, WA, USA). Seawater was sampled for estimating the chlorophyll-a concentration (Chl-*a*) using Niskin water samplers that were attached to the same equipment according to water depths of 0, 10, 20, 30, 50, 75, 100, and 125 m.

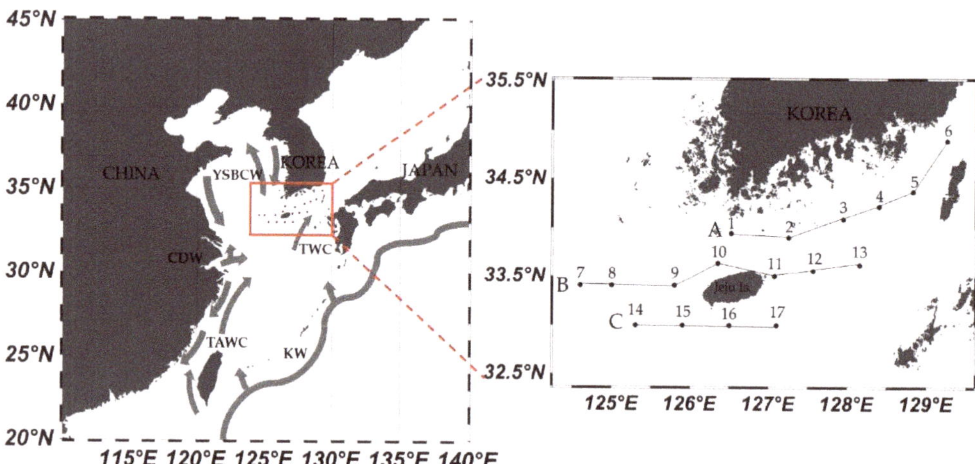

Figure 1. Map of the study sites. Abbreviations: CDW: Changjiang diluted water; Jeju Is.: Jeju Island; TWC: Tsushima Warm Current; TAWC: Taiwan Warm Current water; YSBCW: Yellow Sea bottom cold water; and KW: Kuroshio water.

Table 1. Geographic coordinates of the study sites in the Korea South Sea.

Station	Latitude	Longitude	Bottom Depth (m)
1	33.9	126.5	52
2	33.9	127.3	78
3	34.1	127.9	82
4	34.2	128.4	85
5	34.4	128.8	102
6	34.9	129.3	127
7	33.4	124.6	75
8	33.4	125.0	125
9	33.4	125.8	92
10	33.6	126.4	140
11	33.5	127.1	128
12	33.6	127.6	102
13	33.6	128.2	119
14	33.0	125.3	85
15	33.0	125.9	102
16	33.0	126.5	107
17	33.0	127.1	105

To measure Chl-*a* by size (>20 μm, 3–20 μm, <3 μm), a 20-micrometer membrane filter (Polycarbonate Track Etched, 47 mm; GVS, Sanford, ME, USA), 3-micrometer PC membrane filters (polycarbonate membrane filter, 47 mm; Whatman, Florham Park, NJ, USA), and a 0.45-micrometer membrane filter (polycarbonate membrane filter, 47 mm; Advantec, Tokyo, Japan) were used for sequential filtration by size through a filter holder. Next, the membrane filter was placed in a conical tube (15 mL), wrapped in aluminum foil, and stored at $-80\,^{\circ}$C. All filters were then transferred to frozen storage in our laboratory, Chl-*a* was extracted after solvation in 90% acetone, and they were settled in a dark and cool chamber for 24 h. To filter out particles and extract them from the filter paper, a syringe filter (0.45 μm, polytetrafluoroethylene; Advantec, Florham Park, NJ, USA) was used for filtration and the absorbance was measured using a fluorometer (10-Au; Tuner Designs, San Jose, CA, USA) calibrated with a Chl-*a* standard (Sigma-Aldrich, Darmstadt, Germany). Next, Chl-*a* concentration was calculated from the absorbance measurement using the UNESCO formula [23].

2.2. Zooplankton Sampling

The zooplankton were vertically towed from the bottom 3 m to the surface layer of each station using a conical zooplankton net (mouth size 60 cm, mesh size 200 μm) in February (winter), April (spring), August (summer), and October (autumn) 2018 (Figure 1). The samples were immediately fixed on board at a final concentration of 5–10% with neutralized formaldehyde seawater. A flow meter (model 438115; Hydro-Bios co., Altenholz, Germany) was attached to the inlet of the net to calculate the amount of filtered seawater that passed through the net. Next, the samples were divided into 1:2–1:512 to determine the species composition and abundance of zooplankton using a Folsom-type divider until the final abundance reached approximately 300 or more. Species identification was performed under an optical microscope (SMZ645; Nikon, Tokyo, Japan). When more detailed observation was needed for species identification, the appendages with characteristic features of the species were dissected and observed on a glass slide with an optical microscope (ECLIPSE 80i; Nikon). Next, the zooplankton sample measurements were converted to the number of individuals per unit volume (ind m^{-3}).

2.3. Data Analysis

To understand the characteristics of the water masses in the study area, a T–S diagram (temperature–salinity) was constructed using the water temperature and salinity data (Figure 2). The water masses were classified according to their physical properties as follows: the TWC has a water temperature of 15 °C or higher and salinity of 31–34.5 psu [24]; the YSBCW has a water temperature of 13.2 °C or lower and salinity of 32.6–33.7 psu or lower [25]; and the CDW has a water temperature of 23 °C or higher and salinity of 31 psu or lower [26]. After setting three north-to-south lines (A, B, and C lines), each vertical sectional diagram was drawn using Ocean Data View 5.5.1 software to determine the distribution of the water masses that were affected by the water depth of the study area (Figures 2 and 3). A horizontal cross-sectional diagram for Chl-*a* was also drawn (see Supplementary Material Figure S1).

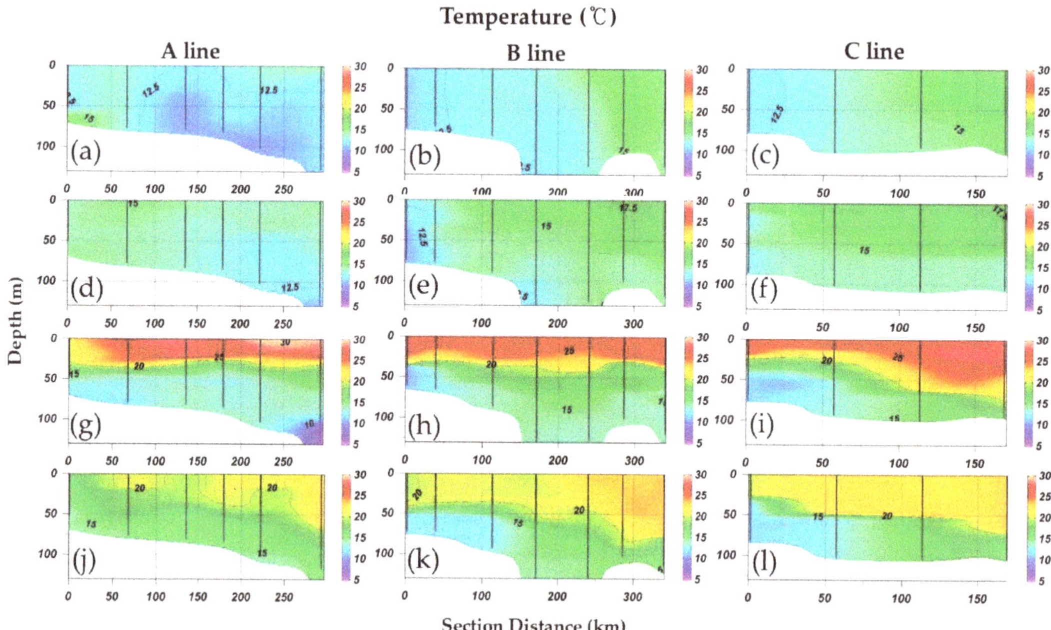

Figure 2. Vertical distribution of water temperature (°C) in (**a–c**) February, (**d–f**) April, (**g–i**) August, and (**j–l**) October 2018. Lines A, B, and C represent north-to-south sampling lines in the study area.

A cluster analysis was performed to compare the similarities between stations based on copepod abundance identified to the species level using PRIMER software (version 6.1.6). After the abundances were converted to Log (x + 1) to reduce the bias in the distribution of the data due to the variation in the abundances, the Bray–Curtis similarity index was calculated. Based on this index, a hierarchical cluster analysis was performed using the unweighted pair group method with arithmetic mean (UPGMA) and compared to the non-metric multidimensional scaling (nMDS) arrangement method. Next, the similarity index for grouping the stations was divided into 50% in February, 55% in April, 53% in August, and 60% in October, and was further subdivided into 55% and 60% in August and October, respectively. The similarity-percentages procedure (SIMPER) analysis was completed to select the major species that affected each classified cluster [27]. In addition, a canonical correspondence analysis (CCA) was conducted using Canonical Community Ordination (CANOCO) software (version 4.5) to examine the relationship between the major species that contributed to the group classification seasonally and the environmental factors (water temperature, salinity, and fractional Chl-*a* by size) [28]. Additionally, a Pearson correlation analysis in the statistical package SPSS (version 12.0) was used to determine the correlation between these major species and the environmental factors (water temperature, salinity, and fractional Chl-*a* by size) of the study area ($p < 0.01$; $p < 0.05$). Data normality was checked before analysis. When data were not normally distributed, a log (x + 1) transformation was applied for data normality.

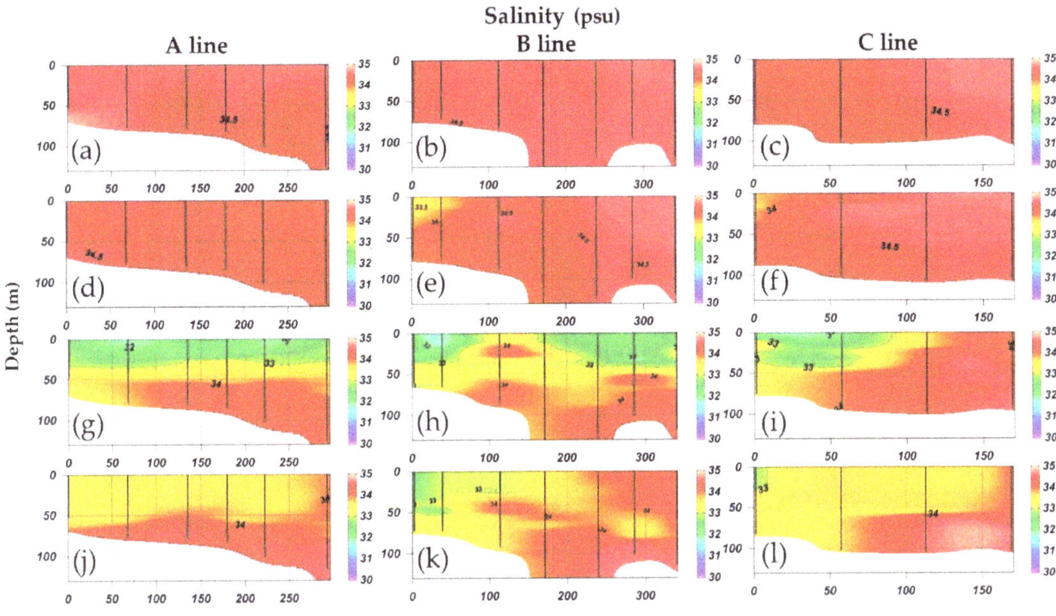

Figure 3. Vertical distribution of salinity (psu) in (**a–c**) February, (**d–f**) April, (**g–i**) August, and (**j–l**) October 2018. Lines A, B, and C represent north-to-south sampling lines in the study area.

3. Results

3.1. Environmental Factors (Water Temperature, Salinity, and Chl-a Concentration)

In February, the water temperature and salinity were in the range of 10.2–16.4 °C and 34.27–34.67 psu, respectively (Figures 2 and 3a–c). The TWC was observed at stations 12 and 13 on the eastern side of Line B and stations 16 and 17 on the eastern side of Line C, and there was a difference of approximately 5 °C in the water temperature when compared to the other stations. Moreover, there was no difference between the surface and bottom layers at the observatory in February (Figures 2 and 3a–c).

In April, the water temperature and salinity were in the range of 10.4–17.6 °C and 33.48–34.67 psu, respectively (Figures 2 and 3d–f). The TWC was observed at stations 9–13 on the eastern side of Line B, and at stations 14 and 15 on the eastern side of Line C. The YSBCW was observed at station 7 on the western side of Line B and station 14 on the western side of Line C, and was located approximately 25 m from the surface (Figures 2 and 3d–f).

In August, the water temperature and salinity were in the range of 6.1–28.3 °C and 31.73–34.59 psu, respectively (Figures 2 and 3g–i). The TWC was observed at all the stations. The YSBCW was observed at stations 7–9 on the western side of Line B and at stations 14 and 15 on the western side of Line C, and it was located at water depths of approximately 50 m or deeper (Figures 2 and 3g–i).

In October, the water temperature and salinity were in the range of 11.7–23.5 °C and 32.42–34.64 psu, respectively (Figures 2 and 3j–l). The TWC was observed at all the stations, as was the case in August. The YSBCW was observed at stations 7–9 west of Line B and stations 14 and 15 of Line C, and was located at water depths of approximately 50 m or deeper (Figures 2 and 3j–l).

The Chl-*a* concentration was high in April and October, whereas it was low in February and August. The stations that were adjacent to the coast had higher Chl-*a* concentrations than those in the open sea (Table 2; see Supplementary Material Figure S1).

Table 2. Range of total chlorophyll-*a* concentrations (µg/L), microplankton (>20 µm), nanoplankton (3–20 µm), and pico-plankton (<3 µm) in the Korea South Sea.

Months	Total		Micro		Nano		Pico	
	Min	Max	Min	Max	Min	Max	Min	Max
2	0.01 (St. 3)	1.64 (St. 17)	0 (St. 2, 3, 5, 10, 13)	0.88 (St. 17)	no data	no data	no data	no data
4	0.52 (St. 15)	4.65 (St. 6)	0.05 (St. 10)	1.17 (St. 13)	0 (except St. 16)	0.01 (St. 16)	0.19 (St. 9)	1.95 (St. 6)
8	0.15 (St. 4)	0.81 (St. 1)	0.02 (St. 13)	0.23 (St. 17)	0.04 (St. 12, 16)	0.11 (St. 7)	0.15 (St. 6)	0.44 (St. 1)
10	0.61 (St. 17)	4.12 (St. 4)	0.02 (St. 1)	1.91 (St. 4)	0.08 (St. 17)	0.25 (St. 1, 8)	0.25 (St. 16)	0.81 (St. 8)

In February, the total Chl-*a* concentration was in the range of 0.01–1.64 µg/L, with the highest value at station 17 and the lowest at station 3. The >20-micrometer Chl-*a* concentration was in the range of 0–0.88 µg/L, with the highest value at station 17, and concentrations of <0.3 µg/L Chl-*a* were observed at the other stations (Table 2; see Supplementary Material Figure S1).

In April, the range of the total Chl-*a* concentration was 0.52–4.65 µg/L, with the highest value at station 6 at 4.65 µg/L and the lowest at 0.52 µg/L at station 15 (Table 2; see Supplementary Material Figure S1). The >20-micrometer Chl-*a* concentration was in the range of 0.05–1.17 µg/L, with the highest value at station 13 and the lowest at station 10. The Chl-*a* concentration measuring between 3 µm and 20 µm ranged from 0 to 0.01 µg/L and was hardly observed at almost all stations. The <3-micrometer Chl-*a* concentration was in the range of 0.28–1.95 µg/L, with the highest value at station 6 and the lowest at station 15. The total Chl-*a* concentration was observed to be high close to the coast of the eastern sea area and further from Jeju Island. The >20-micrometer Chl-*a* was observed to be at a relatively high concentration in the seas near the east of Jeju Island, whereas a relatively high concentration of <3 µm Chl-*a* was observed at the stations that were adjacent to the east coast of the study area (Table 2; see Supplementary Material Figure S1).

In August, the total Chl-*a* ranged from 0.15 to 0.81 µg/L, with the highest value at station 1 and the lowest values at stations 3 and 4. The >20-micrometer Chl-*a* was in the range of 0.02–0.23 µg/L, with the highest value at station 17 and the lowest at station 13. The Chl-*a* measuring between 3 µm and 20 µm was in the range of 0.04–0.11 µg/L, with the highest values at stations 1 and 7, and the lowest at station 16. The Chl-*a* measuring <3 µm or less ranged from 0.15 to 0.38 µg/L, with the highest value at station 7 and the lowest at station 6. This was similar to the <3micrometer Chl-*a* levels in February and was observed to be <1 µg/L (Table 2; see Supplementary Material Figure S1).

In October, the total Chl-*a* ranged from 0.61–4.12 µg/L, with the highest value at station 4, and the lowest at station 17 (Table 2; see Supplementary Material Figure S1). The >20micrometer Chl-*a* concentrations were in the range of 0.02–1.91 µg/L, with the highest value at station 4, and the lowest at station 1. The Chl-*a* concentration measuring between 3 µm and 20 µm was in the range of 0.08–0.25 µg/L, with the highest value at station 1, and the lowest at station 12. The Chl-*a* concentration measuring 3 µm or less ranged from 0.25 to 0.81 µg/L, with the highest value at station 8, and the lowest at station 16. As in April, the Chl-*a* concentration was relatively high close to the east coast and further from Jeju Island. The Chl-*a* measuring 20 µm or more were observed to be at a relatively high concentration at the stations that were adjacent to the east coast of the study area, whereas at the other stations, Chl-*a* occurred at a concentration of less than 1 µg/L (Table 2; see Supplementary Material Figure S1).

3.2. Spatiotemporal Distribution of the Copepods

Supplementary Material Table S1 shows the species list and relative occurrence ratio of copepods in the study area. In total, 106 species of planktonic copepods were observed. The numbers of warm-water oceanic species observed were 39 of the 54 species in February, 34 of the 48 species in April, 66 of the 75 species in August, and 56 of the 66 species in October. The major species that were recorded were as follows: *Calanus sinicus, Ctenocalanus vanus, Oithona longispina, Oithona plumifera, P. aculeatus, P. parvus* s. l., *Scolecithricella longispinosa,* and *S. nicobarica* in the winter; *Ctenocalanus vanus, Ditrichocorycaeus affinis, Oithona longispina, Oithona similis, Paracalanus parvus* s. l. in the spring; *A. pacifica, Calanus sinicus, Canthocalanus pauper, Clausocalanus furcatus, D. affinis, Oncaea venella, Oncaea venusta, Oithona atlantica, Paracalanus parvus* s. l., *Paracalanus aculeatus,* and *Undinula vulgaris* in the summer; and *A. pacifica, Clausocalanus furcatus, Canthocalanus pauper, Clausocalanus minor, D. affinis, Farranula gibbula, Oithona atlantica, Oithona plumifera, Oncaea venella, Oncaea venusta, Paracalanus aculeatus,* and *Paracalanus parvus* s. l in the autumn.

The abundance of the copepods was in the range of 87–800 ind m^{-3} in February, with the lowest value at station 8 and the highest at station 17. In April, it was in the range of 300–3270 ind m^{-3}, with the lowest value at station 13 and the highest at station 7. In August, it was in the range of 258–1647 ind m^{-3}, with the lowest value at station 11 and the highest at station 10. In October, it was in the range of 270–1080 ind m^{-3}, with the lowest value at station 16 and the highest at station 12 (Figure 4).

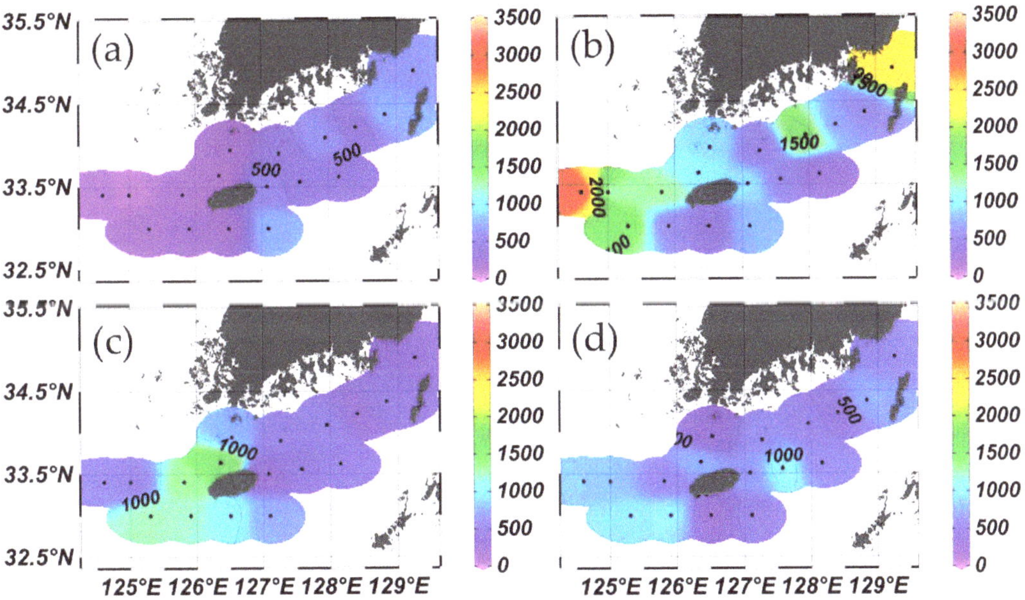

Figure 4. Distribution of copepod abundance (ind m^{-3}) at each station in (**a**) February, (**b**) April, (**c**) August, and (**d**) October 2018.

The Shannon–Wiener species diversity index was in the range of 1.27–3.01 (stations 3 and 16, respectively) in February, 1.52–2.63 (stations 3 and 12, respectively) in April, 2.64–3.36 (stations 5 and 12, respectively) in August, and 2.04–3.25 (stations 1 and 17, respectively) in October (Figure 5).

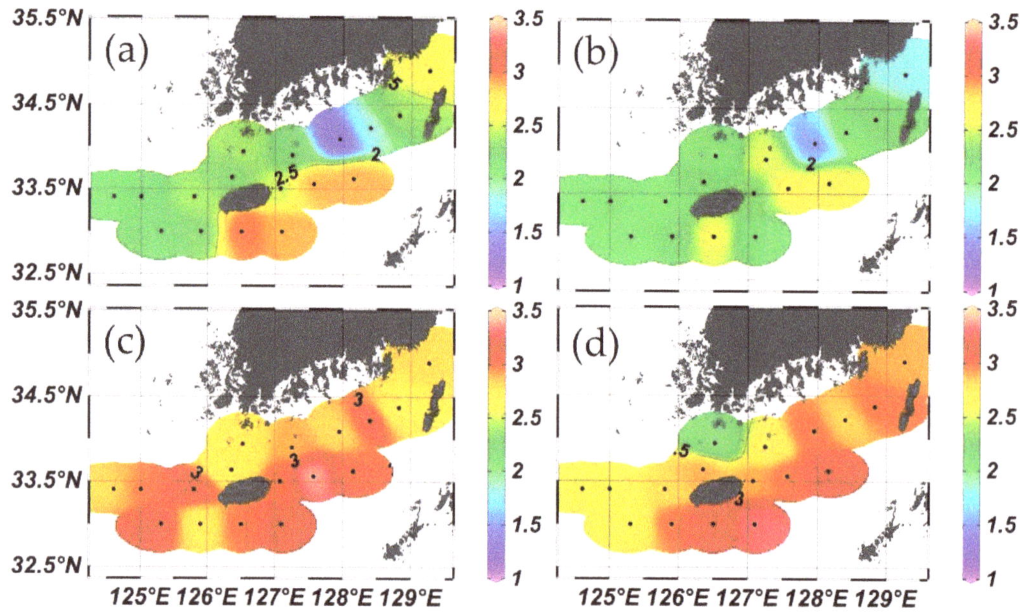

Figure 5. Distribution of the species diversity index based on copepod abundance in (**a**) February, (**b**) April, (**c**) August, and (**d**) October 2018.

3.3. Cluster Analysis

In February, the stations were divided into west (Group A) and east (Group B) with Jeju Island as the reference point and a similarity index of 50%. The contribution of the grouped stations was highest in the following order in Group A: *Paracalanus parvus* s. l., *S. longispinosa*, *Calanus sinicus*, *D. affinis*, and *Oithona plumifera*. It was in the following order in Group B: *Paracalanus parvus* s. l., *Oithona plumifera*, *S. longispinosa*, and *Ctenocalanus vanus* (see Supplementary Material Table S2, Figure 6). In April, the stations were divided into north (Group A) and south (Group B) with Jeju Island as the reference point and a similarity index of 55%. *Paracalanus parvus* s. l., *Oithona similis*, *D. affinis*, *Calanus sinicus*, and *A. omorii* were important contributors in Group A, whereas *Oithona similis*, *Paracalanus parvus* s. l., *Oithona longispina*, *Ctenocalanus vanus*, *D. affinis*, *Calanus sinicus*, *Oncaea scottodicarloi*, and *Oithona plumifera* were important contributors in Group B (see Supplementary Material Table S2, Figure 6). In August, the stations were divided into western offshore stations (Group A) and the rest (Group B), using Jeju Island as the reference point (Figure 6). In Group A, *A. pacifica*, *Paracalanus parvus* s. l., *Calanus sinicus*, *Paracalanus aculeatus*, *Oncaea venella*, *D. affinis*, and *Canthocalanus pauper* were important contributors. By contrast, Group B was divided into Groups B-1 and B-2. *Oncaea venusta*, *Paracalanus parvus* s. l., *Undinula vulgaris*, *Oncaea venella*, *Oithona atlantica*, *D. affinis*, *Clausocalanus furcatus*, and *Paracalanus aculeatus* made important contributions to Group B-1, whereas in Group B-2, the important contributors were *Paracalanus aculeatus*, *Oncaea venusta*, *Oncaea venella*, *Oithona plumifera*, *Clausocalanus furcatus*, and *Paracalanus parvus* s. l. (see Supplementary Material Table S2, Figure 6). In October, the stations were divided into two groups (Groups A and B) with a similarity index of approximately 60%. Group A was divided into Groups B-1 and B-2. The highest contributors to the B-1 group were in the following order: *Paracalanus parvus* s. l., *Paracalanus aculeatus*, *Oncaea venusta*, *D. affinis*, and *A. pacifica*. The highest contributors to Group B-2 were in the following order: *Paracalanus aculeatus*, *Oncaea venusta*, and *Oithona plumifera* (see Supplementary Material Table S2, Figure 6).

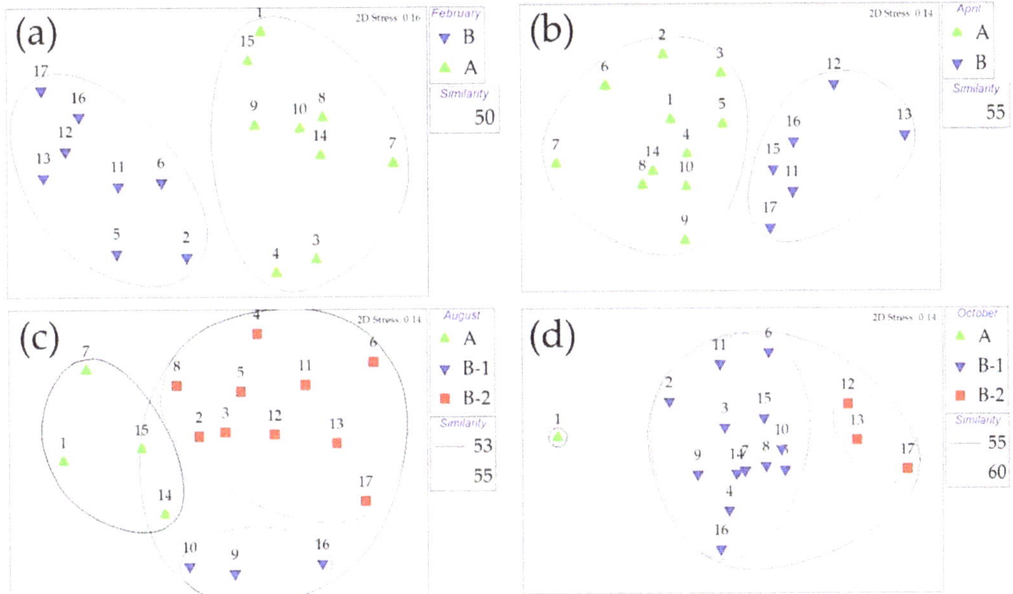

Figure 6. Dendrogram from the cluster analysis based on the Bray–Curtis similarities that were estimated using copepod abundance in the Korea South Sea and a non-metric multidimensional scaling (nMDS) analysis in (**a**) February, (**b**) April, (**c**) August, and (**d**) October 2018.

3.4. Correlation between the Environmental Factors and Major Copepods

Among the species that played an important role in the grouping of similar community structures in February, *Calanus sinicus*, in Group A, showed a negative correlation with water temperature ($p < 0.01$), and *Oithona plumifera*, in, Group B showed a positive correlation with water temperature and salinity ($p < 0.05$; Table 3, see Supplementary Material Table S3 and Figure S2a). In April, *A. omorii* and *Calanus sinicus*, in Group A, showed a negative correlation with water temperature and salinity, respectively ($p < 0.05$; Table 3, see Supplementary Material Figure S2b). In August, *Oncaea venella*, in Group B-1, showed a positive correlation with salinity and Chl-*a* (> 20 μm; $p < 0.01$), whereas *Paracalanus aculeatus* showed a positive correlation with salinity ($p < 0.01$) and Chl-*a* (> 20 μm) ($p < 0.05$). *Oncaea venusta*, in Group B-2, showed a positive correlation with the total Chl-*a* ($p < 0.01$; Table 3, see Supplementary Materials Table S3 and Figure S2c). In October, *Oncaea venusta*, in Group B-1, showed a positive correlation with salinity, and *A. pacifica* showed a negative correlation with salinity ($p < 0.05$). Furthermore, *Paracalanus aculeatus*, in Group B-2, showed a positive correlation with Chl-*a* (>20 μm; $p < 0.05$; Table 3, see Supplementary Material Table S3).

Table 3. Correlation analysis between species and environmental factors in February (Feb.), April (Apr.), August (Aug.), and October (Oct.) 2018. (** correlation is significant at the 0.01 level, * correlation is significant at the 0.05 level).

	Group	Species	T	S	Chlorophyll−a			
					Total	>20 μm	3–20 μm	<3 μm
Feb.	A	*Paracalanus parvus* s. l.	−0.264	0.350	−0.716 *	−0.252	no data	no data
		Scolecithricella longispinosa	0.093	−0.475	−0.310	−0.177	no data	no data
		Calanus sinicus	−0.880 **	−0.207	−0.197	0.212	no data	no data
		Ditrichocorycaeus affinis	0.460	−0.179	0.312	−0.008	no data	no data
		Oithona plumifera	0.435	0.571	−0.056	−0.456	no data	no data
	B	*Paracalanus parvus* s. l.	−0.572	−0.234	−0.110	0.050	no data	no data
		Oithona plumifera	0.769 *	0.801 *	0.412	0.409	no data	no data
		Scolecithricella longispinosa	0.589	0.270	0.280	0.229	no data	no data
		Ctenocalanus vanus	−0.197	−0.639	−0.364	−0.550	no data	no data
Apr.	A	*Paracalanus parvus* s. l.	−0.163	−0.376	0.094	−0.488	0	0.063
		Oithona similis	−0.479	−0.495	0.226	−0.227	0	0.275
		Ditrichocorycaeus affinis	−0.189	0.139	0.323	0.082	0	0.368
		Calanus sinicus	−0.586	−0.713 *	−0.123	−0.291	0	−0.065
		Acartia omorii	−0.633 *	−0.570	0.104	−0.220	0	0.181
	B	*Oithona similis*	−0.314	0.108	0.207	−0.081	−0.328	−0.192
		Paracalanus parvus s. l.	−0.779	0.043	0.419	−0.021	0.137	0.310
		Oithona longispina	−0.016	0.003	−0.101	−0.200	−0.426	−0.407
		Ctenocalanus vanus	0.302	−0.147	−0.749	−0.587	0.313	0.027
		Ditrichocorycaeus affinis	−0.093	−0.356	−0.247	−0.432	0.197	0.053
		Calanus sinicus	0.224	0.139	−0.374	−0.235	0.554	0.403
		Oncaea scottodicarloi	−0.513	−0.470	0.072	−0.392	−0.060	−0.076
		Oithona plumifera	−0.786	−0.107	0.282	−0.131	0.152	0.261
Aug.	A	*Acartia pacifica*	−0.898	−0.149	0.672	0.712	−0.118	0.219
		Paracalanus parvus s. l.	−0.657	−0.289	0.574	0.564	−0.437	−0.044
		Paracalanus aculeatus	0.851	0.365	−0.884	−0.235	−0.670	−0.884
		Oncaea venella	−0.204	0.824	−0.501	0.842	−0.540	−0.650
		Calanus sinicus	0.049	−0.789	0.455	−0.276	−0.491	−0.107
		Ditrichocorycaeus affinis	−0.563	0.339	0.062	0.893	−0.588	−0.411
		Canthocalanus pauper	0.824	0.246	−0.679	−0.639	0.220	−0.154
	B−1	*Oncaea venusta*	−0.207	0.425	0.599	0.087	−0.029	−0.302
		Paracalanus parvus s. l.	−0.190	−0.379	−0.135	−0.165	0.023	0.237
		Undinula vulgaris	0.619	0.388	−0.029	0.430	−0.240	−0.709 *
		Oncaea venella	0.345	0.780 **	−0.061	0.788 **	0.455	−0.332
		Oithona atlantica	−0.231	−0.352	0.190	−0.372	−0.503	−0.041
		Ditrichocorycaeus affinis	0.138	−0.087	−0.530	0.018	−0.420	−0.407
		Clausocalanus furcatus	0.129	0.351	0.211	−0.082	0.066	−0.173
		Paracalanus aculeatus	0.076	0.822 **	0.309	0.726 *	0.525	−0.151
	B−2	*Paracalanus aculeatus*	−0.981	−0.971	0.875	−0.191	0.986	−0.139
		Oncaea venusta	−0.766	−0.736	1.000 **	0.306	0.785	0.356
		Oncaea venella	0.529	0.490	−0.951	−0.585	−0.555	−0.627
		Oithona plumifera	0.813	0.838	−0.246	0.846	−0.795	0.817
		Clausocalanus furcatus	0.697	0.664	−0.995	−0.400	−0.718	−0.448
		Paracalanus parvus s. l.	−0.957	−0.969	0.546	−0.630	0.948	−0.588
Oct.	B−1	*Paracalanus parvus* s. l.	0.067	−0.473	0.006	0.027	0.062	0.414
		Paracalanus aculeatus	0.074	−0.320	−0.089	−0.190	−0.118	0.444
		Oncaea venusta	0.442	0.654 *	0.236	0.207	0.096	−0.012
		Ditrichocorycaeus affinis	−0.176	0.113	−0.301	−0.325	−0.249	0.012
		Acartia pacifica	−0.004	−0.614 *	−0.266	−0.376	−0.375	0.385
	B−2	*Paracalanus aculeatus*	−0.890	−0.729	0.892	0.999 *	0.139	0.996
		Oncaea venusta	−0.712	−0.489	0.987	0.968	0.431	0.977
		Oithona plumifera	0.687	0.459	−0.992	−0.958	−0.462	−0.970

T, water temperature; S, salinity.

4. Discussion

Over the past several decades, many studies have revealed that zooplankton species, in particular copepod species, are closely related to large-scale physical processes, such as the transport of water masses by ocean currents [28–36]. In this study, 106 species were found to occur and among them, 85 were warm-water oceanic species. However, there was a seasonal difference: the warm-water oceanic species were most prevalent in summer, and few species appeared in spring (see Supplementary Material Table S1). By contrast, Kang and Hong [17] showed that the number of warm-water oceanic species was the highest in autumn, and they explained that this was because the force of the TWC increases as autumn approaches [37]. For the past three decades, the warm-water oceanic species from Korean waters have been introduced along the TWC, which branches from the KW [11,17,22,38]. Nevertheless, Cho et al. [39] argued that the TAWC may have a greater effect on the study area than the TWC in summer. In this study, warm-water oceanic calanoid copepods (59 species) occurred in the Korea South Sea, including species from the TAWC and KW. Among these species, *A. danae, Centropages gracilis, Nannocalanus minor, Euchaeta media, Heterorhabdus subspinifrons, Labidocera acuta, Pontellina morii, Rhincalanus nasutus, S. nicobarica,* and *Temoropia mayumbaensis* occurred in the Taiwanese waters, whereas *Clausocalanus parapergens, Haloptilus longicornis, Rhincalanus cornutus, Pleuromamma piseki,* and *S. longispinosa* appeared in the KW. Although the data need to be supplemented by further extensive surveys in the KW, this occurrence pattern seems to support Cho et al. [39]. Moreover, since the study conducted by Kang and Hong [17], the oceanic warm-water species appear to have continuously increased in Korean waters (see Supplementary Material Table S4), likely due to global warming, which could cause changes in the community structure of the zooplankton in the marine ecosystem.

The cluster analysis indicated that the stations were divided into two or three groups (Figure 6). The contributing species for the grouped stations are shown in Table 3. Among them, in winter, the neritic species, *Paracalanus parvus* s. l., showed a significantly negative relationship with the total Chl-*a* ($p < 0.05$) in Group A, but in the other seasons, it was not significantly related to the other environmental factors. The warm-water oceanic species, *Oithona plumifera,* had a significantly positive relationship with temperature and salinity ($p < 0.05$) in Group B in winter. The neritic species *A. omorii* had a significantly negative relationship with temperature ($p < 0.05$) in Group A in spring. The warm-water oceanic species *Oncaea venella* and *Paracalanus aculeatus* were significantly positively related with salinity and Chl-*a* (>20 μm) in Group B in summer. Additionally, the warm-water oceanic species *Oncaea venusta* had a positive relationship with salinity ($p < 0.05$), whereas the neritic species *A. pacifica* had a negative relationship with salinity ($p < 0.05$) in Group B in autumn. Additionally, in the correlation analysis between the environmental factors and dominant copepods, *Oithona plumifera* (February, Group B), *Paracalanus aculeatus* (August, Group B-1), *Oncaea venusta* (October, B Group -1), and *Oncaea venella* (August, Group B-1) showed positive correlations with water temperature or salinity, indicating that these species are possible indicators of the TWC. Many studies [3,6–8,40] suggest that *Clausocalanus furcatus, Oithona plumifera, Paracalanus aculeatus, Oncaea venusta, Oncaea venella, Oncaea mediterranea, Oncaea media, Triconia conifera, Nannocalanus minor, Canthocalanus pauper, Scolecithrix danae, Temora turbinata,* and *Calocalanus plumulosus* are abundant in the KW, which has a high temperature and high saltwater masses, and their occurrence indicates inflow from the TWC (see Supplementary Material Table S4).

In the present study, as a result of examining the correlation between the dominant species and the phytoplankton, three species, namely *Paracalanus aculeatus, Oncaea venusta,* and *Oncaea venella,* showed a positive correlation with >20-micrometer phytoplankton. It is easy for zooplankton to find and capture large prey [41]. Although analyzing the relationship with phytoplankton according to the size of the zooplankton is ideal, a limitation of the present study is its understanding of the food web from phytoplankton to zooplankton, because size was not measured.

By contrast, the neritic cold-water species, *Calanus sinicus*, was a key species in Group A in summer. *Calanus sinicus* was well distributed in the YSBCW in summer [9,10]. The YSBCW is formed in the winter and remains in the bottom layer, and as the seasons change, and the surface water temperature rises under the influence of the rising temperature of the seasons; consequently, a strong stratification with the bottom water is created, and it flows south to the Korea South Sea in all seasons except winter [42,43]. In April, August, and October, the YSBCW was found in the bottom layers at <50 m in the western sea area (stations 1, 7, 8, 9, 14, and 15) based on Jeju Island. In August and October, it extended to the western area of the study area, and the extension range was the narrowest in April (stations 7 and 14). *Calanus sinicus*, which appeared in the areas where the YSBCW was observed between April and October, had a higher average abundance, including immature individuals, than in other waters: 842 ind m^{-3} in April, 200 ind m^{-3} in August, and 57 ind m^{-3} in October. Furthermore, in April, the high abundance in the western area (station 7) was due to the high abundance of the copepods stage of the *Calanus sinicus*. These results suggest that the southern flow of the YSBCW can affect the copepod community of the Korea South Sea. However, the geographical distribution of *Calanus sinicus* extends southward to northeastern Taiwan in the marginal seas of the northwestern Pacific Ocean [3,33,35,43]. Moreover, *Calanus sinicus* retains its population in cold water at <20 °C in the Yeosu Strait of Gwangyang Bay, located in the central region of the southern Korea South Sea in summer [11]. This indicates that *Calanus sinicus* can have variable sites for its oversummering period [9,10].

The species diversity index could also have responded to the effect of the inclusion of the warm currents. Tseng et al. [5] suggested that when comparing the species diversity indices in seas where various water masses exist, the species diversity index is usually higher in sea areas that are affected by warm currents. In this study, most of the stations that were affected by the TWC showed a high species diversity index, and it seems that the inflow of the oceanic warm currents, such as the TWC, could have affected the species diversity index. In August, the TWC and YSBCW flowed together in the waters near the western part of Jeju Island (stations 8 and 9), indicating that these water masses may have contributed to the increase in the species diversity index. However, the CDW, which is known to affect the Korea South Sea in the summer, was not recorded during the study period. This may be because the inflow of the CDW in 2018 (<45,000 m^3 s^{-1}) was significantly lower than in 2016, 2017, 2019, and 2020 (approximately 60,000–80,000 m^3 s^{-1}) [https://www.nifs.go.kr/bbs?id=insmaterial&flag=pre&boardIdx=3861&site=&gubun=A&sc=&sv=&cPage=1&startDate=&endDate=], accessed on 20 April 2022.

5. Conclusions

There are various water masses or currents in the Korea South Sea, and the spatiotemporal occurrence pattern of the planktonic copepods seems to be affected by the seasonal fluctuations in these water masses. In particular, some of the following are possible indicator species: *Oithona plumifera*, *Oncaea venusta*, and *Paracalanus aculeatus* for the TWC, *Calanus sinicus* for the YSBCW, and *A. pacifica* for coastal waters. Moreover, as global warming intensifies, studies on zooplankton diversity are expected to play an important role in determining the fluctuations in various water masses.

Supplementary Materials: The following supporting information can be downloaded at: https://www.mdpi.com/article/10.3390/jmse10060754/s1. Table S1: Abundance of the copepods; Table S2: Similarity-percentages (SIMPER) analysis; Table S3: Canonical correspondence analysis (CCA) table; Table S4: Warm-water species; Figure S1: Distribution of chlorophyll-a concentration; Figure S2: Canonical correspondence analysis (CCA) figure. References [44–47] are cited in Supplementary Materials.

J. Mar. Sci. Eng. **2022**, *10*, 754

Author Contributions: Conceptualization, S.S.S., S.Y.C., M.H.S., S.J.L., H.Y.S. and S.H.Y.; methodology, S.S.S. and H.Y.S.; software, S.S.S. and S.Y.C.; validation, H.Y.S.; investigation, S.H.Y.; resources, S.H.Y.; data curation, S.S.S., M.H.S., S.J.L. and H.Y.S.; writing—original draft preparation, S.S.S., S.Y.C. and H.Y.S.; writing—review and editing, H.Y.S.; visualization, S.S.S. and S.Y.C.; supervision, S.H.Y.; project administration, H.Y.S.; funding acquisition, H.Y.S. All authors have read and agreed to the published version of the manuscript.

Funding: This research was funded by the National Institute of Fisheries Science (NIFS), Korea, effects of typhoons on the marine ecosystem in Korea, grant number R2022058.

Institutional Review Board Statement: Not applicable.

Data Availability Statement: Not applicable.

Acknowledgments: We thank the anonymous reviewers and the editor, who made constructive and invaluable suggestions and comments.

Conflicts of Interest: The authors declare no conflict of interest.

References

1. Haury, L.R.; Yamazaki, H.; Fey, C.L. Simultaneous measurements of small-scale physical dynamics and zooplankton distributions. *J. Plankton Res.* **1992**, *14*, 513–530. [CrossRef]
2. Uye, S.; Shimazu, T.; Yamamuro, M.; Ishitobi, Y.; Kamiya, H. Geographical and seasonal variations in mesozooplankton abundance and biomass in relation to environmental parameters in Lake Shinji-Ohashi River-Lake Nakaumi brackish-water system. *J. Mar. Syst.* **2000**, *26*, 193–207. [CrossRef]
3. Hwang, J.S.; Wong, C.K. The China coastal current as a driving force for transporting Calanus sinicus (Copepoda: Calanoida) from its population centers to waters of Taiwan and Hong Kong during the NE monsoon period in winter. *J. Plankton Res.* **2005**, *27*, 205–210. [CrossRef]
4. Hwang, J.S.; Souissi, S.; Tseng, L.C.; Seuront, L.; Schmitt, F.G.; Fang, L.S.; Peng, S.H.; Wu, C.H.; Hsiao, S.H.; Twan, W.H.; et al. A 5-year study of the influence of the northeast and southwest monsoons on copepod assemblages in the boundary coastal waters between the East China Sea and the Taiwan Strait. *J. Plankton Res.* **2006**, *28*, 943–958. [CrossRef]
5. Tseng, L.C.; Souissi, S.; Dahms, H.U.; Chen, Q.C.; Hwang, J.S. Copepod communities related to water masses in the southwest East China Sea. *Helgol. Mar. Res.* **2008**, *62*, 153–165. [CrossRef]
6. Hsieh, C.H.; Chiu, T.S.; Shih, C.T. Copepod diversity and compositions as indicators of intrusion of the Kuroshio branch current into the Northern Taiwan Strait in spring 2000. *Zool. Stud.* **2004**, *43*, 393–403.
7. Liao, C.H.; Chang, W.J.; Lee, M.A.; Lee, K.T. Summer distribution and diversity of copepods in the upwelling waters of the southeastern East China Sea. *Zool. Stud.* **2006**, *45*, 378–394.
8. Lan, Y.C.; Lee, M.A.; Liao, C.H.; Lee, K.T. Copepod community structure of the winter frontal zone induced by the Kuroshio Branch Current and the China Coastal Current in the Taiwan Strait. *J. Mar. Sci. Technol.* **2009**, *17*, 1–6. [CrossRef]
9. Wang, R.; Zuo, T.; Wang, K. The Yellow Sea cold bottom water-an oversummering site for Calanus sinicus (Copepoda, Crustacea). *J. Plankton Res.* **2003**, *25*, 169–183. [CrossRef]
10. Pu, X.M.; Sun, S.; Yang, B.; Zhang, G.T.; Zhang, F. Life history strategies of Calanus sinicus in the southern Yellow Sea in summer. *J. Plankton Res.* **2004**, *26*, 1059–1068. [CrossRef]
11. Jang, M.C.; Jang, P.G.; Shin, K.S.; Park, D.W.; Jang, M. Seasonal variation of zooplankton community in Gwangyang Bay. *Korean J. Environ. Biol.* **2004**, *22*, 11–29.
12. Moon, S.Y.; Oh, H.J.; Soh, H.Y. Seasonal variation of zooplankton communities in the southern coastal waters of Korea. *Ocean Polar Res.* **2010**, *32*, 411–426. [CrossRef]
13. Cho, Y.K.; Kim, K. Characteristics and origin of the cold water in the South Sea of Korea in summer. *Korean Soc. Oceanogr.* **1994**, *29*, 414–421.
14. Eisner, L.; Hillgruber, N.; Martinson, E.; Maselko, J. Pelagic fish and zooplankton species assemblages in relation to water mass characteristics in the northern Bering and southeast Chukchi seas. *Polar Biol.* **2013**, *36*, 87–113. [CrossRef]
15. Park, C.; Lee, C.R.; Kim, J.C. Zooplankton community in the front zone of the East Sea (Sea of Japan), Korea: 2. Relationship between abundance distribution and seawater temperature. *Korean J. Fish. Aquat. Sci.* **1998**, *31*, 749–759.
16. Park, J.S.; Lee, S.S.; Kang, Y.S.; Lee, B.D.; Huh, S.H. The distributions of copepods and chaetognaths in the southern waters of Korea and their relationship to the characteristics of water masses. *Korean J. Fish. Aquat. Sci.* **1990**, *23*, 245–252.
17. Kang, Y.S.; Hong, S.Y. Occurrences of oceanic warm-water calanoid copepods and their relationship to hydrographic conditions in Korean waters. *Bull. Plankton Soc. Jpn. Hiroshima* **1995**, *42*, 29–41.
18. Lee, C.R.; Lee, P.G.; Park, C. Seasonal and vertical distribution of planktonic copepods in the Korea Strait. *Korean J. Fish. Aquat. Sci.* **1999**, *32*, 525–533.

19. Choi, S.Y. Spatial and Temporal Distribution of Acartia (Crustacea, Copepoda, Calanoida) Species in the Southern Coastal Waters of Korea and the Physio-Ecological Characteristics of Acartia eryhraea. Ph.D. Thesis, Chonnam National University, Yeosu, Korea, February 2020.
20. Kondo, M. Oceanographic investigations of fishing grounds in the East China Sea and the Yellow Sea-I. Characteristics of the mean temperature and salinity distributions measured at 50 m and near the bottom. *Bull. Seikai Reg. Fish. Res. Lab.* **1985**, *62*, 19–66.
21. Lie, H.J.; Cho, C.H. Seasonal circulation patterns of the Yellow and East China Seas derived from satellite-tracked drifter trajectories and hydrographic observations. *Prog. Oceanogr.* **2016**, *146*, 121–141. [CrossRef]
22. Park, J.S.; Lee, S.S.; Kang, Y.S.; Huh, S.H. Distribution of indicator species of copepods and chaetognaths in the middle East Sea of Korea and their relationships to the characteristics of water masses. *Korean J. Fish. Aquat. Sci.* **1991**, *24*, 203–212.
23. Vohra, D.F. *Determination of Photosynthetic Pigments in SeaWater. Monographs on Oceanographic Methodology*; UNESCO, Ed.; UNESCO: Paris, France, 1996; pp. 10–18.
24. Seung, Y.H. Water masses and circulations around Korean Peninsula. *J. Korean Soc. Oceanogr.* **1992**, *27*, 324–331.
25. Jang, S.T.; Lee, J.H.; Kim, C.H.; Jang, C.J.; Jang, Y.S. Movement of cold water mass in the northern East China Sea in summer. *J. Korea Soc. Oceanogr.* **2011**, *16*, 1–13.
26. Gong, G.C.; Chen, Y.L.L.; Liu, K.K. Chemical hydrography and chlorophyll-a distribution in the East China Sea in summer: Implications in nutrient dynamics. *Cont. Shelf Res.* **1996**, *16*, 1561–1590. [CrossRef]
27. Clarke, K.R.; Gorley, R.N. *Primer v6: User Manual/Tutorial*; PRIMER-E: Plymouth, UK, 2006; p. 866.
28. Ter Braak, C.J.F.; Milauer, P. *CANOCO Reference Manual and CanoDraw for Windows User's Guide: Software for Canonical Community Ordination (Version 4.5)*; Microcomputer Power: Ithaca, NY, USA, 2002.
29. Haury, L.R.; Pieper, R.E. *Zooplankton: Scales of biological and physical events. Marine Organisms as Indicators*; Springer: New York, NY, USA, 1988; pp. 35–72.
30. Yuhui, L.; Nakamura, Y. Distribution of planktonic copepods in the Kuroshio area of the East Chine Sea to the southwest of Kyushu in Japan. *Acta Oceanol. Sin.* **1993**, *3*, 445–456.
31. Beaugrand, G.; Reid, P.C.; Ibanez, F.; Lindley, J.A.; Edwards, M. Reorganization of North Atlantic marine copepod biodiversity and climate. *Science* **2002**, *296*, 1692–1694. [CrossRef] [PubMed]
32. Peterson, W.T.; Keister, J.E. Interannual variability in copepod community composition at a coastal station in the northern California Current: A multivariate approach. *Deep Sea Res. Part II Top. Stud. Oceanogr.* **2003**, *50*, 2499–2517. [CrossRef]
33. Dur, G.; Hwang, J.S.; Souissi, S.; Tseng, L.C.; Wu, C.H.; Hsiao, S.H.; Chen, Q.C. An overview of the influence of hydrodynamics on the spatial and temporal patterns of calanoid copepod communities around Taiwan. *J. Plankton Res.* **2007**, *29*, 97–116. [CrossRef]
34. Hwang, J.S.; Dahms, H.U.; Tseng, L.C.; Chen, Q.C. Intrusions of the Kuroshio Current in the northern South China Sea affect copepod assemblages of the Luzon Strait. *J. Exp. Mar. Biol. Ecol.* **2007**, *352*, 12–27. [CrossRef]
35. Hsiao, S.H.; Kâ, S.; Fang, T.H.; Hwang, J.S. Zooplankton assemblages as indicators of seasonal changes in water masses in the boundary waters between the East China Sea and the Taiwan Strait. *Hydrobiologia* **2011**, *666*, 317–330. [CrossRef]
36. Sogawa, S.; Kidachi, T.; Nagayama, M.; Ichikawa, T.; Hidaka, K.; Ono, T.; Shimizu, Y. Short-term variation in copepod community and physical environment in the waters adjacent to the Kuroshio Current. *J. Oceanogr.* **2017**, *73*, 603–622. [CrossRef]
37. Teague, W.J.; Jacobs, G.A.; Perkins, H.T.; Book, J.W.; Chang, K.I.; Suk, M.S. Low-frequency current observation in the Korea/Tsushima Strait. *J. Phys. Oceanogr.* **2002**, *32*, 1621–1641. [CrossRef]
38. Kim, W.S.; Yoo, J.M.; Myung, C.S. A review on the copepods in the South Sea of Korea. *Korean J. Fish. Aquat. Sci.* **1993**, *26*, 266–278.
39. Cho, Y.K.; Seo, G.H.; Choi, B.J.; Kim, S.; Kim, Y.G.; Youn, Y.H.; Dever, E.P. Connectivity among straits of the northwest Pacific marginal seas. *J. Geophys. Res. Ocean.* **2009**, *114*, C06018. [CrossRef]
40. Lee, C.Y.; Liu, D.C.; Su, W.C. Seasonal and spatial variations in the planktonic copepod community of Ilan Bay and adjacent Kuroshio waters off northeastern Taiwan. *Zool. Stud.* **2009**, *48*, 151–161.
41. Stoecker, D.K.; Egloff, D.A. Predation by Acartia tonsa Dana on planktonic ciliates and rotifers. *J. Exp. Mar. Biol. Ecol.* **1987**, *110*, 53–68. [CrossRef]
42. Yun, Y.H.; Park, Y.H.; Bong, J.H. Enlightenment of the characteristics of the Yellow Sea bottom cold water and its southward extension. *J. Korean Earth Sci. Soc.* **1991**, *12*, 25.
43. Choi, Y.C. The Characteristics of Yellow Sea Bottom Cold Water in September, 2006. *J. Fish. Mar. Sci. Educ.* **2011**, *23*, 425–432.
44. Nakata, K.; Itoh, H.; Ichikawa, T.; Sasaki, K. Seasonal changes in the reproduction of three oncaeid copepods in the surface layer of the Kuroshio Extension. *Fish. Oceanogr.* **2004**, *13*, 21–33. [CrossRef]
45. Ka, S.; Hwang, J.S. Mesozooplankton distribution and composition on the northeastern coast of Taiwan during autumn: Effects of the Kuroshio Current and hydrothermal vents. *Zool. Stud.* **2011**, *50*, 155–163.
46. Tseng, L.C.; Hung, J.J.; Chen, Q.C.; Hwang, J.S. Seasonality of the copepod assemblages associated with interplay waters off northeastern Taiwan. *Helgol. Mar. Res.* **2013**, *67*, 507–520. [CrossRef]
47. Miyamoto, H.; Itoh, H.; Okazaki, Y. Temporal and spatial changes in the copepod community during the 1974–1998 spring seasons in the Kuroshio region; a time period of profound changes in pelagic fish populations. *Deep Sea Res. Part I Oceanogr. Res. Pap.* **2017**, *128*, 131–140. [CrossRef]

Journal of
Marine Science and Engineering

MDPI

Article

Spatial Distribution of Sound Scattering Layer and Density Estimation of *Euphausia pacifica* in the Center of the Yellow Sea Bottom Cold Water Determined by Hydroacoustic Surveying

Hansoo Kim [1], Garam Kim [2], Mira Kim [1] and Donhyug Kang [1,*]

[1] Maritime Security and Safety Research Center, Korea Institute of Ocean Science and Technology (KIOST), Busan 49111, Korea; hskim@kiost.ac.kr (H.K.); mrkim0825@kiost.ac.kr (M.K.)
[2] Marine Ecosystem Research Center, Korea Institute of Ocean Science and Technology (KIOST), Busan 49111, Korea; garamkim@kiost.ac.kr
* Correspondence: dhkang@kiost.ac.kr; Tel.: +82-51-664-3650

Abstract: The Yellow Sea Bottom Cold Water (YSBCW) refers to seawater with a water temperature of 10 °C or less found at the bottom of the center of the Yellow Sea. The spatiotemporal variability of the YSBCW directly affects the distribution of organisms in the marine ecosystem. In this study, hydroacoustic and net surveys were conducted in April (spring) to understand the spatial distribution of the sound scattering layer (SSL) and estimate the density of *Euphausia pacifica* (*E. pacifica*) in the YSBCW. Despite the shallow water in the YSBCW region, *E. pacifica* formed an SSL, which was distributed near the bottom during the daytime; it showed a diel vertical migration (DVM) pattern of movement toward the surface during the nighttime. The mean upward and downward swimming speeds around sunset and sunrise were approximately 0.6 and 0.3–0.4 m/min, respectively. The *E. pacifica* density was estimated in the central, western, and eastern regions; the results were approximately 15.8, 1.3, and 10.3 g/m^2, respectively, indicating significant differences according to region. The results revealed high-density distributions in the central and eastern regions related to the water temperature structure, which differs regionally in the YSBCW area. Additional studies are needed regarding the spatial distribution of *E. pacifica* in the YSBCW and its relationship with various ocean environmental parameters according to season. The results of this study contribute to a greater understanding of the structure of the marine ecosystem in the YSBCW.

Keywords: diel vertical migration; sound scattering layer; spatial and regional distributions; Yellow Sea Bottom Cold Water

Citation: Kim, H.; Kim, G.; Kim, M.; Kang, D. Spatial Distribution of Sound Scattering Layer and Density Estimation of *Euphausia pacifica* in the Center of the Yellow Sea Bottom Cold Water Determined by Hydroacoustic Surveying. *J. Mar. Sci. Eng.* 2022, 10, 56. https://doi.org/10.3390/jmse 10010056

Academic Editor: Giuseppa Buscaino

Received: 24 November 2021
Accepted: 28 December 2021
Published: 4 January 2022

Publisher's Note: MDPI stays neutral with regard to jurisdictional claims in published maps and institutional affiliations.

1. Introduction

The Yellow Sea, a semi-enclosed marginal sea located between the Korean peninsula and the China mainland, is connected to the northwestern Pacific Ocean and has a mean depth of approximately 44 m [1]. The maximum water depth in the central part of the sea, which has the characteristics of a continental shelf, is less than approximately 100 m [2]. The Yellow Sea Bottom Cold Water (YSBCW), an important oceanic phenomenon in the Yellow Sea, occurs from spring to autumn at the bottom of the central area [3,4]. The YSBCW, which occupies more than 70% of the total area of the Yellow Sea, has a low water temperature of less than 10 °C and a high salinity of 32–33 practical salinity unit (psu) [5,6].

With catches exceeding 3 million tons of marine organisms per year, the Yellow Sea is an important source of aquatic resources [7,8]. However, its fishery resources are decreasing rapidly because of overfishing, as well as increases in temperature and marine–environmental pollution related to global change [9]. The zooplankton species *Euphausia pacifica* (*E. pacifica*) is a key species in the Yellow Sea. The YSBCW provides a cold-water refuge for this species, allowing it to survive summer and autumn when surface temperatures are high. *E. pacifica* is a major food organism for fish and marine mammals; it has an important role in the food web, where it links predators and phytoplankton [10,11]. Thus, *E. pacifica* is one of the most important biological factors that affect marine ecosystems and fishery resources in the Yellow Sea [12]. Because the temporal and spatial variabilities of the YSBCW greatly affect the distribution of marine life, there is a need to study the characteristics of marine ecology.

Various zooplankton (Mesozooplankton, Macrozooplankton), small mesopelagic fish and other juvenile fishes generally tend to form a community at a specific depth in the ocean, defined as the sound scattering layer (SSL), or deep scattering layer (DSL), which is of several meters vertical extent [13,14]. The distribution of the SSL varies depending on the marine environment of the specific ocean; it moves toward the surface at night and the bottom during the day because of diel vertical migration (DVM) [14,15]. The DVM of zooplankton changes the distribution of SSLs in accordance with changes in illumination, the constituent organisms, marine environmental conditions, seasons, prey, and escape [16,17]. *E. pacifica* migrates to the surface at night to avoid predators and moves to the bottom during the day to feed [18,19]. Identifying the distribution of the important zooplankton constituting the SSL is an important consideration for understanding the characteristics of marine ecosystems and relationships within food chains; thus, it requires investigation in aquatic resources research.

To determine the characteristics and structure of marine ecology in the YSBCW, previous studies have mainly examined spatial distribution, individual structure, reproduction, vertical distribution, seasonal volume, and regional appearance distribution via biological sampling [20–25]. These previous studies provided qualitative data that are useful for understanding these characteristics, but they were often limited in their abilities to provide quantitative characteristics because of the zooplankton-detection problem associated with the sampling process. When the zooplankton density is determined by net survey alone, there is a possibility that it will be underestimated. However, hydroacoustic techniques are used in various ways to evaluate the spatiotemporal distribution, quantitative distribution, and biomass of zooplankton [26]. In the YSBCW, only the characteristics of circadian changes related to changes in temperature structure have been reported, and few studies have been conducted [12,27]. There have also been studies regarding the distribution of the SSL in the southern Yellow Sea [28] and the estimation of the *E. pacifica* density in the East China Sea [29]; however, the biomass and spatiotemporal distribution of *E. pacifica* have not been previously assessed in the YSBCW, and the optimal management strategy for this resource remains unclear. Therefore, because this species is an important biological factor that regulates the total biomass of the marine ecosystem and the fishery resources in the YSBCW area, qualitative and quantitative research regarding *E. pacifica* is continuously needed for effective resource management.

The goals of this study were to understand the spatial distribution of the SSL and density estimation of the *E. pacifica* density in the YSBCW. We used hydroacoustic data, net sampling data, and water temperature data gathered near the Yellow Sea during the spring season. The results of this study will enhance understanding of the marine ecosystem structure in the YSBCW.

2. Materials and Methods

2.1. Survey Area

The survey area covered the central Yellow Sea and was divided into the Korea–China provisional water zone (at latitudes of 35° to 36°) and the exclusive economic zone (EEZ) of Korea. YSBCW distribution occurs in the survey area. A research vessel *R/V Onnuri* was used to conduct the hydroacoustic survey, biological net sampling, and conductivity–temperature–depth (CTD) castings in Figure 1 from 19 to 26 April 2019. Table 1 provides the region, time, latitude, longitude, range, and sea-depth data of each transect.

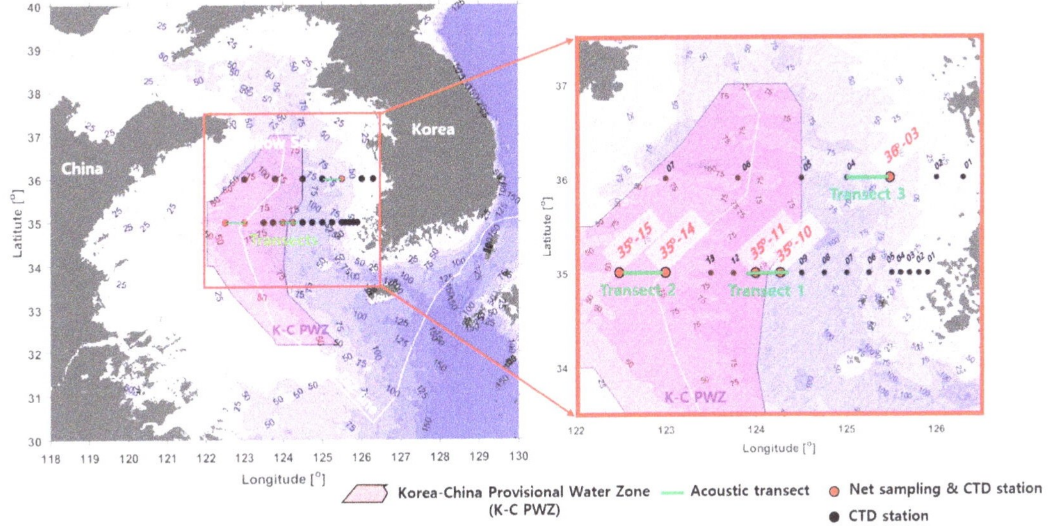

Figure 1. Map of acoustic transects, net sampling, and CTD stations used for evaluation of the sound scattering layer (SSL) including *Euphausia pacifica* in the Yellow Sea Bottom Cold Water (YSBCW).

Table 1. Transect, region, time, start/end position, range, and sea depth information of the acoustic measurements.

Transect	Region		Time (Local)	Start lat./lon. (°)	End lat./lon. (°)	Range (Nautical Miles)	Sea Depth (m)
Transect 1	YSBCW Central	K-C PWZ [1] (Korea)	04:00–07:00 Sunrise	35.00° N 124.25° E	35.00° N 124.00° E	16	78–85
Transect 2	YSBCW Western	K-C PWZ [1] (China)	16:00–20:30 Sunset	35.00° N 123.00° E	35.00° N 122.50° E	22	65–75
Transect 3	YSBCW Eastern	EEZ (Korea)	17:50–20:00 Sunset	36.00° N 125.00° E	36.00° N 125.50° E	12	70–80

[1] K-C PWZ: Korea-China Provisional Water Zone.

2.2. Data Collection

Acoustic measurements were performed using a scientific echosounder system (DT-X Extreme, BioSonics, Inc., Seattle, WA, USA) frequencies of 38 and 200 kHz [30]. The measurement of *E. pacifica* in the YSBCW was determined using the 38 and 200 kHz. However, the frequencies were mainly used in previous studies of the Yellow Sea [28,29]. Because this study was performed in spring, and given the proliferation of phytoplankton as prey, it was presumed that large numbers of zooplankton would appear in this season; thus, the present study used the frequency of 200 kHz for good resolution in the field acoustic survey.

Before the measurements, the transducers of the echosounder were calibrated in a water tank using 38 mm (38 kHz) and 36 mm (200 kHz) tungsten-carbide calibration spheres and standard procedures [31,32]. After 38 and 200 kHz transducers had been deployed to a small-towed body, it was towed by the research vessel. In order to minimize the effect of ship or surface bubbles by wind, the towed body was located at a depth of roughly 7–9 m from the sea surface to obtain measurements for each transect line (Figure 2, top).

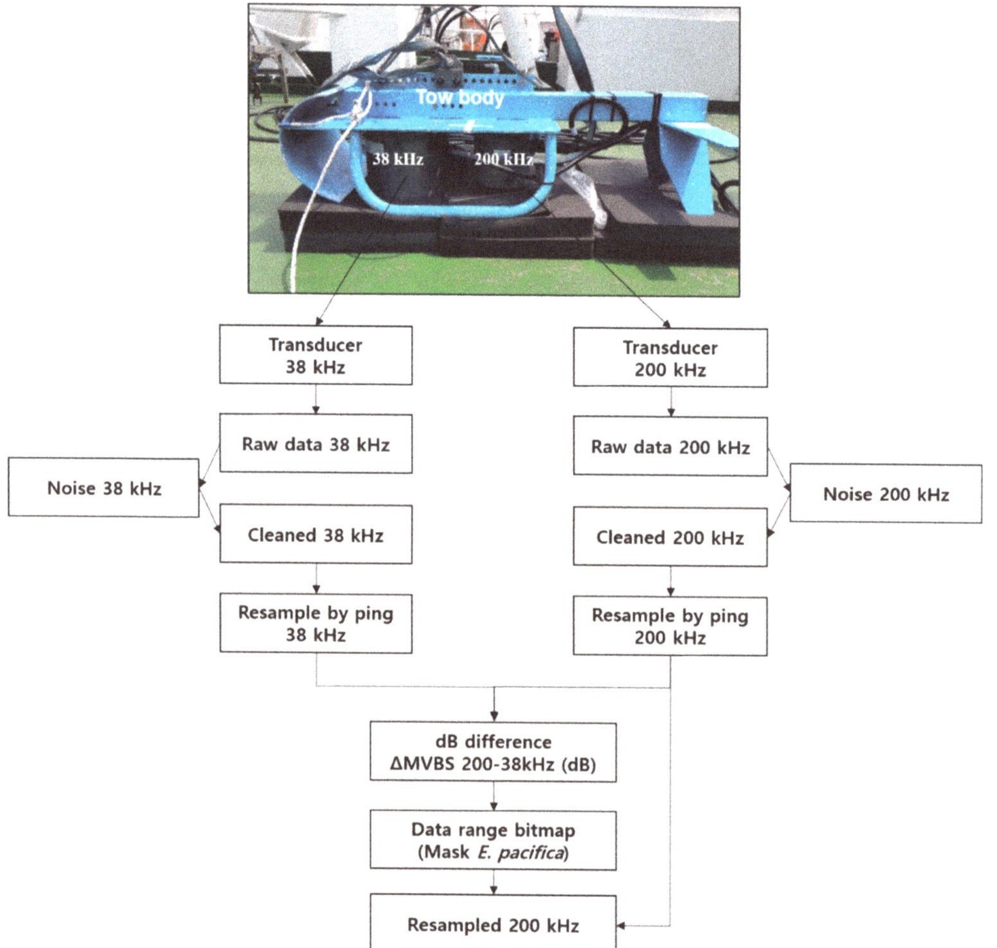

Figure 2. Flowchart of acoustic data processing using the dB difference method at 38 and 200 kHz.

To identify the regional distribution of *E. pacifica*, measurements were conducted for three transects at latitudes 35° and 36° in the central, western, and eastern regions (approximately 4 h of measurements for each transect).

Centered on the imaginary midline in the Korea–China provisional water zone (K-C PWZ), the acoustic measurement areas were divided into the central Korea ocean (Transect 1), western China ocean (Transect 2), and eastern Korea exclusive economic zone (EEZ) ocean (Transect 3). The acoustic measurements were performed at sunrise or sunset times for each transect to confirm the DVM pattern of the SSL. Both the 38 and 200 kHz transducers transmitted and received 2 pings/s; the pulse width was 0.5 ms, and the detection range was set at 0–120 m water depth. The threshold level of the received signal was fixed at −100 dB to determine the influence of small organism or zooplankton layers on the volume backscattering strength (*Sv*). Location information was collected from the differential global positioning system within the echosounder while traveling along the transect lines at a speed of 5–7 knots. The system and environmental parameters of the acoustic measurements are detailed in Table 2.

Table 2. System parameters of the scientific echosounder used for acoustic surveys in the Yellow Sea Bottom Cold Water.

Parameter	38 kHz	200 kHz
Source Level (dB/μPa)	212.6	220.5
Pulse Length (ms)	0.5	0.5
Beam Width (degree)	9.0	6.9
Ping Rate (ping/s)	2	
Absorption Coefficient (dB/m)	0.00904	0.05259
Threshold (dB)	-100	
Sound Speed (m/s)	1489.86	
Collection Range (m)	0–120	
Sensor Depth (m)	7–9	

The ocean environment and net surveys were carried out to understand the ecological characteristics of *E. pacifica* in the ocean area. Vertical profiles of water temperature were measured at 22 stations (35°, 01–15; 36°, 01–07) using a CTD system (SBE 911plus, Sea-Bird Electronics, Bellevue, WA, USA) from the surface to 5 m above the bottom (Figure 1). The net survey was conducted at five stations: two stations (35°, 10 and 11) in the central region, two stations (35°, 14 and 15) in the western region, and one station (36°, 03) in the eastern region (Figure 1). The net survey was conducted before or after the acoustic transects. The bongo net used for the net survey was a conical net with an attached flowmeter. The bongo net's diameter, length, and mesh size are 60 cm, 3 m, 200 μm, respectively.

Slope net sampling from the bottom to the surface was performed for approximately 20 min at a net-specific towing speed of 2–3 knots. The samples were fixed immediately using 5% formalin.

After the field surveys, the samples were fixed in formalin and moved to the laboratory. One hundred individuals were randomly selected from each net sampling station, and standard length (L) was measured for 500 specimens obtained through the net surveys at five stations. Because only L was measured and analyzed for these net sampling data, the wet weight (W) was calculated using the W–L relationship of Equation (1) proposed by Kang et al. (2003) [29],

$$W \ (\text{mg}) = 0.0054 \cdot L \ (\text{mm})^{3.15} \tag{1}$$

2.3. Acoustic Data Analysis and E. pacifica Identification

The acoustic data collected from the transects were analyzed using acoustic analysis and processing software (Echoview v 9.0; Echoview Ltd., Hobart, Tasmania, Australia). From the acquired Sv of the acoustic measurements, a virtual echogram was generated by removing unnecessary data (e.g., sensor depth, sensor near-field effects, air bubbles in the surface and bottom signals) [33,34]. To understand the DVM of the SSL and identify *E. pacifica*, the acoustic data were compressed by the horizontal resolution (3.5 s) and the vertical resolution (0.5 m). An analysis and processing flowchart of the acquired acoustic data is shown in Figure 2.

To extract the *E. pacifica* echoes, this species identification was performed using the difference of the mean volume backscattering strength ($\Delta MVBS$) method, which is also known as the dB difference method [35–38]. The dB difference method depends on the frequency characteristics of the SSL attributable to marine organisms. Generally, zooplankton target strength (TS) is characterized by fluctuations between the low frequency of 38 kHz and the high frequency of 200 kHz [35,39]. Thus, the Sv difference between 38 and 200 kHz is large, providing a good method for species classification. The $\Delta MVBS$ window can be used to identify *E. pacifica* echo signals. By applying the $\Delta MVBS$, other zooplankton (ex. Copepoda) and fish signals were excluded from the net sampling data. The echoes were selected using the data range bitmap algorithm to eliminate noise smaller than the minimum Sv and larger than the maximum Sv of *E. pacifica* and then masking it by applying the $\Delta MVBS$. *E. pacifica* were used as the true signals, and all other values were not-selected signals from the data range bitmap (Figure 2). In this study, the range of the $\Delta MVBS$ applied (on the basis of biological net sampling data) was 14–19 dB. The range of dB differences used to identify *E. pacifica* was the recommended range of Sv difference based on the size distribution of *E. pacifica* by net sampling data.

For the acoustic characteristics of *E. pacifica*, the TS at 200 kHz was estimated using the distorted wave Born approximation (DWBA) model, a widely used acoustic model for estimating the TS of zooplankton, as shown in Equation (2) [40–42],

$$TS_{200 \text{ kHz}} = 40.42 \cdot log_{10}(L) - 131.9 \tag{2}$$

The $\Delta MVBS$ value was obtained by separating the 38 kHz and 200 kHz acoustic signals, using Equation (3),

$$\Delta MVBS = TS(200 \text{ kHz}) - TS(38 \text{ kHz}) = Sv(200 \text{ kHz}) - Sv(38 \text{ kHz}) \tag{3}$$

The swimming speed of DVM of the SSL was estimated by using the echo signal of Sv, to which the $\Delta MVBS$ was applied. To calculate the swimming speed, the center of the SSL of each compressed signal was calculated as a function of time $D(t)$. Then, a threshold of -90 dB of Sv was set to accurately extract the center of the SSL. The upward and downward swimming speeds (Vs) were calculated from the rate of change in the SSL as the derivative of $D(t)$ with the time(t) [28],

$$Vs = dD(t)/dt \tag{4}$$

2.4. Density Estimation of E. pacifica

To estimate the spatially distributed density of *E. pacifica*, the SSL was separated by excluding fish signals and other zooplankton signals from the acoustic data. The densities by transect of *E. pacifica* in the SSL were calculated using the $\Delta MVBS$, the Sv, and the nautical area scattering cross-section ($NASC$) of the 1 nautical mile (n·mile) interval of the elementary distance sampling unit. The density of *E. pacifica* was calculated via Equation (4) using the L with frequency of L, W–L relationship, and the $NASC$ values obtained from the net sampling data [26,43–45],

$$\rho\left(\text{g/m}^2\right) = a \cdot \sum_{i=1}^{n} f_i(L_i)^b * NASC \tag{5}$$

where f_i is the relative frequency value of *E. pacifica* concerning the standard length (L_i), and the sum is 1; a and b are constants calculated from the conversion factor, considering the $NASC$ and W–L relationship. To calculate the mean density ($\bar{\rho}$) in the survey area, the weighted mean of the data obtained from each transect was used, as in Equation (6):

$$\bar{\rho}\left(\text{g/m}^2\right) = \frac{\sum_{i=1}^{N} \bar{\rho}_i \cdot n_i}{\sum_{i=1}^{N} n_i} \tag{6}$$

where N is the number of transect lines, $\bar{\rho}_i$ is the mean density of the i-th transect line, and n_i is the distance obtained by converting the i-th transect line into 1 n·mile.

3. Results

3.1. Zooplankton Community Analysis and Size Distribution

According to an analysis of the net sampling data at the five stations, Copepoda was a dominant class with an average of 86%. Euphausiacea and etc. comprised 11% and under 3%, respectively (Figure 3a). It was confirmed that Copepoda consisted of individuals less than 2 mm in L, and Euphausiacea mainly consisted of juvenile and adults of *E. pacifica*.

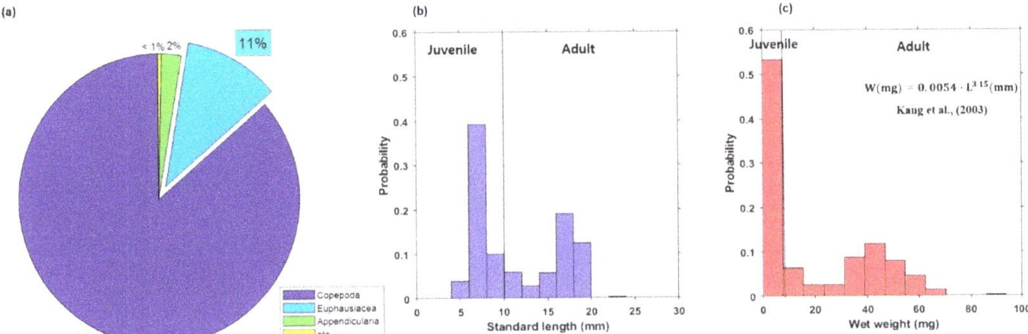

Figure 3. (**a**) Zooplankton species composition from net sampling data and the histograms of the (**b**) standard length (*L*) frequency distribution and (**c**) wet weight (*W*) frequency distribution from all net surveys of *E. pacifica* in the SSL.

Analysis of the size and weight distribution of *E. pacifica* was performed using the net survey data; the findings indicated that the juveniles and adults were mixed and distributed concurrently. In the central region, the adult mean *L* and mean *W* were 15.1 mm and 33.4 mg, respectively. In contrast, the juvenile mean *L* and mean *W* were 6.6 mm and 2.1 mg, respectively. In the western region, the juvenile and adult of mean *L* and *W* were 7.2 mm and 2.8 mg, respectively, whereas the juvenile and adult of mean *L* and *W* in the eastern region were 17.5 mm and 45.5 mg, respectively. Hence, larger individuals of *E. pacifica* were caught in the eastern region than at the other net sampling stations. Among five net surveys performed at all net sampling stations, the mean *L* of *E. pacifica* was 11.4 mm and the mean *W* was 19.4 mg.

Large individuals of *E. pacifica* were found in the eastern region and the Korean oceans; small individuals inhabited the western region. A histogram analysis, performed by dividing the distribution of *L* of *E. pacifica* from all net sampling stations by 2 mm, revealed a bimodal distribution ranging from 6 to 8 mm and from 16 to 18 mm (Figure 3b). The highest frequency of *W* was 0–8 mg (Figure 3c). *L* and *W* information obtained via net sampling was used to estimate the density of *E. pacifica* from the acoustic data.

3.2. DVM of SSL

Figure 4 shows the compressed sample echogram received from the 38 (Figure 4a,c,e) and 200 (Figure 4b,d,f) kHz transducers. The 38 kHz echogram was different from that of the 200 kHz echogram. Despite the shallowness of the ocean, an SSL was formed from *E. pacifica* and distributed in the YSBCW. Analysis of acoustic data obtained before and after sunrise and sunset revealed that the SSL was distributed at the surface during the nighttime; the DVM findings were characteristic of swimming toward the bottom during the daytime.

In Transect 1 at 200 kHz of the central region, the biological community of the SSL during the nighttime was divided into two layers. At nighttime, an SSL located in the upper layer of the water column was distributed from the surface to a depth of approximately 50 m; an SSL in the lower layer was distributed from a depth of approximately 60 m from the bottom. The upper layer (SSL 1), located at a depth of approximately 25 m, began to swim toward the bottom with a swimming speed of 0.63 m/min from 05:40. The distribution was mixed, with the lower layer (SSL 2) located at a depth of approximately 75 m (Figure 4b).

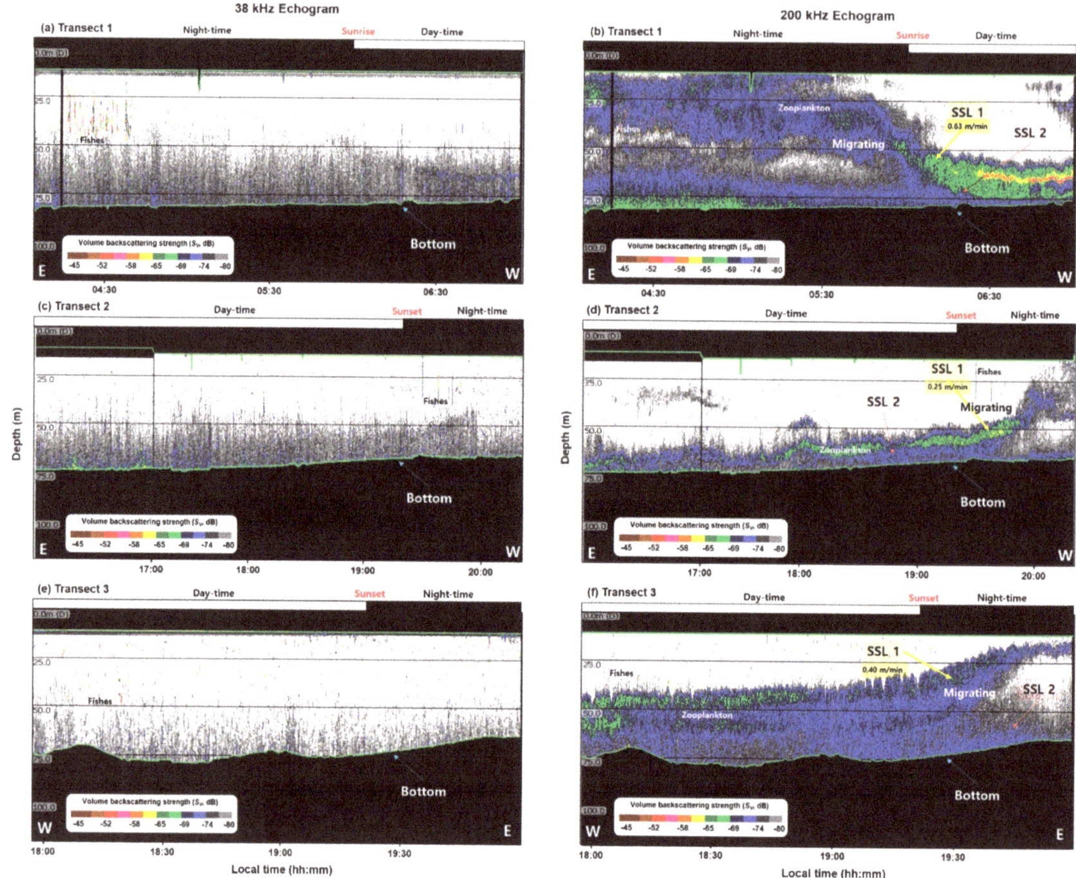

Figure 4. Vertical and spatial distributions of the 38 kHz (**a**,**c**,**e**) and 200 kHz (**b**,**d**,**f**) sample echograms for (**a**,**b**) Transect 1, (**c**,**d**) Transect 2, and (**e**,**f**) Transect 3. In the figure, E is east, W is west, black regions are bottom, and depth are divided by 25 m. Daytime and nighttime at the top are divided based on the on-site sunrise and sunset, and the green line is the measured and calculated regions considering the transducer depth, near-field effects, and bottom.

In Transect 2 at 200 kHz in the western region, the SSL was densely distributed near the bottom during the daytime, but it then swam slowly to the surface at the swimming speed of 0.25 m/min. At 20:00, after sunset, the SSL was distributed from the surface layer to a depth of 40 m. As in Transect 1, the SSL in Transect 2 tended to be divided and distributed into two layers (Figure 4d).

In contrast, Transect 3 at 200 kHz in the eastern region exhibited SSL distribution from 40 to 65 m depth during the daytime. From 19:00, before sunset, the SSL moved toward the surface at the swimming speed of 0.40 m/min; it rapidly became densely located at a depth of approximately 20 m after sunset (Figure 4f). The *E. pacifica* results confirmed DVM during the daytime and nighttime; the downward swimming speed was higher than the upward speed.

3.3. DVM of SSL Vertical Distribution of E. pacifica

The results of the 200 kHz echo signal were comparatively analyzed (Figure 5). In Transect 1, the acoustic signal was highest (−60 dB) at a depth of 60–70 m during the daytime; it was high at a depth of approximately 15–30 m during the nighttime (Figure 5a). In Transect 2, the *Sv* was high at a depth of approximately 70 m during the daytime, as well as at depths of 30–35 m and 45–60 m during the nighttime (Figure 5b). Conversely, in Transect 3, the *Sv* was high at a depth of 10–50 m during the nighttime and 45–65 m during the daytime (Figure 5c).

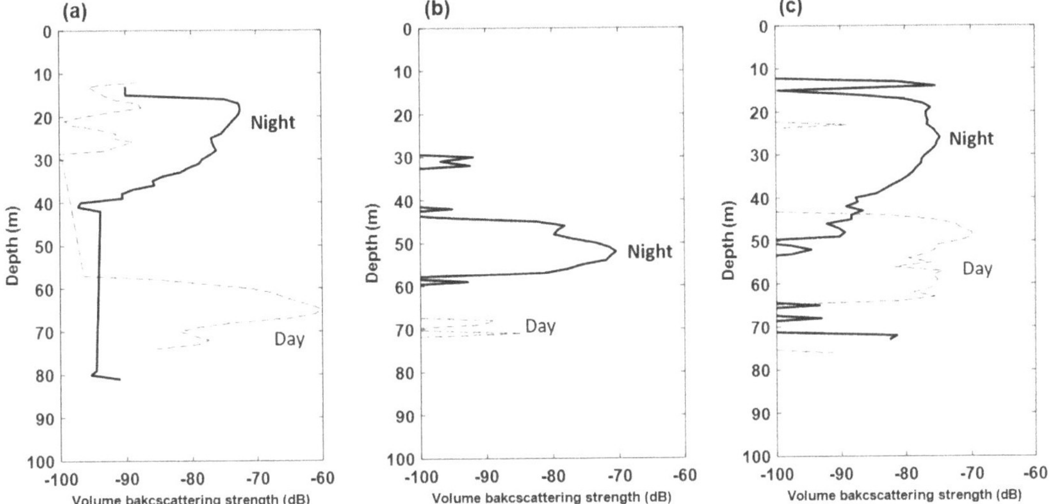

Figure 5. Variations in mean volume backscattering strength ($\Delta MVBS$) of the 200 kHz during daytime and nighttime in the YSBCW at (**a**) Transect 1, (**b**) Transect 2, and (**c**) Transect 3.

The distribution depth of *E. pacifica* during the nighttime was similar from the acoustic data of Transect 1 and Transect 3. However, the Transect 2 data were densely distributed at a low water depth. The data implied that *E. pacifica* was densely distributed in the lower layer during the daytime, whereas it was widely distributed from the surface layer to the middle layer during the nighttime. The acoustic data confirmed that the vertical distribution of *E. pacifica* differed according to the day and night DVM characteristics, as well as the region.

3.4. Density and Spatial Distribution of E. pacifica

Using the dB difference method with the acoustic and net sampling results, DVM patterns in the surface and bottom waters were revealed. The echo signals for Copepoda and fish contained in zooplankton were removed through the $\Delta MVBS$ method, and only *E. pacifica* was extracted. Figure 6 shows the mean density along each transect line. The density was similar when using *Sv* or *NASC*. Analysis of the *E. pacifica* density according to section showed that the *E. pacifica* density was especially high in the central region, where it was 1.60–55.96 g/m². In the western and eastern regions, the overall densities were 0.01–9.50 g/m² and 2.38–25.81 g/m², respectively. The *E. pacifica* density in the YSBCW was noticeably higher in the central and eastern regions than in the western region.

The mean *E. pacifica* densities were 15.78 g/m², 1.33 g/m², and 10.26 g/m², and the coefficients of variation (*CV*) were 1.14, 1.97, 0.61, respectively, in the central, western, and eastern regions. The overall mean density was 8.33 g/m². Thus, the SSL had a high-density distribution in the central and eastern regions of the Korean oceans, while it exhibited a low-density distribution in the western region of the YSBCW.

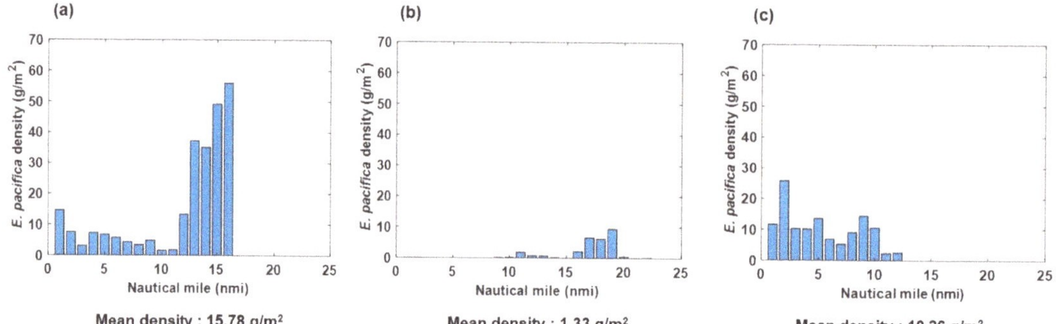

Figure 6. Calculated densities at each nautical mile, and mean density of *E. pacifica* in the YSBCW at (**a**) Transect 1, (**b**) Transect 2, and (**c**) Transect 3.

4. Discussion

The YSBCW is formed during the previous winter but remains in the lower layer from spring to late autumn [1,5]. In summer, high temperature and low salinity are distributed at the surface, but low temperature and high salinity are distributed at the bottom in the YSBCW [1]. Therefore, the spatiotemporal distribution of marine organisms in the Yellow Sea is greatly affected by changes in the physical environment, which are related to seasonal changes. In a previous study, *E. pacifica* exhibited DVM patterns in all water columns from the surface to the bottom in the spring season; it moved to only the lower water column where the thermocline occurred in summer [12]. During the spring season, when the present study was conducted, the water temperature structure of the YSBCW showed a mixed layer from the surface to the bottom; the observed biological characteristics implied SSL movement to the surface and to the bottom through DVM. By applying dB difference, the SSL showing DVM pattern was found to be *E. pacifica*, and the SSL showing no DVM pattern was removed.

In general, water temperature, salinity, and chlorophyll-a concentration are considered major environmental factors that affect the distribution of *E. pacifica* [20–22]. In the Yellow Sea, *E. pacifica* is mainly distributed at temperatures lower than 20 °C during the summer [46]; the spatial distribution of *E. pacifica* is determined primarily by temperature, with a tendency to inhabit cold water (8–16 °C) [23]. Although the net sampling results could not demonstrate a high-resolution SSL pattern (because of the discrete sampling method and net avoidance by *E. pacifica*), the sampling data provide valuable information for understanding the species composition of the SSL. Therefore, we compared the spatial distribution of temperature with the density of *E. pacifica* to understand the regional distribution of the YSBCW (Figure 7). The spatial temperature structure obtained from the CTD survey revealed that the difference in temperature between the surface and bottom was within approximately 1 to 4 °C during this period, indicating that all water columns were mixed without significant differences in temperature. The mean temperature was less than approximately 10 °C in the central region (Figure 7b, Transect 1) and less than approximately 8 °C in the eastern region (Figure 7a, Transect 3), except for a portion at the surface; in contrast, a high temperature of approximately 10 °C or higher existed in the western region (Figure 7b, Transect 2). Based on the distribution of water temperature, we concluded that the density of *E. pacifica* was high in the central and eastern regions because the temperature was lower there than in the western region (less than 10 °C). Therefore, it is assumed that the density of *E. pacifica* was related to the water temperature.

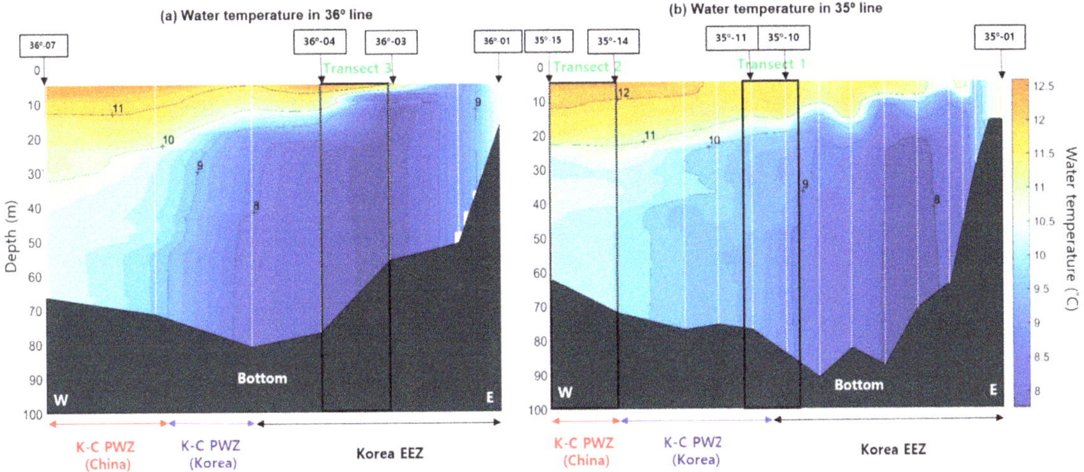

Figure 7. Spatial distribution of water temperature in the (**a**) 36° and (**b**) 35° lines from CTD stations in the YSBCW. In the figure, W is west, E is east, and the depth is divided by 10 m. White dotted lines and grey regions are CTD stations and bottom, respectively.

Thus far, various studies have reported the density or biomass of *E. pacifica* in the YSBCW, as well as its spatial distribution. In addition to the spatial and temporal distributions of *E. pacifica*, the estimated density distributions obtained through acoustic and net surveys have been compared with the densities of *E. pacifica* reported in previous studies. In August 1997 and February 1998, the densities in the northwestern sea of Jeju Island and the central Yellow Sea obtained through net surveys in summer and winter, respectively, were in the range of 39.5–314.7 mg/m^3 [20]. In spring 1998, the density of *E. pacifica* was in the range of 16.5–129.1 mg/m^3 [21]. In summer 2002, the density in the East China Sea obtained through acoustic and net surveys was in the range of 20.4–221.4 mg/m^3 [29], similar to our results. According to the regular ocean observations of zooplankton biomass conducted from 1978 to 2010 using the net sampling method of the National Institute of Fisheries Science of the Korean government at Korea Oceanographic Data Center stations, the density of *E. pacifica* in spring was in the range of 10.4–913.9 mg/m^3 [47]. Thus, the density shows great variability according to season and year. In an examination of the spatial distribution pattern during the spring of 2006, the mean concentrations of individual eggs, larvae, adults, and juveniles obtained through net surveys in the Yellow Sea were 39.9, 3.58, 0.28, and 0.46 ind./m^3 [23], resulting in density range of 0.5–69.2 mg/m^3. In that study, egg abundance was positively correlated with chlorophyll-a concentration to a significant degree; the locations of adults were closely correlated with water temperature. In the late spring (May) of 2010, the density of *E. pacifica* obtained through net surveys in the Yellow Sea was 13.1 ± 14.3 mg/m^3 [24]. Kim and Kang (2020) analyzed the density of *E. pacifica* by net survey in the YSBCW during spring [25]. They reported that the densities in central, western, and eastern regions were in the range of 12.7–495.6 mg/m^3, thus revealing some regional differences. According to the acoustic and net surveys in the present study, the density of *E. pacifica* was in the range of 17.7–210.4 mg/m^3. At this time, the density unit of g/m^2 was converted to mg/m^3 considering the mean depth of 75 m in the YSBCW, and unit of ind./m^3 was converted to mg/m^3 considering the mean W of *E. pacifica*.

The density results calculated from the acoustic and net surveys of previous studies provided the same range of values, but some previous results showed a larger variation compared with the density results in the present study. The acoustic survey revealed a high-density distribution of *E. pacifica*, but quantitative net sampling was difficult because of the strong swimming ability of this species; thus, relative underestimation by net sampling is likely [29]. Table 3 summarizes the density values of *E. pacifica* according to region, season, and year in the Yellow Sea and the East China Sea, as recorded in previous studies and the present study.

Table 3. Comparison of *E. pacifica* density results between previous studies and the present study.

Region	Season (Month)	Survey Year	Survey Method	*E. pacifica* Density (mg/m^3)	References
Yellow Sea	Summer & Winter (August, February)	1997–1998	Sampling	39.5–314.7	[20]
	Spring (April)	1998		16.5–129.1	[21]
East China Sea	Summer (July)	2002	Acoustic	20.4–221.4	[29]
Yellow Sea	Spring (April)	1978–2006	Sampling	10.4–913.9	[47]
East China Sea Yellow Sea	Spring (April)	2007	Sampling	0.5–69.2	[23]
Yellow Sea	Spring (May)	2010	Sampling	13.1	[24]
Yellow Sea	Spring (April)	2019	Sampling	12.7–495.6	[25]
Yellow Sea	Spring (April)	2019	Acoustic	17.7–210.4	This study

Further studies on more accurate *E. pacifica* density will be obtained by a multi-frequency technique using 38, 120, 200, and 420 kHz with the dB difference method. If the dB difference method is applied by simultaneously operating various frequencies, it is possible to estimate the density more accurately through the application of the acoustic characteristics for each frequency of *E. pacifica*. It is planned to estimate the density through acoustic data analysis only during the daytime when the SSL is stably located near the bottom.

On the other hand, the DVM pattern is influenced by environmental and biological factors such as salinity, chlorophyll, suspended solid concentration, nutrient, light intensity, and predator relationship, but we only focused on the water temperature. Therefore, in order to fully understand the DVM mechanism and the characteristics of the regional SSL in the whole YSBCW area, other environmental and biological factors must be investigated in future study.

5. Conclusions

This study proposes the DVM characteristics of the SSL during daytime and nighttime in the YSBCW using hydroacoustic techniques; density characteristics of *E. pacifica* attributable to regional differences were also identified. In the spring season, *E. pacifica* showed DVM between the surface and the bottom; upward and downward swimming speeds in the YSBCW were in the range of 0.25–0.63 m/min. The central and eastern regions exhibited a high-density distribution of *E. pacifica*, while the western region showed a low-density distribution, indicating regional differences. The regional distribution of *E. pacifica* was related to the water temperature. These results could be used to enhance understanding of the marine ecosystem structure of the YSBCW.

Author Contributions: Conceptualization, H.K. and D.K.; methodology, H.K. and D.K.; formal analysis, H.K., G.K. and M.K.; investigation, H.K., G.K. and D.K.; writing—original draft preparation, H.K. and D.K. All authors have read and agreed to the published version of the manuscript.

Funding: This research was conducted with the support of the research project of "Research based on Understanding and Response to Changes in Marine Ecosystems in the Adjacent Sea of Korea (PE99913)" funded by the Korea Institute of Ocean Science and Technology (KIOST) and "Development of Marine Science Exploration Technology in Coastal areas (PM62603)" funded by the Ministry of Oceans and Fisheries (MOF), Korea.

Institutional Review Board Statement: Not applicable.

Informed Consent Statement: Not applicable.

Data Availability Statement: The data presented in this study are available on request from the corresponding author. The data are not publicly available since the data also forms part of an ongoing study.

Acknowledgments: We are grateful to Seok Lee, other scientists, and crew members of "*R/V Onnuri*" in the Korea Institute of Ocean Science and Technology (KIOST).

Conflicts of Interest: The authors declare no conflict of interest.

References

1. Lie, H.J. A note on water masses and general circulation in the Yellow Sea. *J. Oceanol. Soc. Korea* **1984**, *19*, 187–194. (In Korean)
2. Teague, W.J.; Jacobs, G.A. Current observations on the development of the Yellow Sea warm current. *J. Geophys. Res.* **2000**, *105*, 3401–3411. [CrossRef]
3. Park, Y.H. Water characteristics and movements of the Yellow Sea Warm Current in summer. *Prog. Oceanogr.* **1986**, *17*, 243–254. [CrossRef]
4. Lie, H.J.; Cho, C.H.; Lee, S. Tongue-shaped frontal structure and warm water intrusion in the southern Yellow Sea in winter. *J. Geophys. Res. Ocean.* **2009**, *114*, C01003. [CrossRef]
5. Hur, H.B.; Jacobs, G.A.; Teaque, W.J. Monthly variations of water masses in the Yellow and East China seas. *J. Oceanogr.* **1999**, *55*, 171–184. [CrossRef]
6. Jang, S.T.; Lee, J.H.; Kim, C.H.; Jang, C.J.; Jang, Y.S. Movement of cold water mass in the Northern East China Sea in summer. *Sea J. Korean Soc. Oceanogr.* **2011**, *16*, 1–13. (In Korean)
7. Food and Agriculture Organization (FAO). Fishery Production Statistics. 2020. Available online: http://www.fao.org/home/en/ (accessed on 1 March 2021).
8. Ministry of Oceans and Fisheries (MOF). Fisheries Production Statistics, Ministry of Oceans and Fisheries. 2020. Available online: https://www.mof.go.kr/en/index.do/ (accessed on 17 October 2020).
9. Jang, M.; Jo, H.; Kweon, D.; Cha, B.; Hwang, J.; Han, K.; Im, Y. Geographical distribution and catch fluctuations of Mottled skate, Beringraja pulchra in the eastern Yellow sea. *Korean J. Ichthyol.* **2014**, *26*, 295–302. (In Korean)
10. Everson, I. The Southern Ocean. In *Krill Biology, Ecology and Fisheries*; Everson, I., Ed.; Blackwell Science Oxford: London, UK, 2008; pp. 63–79, ISBN 9780470999486.
11. Kim, Y.S.; Choi, J.H.; Kim, J.N.; Oh, T.Y.; Choi, K.H.; Lee, D.W.; Cha, H.K. Seasonal variation of fish assemblage in Sacheon marine ranching, the southern coast of Korea. *J. Korean Soc. Fish Technol.* **2010**, *46*, 335–345. [CrossRef]
12. Lee, H.; Cho, S.; Kim, W.; Kang, D. The diel vertical migration of the sound-scattering layer in the Yellow Sea Bottom Cold Water of the southeastern Yellow sea: Focus on its relationship with a temperature structure. *Acta Oceanol. Sin.* **2013**, *32*, 44–49. [CrossRef]
13. Zaret, T.M.; Suffern, J.S. Vertical migration in zooplankton as a predator avoidance mechanism. *Limnol. Oceanogr.* **1976**, *21*, 804–813. [CrossRef]
14. Forward, R.B. Diel vertical migration: Zooplankton photobiology and behaviour. *Oceanogr. Mar. Biol. Annu. Rev.* **1988**, *26*, 361–393.
15. Berge, J.; Cottier, F.; Varpe, Ø.; Renaud, P.E.; Falk-Petersen, S.; Kwasniewski, S.; Griffiths, C.; Søreide, J.E.; Johnsen, G.; Aubert, A.; et al. Arctic complexity: A case study on diel vertical migration of zooplankton. *J. Plankton Res.* **2014**, *36*, 1279–1297. [CrossRef] [PubMed]
16. Lampert, W. The adaptive significance of diel vertical migration of zooplankton. *Funct. Ecol.* **1989**, *3*, 21–27. [CrossRef]
17. Murphy, H.M.; Jenkins, G.P.; Hamer, P.A.; Swearer, S.E. Diel vertical migration related to foraging success in snapper Chrysophrys auratus larvae. *Mar. Ecol. Prog. Ser.* **2011**, *433*, 185–194. [CrossRef]
18. Zhou, M.; Dorland, R.D. Aggregation and vertical migration behavior of *Euphausia superba*. *Deep Sea Res. Part II Top. Stud. Oceanogr.* **2004**, *51*, 2119–2137. [CrossRef]
19. Gaten, E.; Tarling, G.; Dowse, H.; Kyriacou, C.; Rosato, E. Is vertical migration in Antarctic krill (*Euphausia superba*) influenced by an underlying circadian rhythm? *J. Genet.* **2008**, *87*, 473–483. [CrossRef] [PubMed]
20. Yoon, W.D.; Cho, S.H.; Lim, D.H.; Choi, Y.K.; Lee, Y. Spatial distribution of *Euphausia pacifica* (Euphausiacea: Crustacea) in the Yellow Sea. *J. Plankton Res.* **2000**, *22*, 939–949. [CrossRef]
21. Yoon, W.D.; Yang, J.Y.; Lim, D.; Cho, S.H.; Park, G.S. Species composition and spatial distribution of euphausiids of the Yellow Sea and relationships with environmental factors. *Ocean Sci. J.* **2006**, *41*, 11–29. [CrossRef]

22. Liu, H.L.; Sun, S. Diel vertical distribution and migration of a euphausiid *Euphausia pacifica* in the Southern Yellow Sea. *Deep Sea Res. Part II Top. Stud. Oceanogr.* **2010**, *57*, 594–605. [CrossRef]

23. Sun, S.; Tao, Z.; Li, C. Spatial distribution and population structure of *Euphausia pacifica* in the Yellow Sea (2006–2007). *J. Plankton Res.* **2011**, *33*, 873–889. [CrossRef]

24. Zuo, T.; Liu, H. Seasonal size composition and abundance distribution of *Euphausia pacifica* in relation to environmental factors in the southern Yellow Sea. *Acta Oceanol. Sin.* **2019**, *38*, 70–77. [CrossRef]

25. Kim, G.; Kang, H.K. Mesozooplankton community structure in the Yellow Sea in Spring. *Ocean Polar Res.* **2020**, *42*, 271–281. (In Korean) [CrossRef]

26. Simmonds, E.J.; MacLennan, D.N. *Fisheries Acoustics: Theory and Practice*, 2nd ed.; Blackwell Science: Oxford, UK, 2005; pp. 1–437, ISBN 9780632059942.

27. Lü, L.-G.; Liu, J.; Yu, F.; Wu, W.; Yang, X. Vertical migration of sound scatterers in the southern Yellow Sea in summer. *Ocean Sci. J.* **2007**, *42*, 1–8. [CrossRef]

28. Lee, M.A.; Chao, M.H.; Weng, J.S.; Lan, Y.C.; Lu, H.J. Diel distribution and movement of sound scattering layer in Kuroshio waters, northeastern Taiwan. *J. Mar. Sci. Technol.* **2011**, *19*, 253–258. [CrossRef]

29. Kang, D.; Hwang, D.J.; Soh, H.Y.; Yoon, Y.H.; Suh, H.L.; Kim, Y.J.; Shin, H.C.; Iida, K. Density estimation of Euphausiid (*Euphausia pacifica*) in the sound scattering layer of the East China sea. *J. Korean Fish. Soc.* **2003**, *36*, 749–756. (In Korean)

30. BioSonics. *Visual Acquisition DT-X User Guide, Version 3.37*; BioSonics: Seattle, WA, USA, 2019; pp. 1–64.

31. BioSonics. *Calibration Manual (Wireless Scienfitic Echo Sounder DT-X Series), Version 2*; BioSonics: Seattle, WA, USA, 2004; pp. 1–23.

32. Svein, V.; Kenneth, F.; Mark, T.; David, F. A technique for calibration of monostatic echosounder systems. *IEEE J. Ocean. Eng.* **1996**, *21*, 298–305.

33. De Robertis, A.; Higginbottom, I. A post-processing technique for estimation of signal-to-noise ratio and removal of echosounder background noise. *ICES J. Mar. Sci.* **2007**, *64*, 1282–1291. [CrossRef]

34. Ryan, T.E.; Downie, R.A.; Kloser, R.J.; Keith, G. Reducing bias due to noise and attenuation in open-ocean echo integration. *ICES J. Mar. Sci.* **2015**, *72*, 2482–2493. [CrossRef]

35. Miyashita, K.; Aoki, I.; Seno, K.; Taki, K.; Ogishima, T. Acoustic identification of isada krill, *Euphausia pacifica* Hansen, off the Sanriku coast, north-eastern Japan. *Fish. Oceanogr.* **1997**, *6*, 266–271. [CrossRef]

36. Kang, M.H.; Furusawa, M.; Miyashita, K. Effective and accurate use of difference in mean volume backscattering strength to identify fish and plankton. *ICES J. Mar. Sci.* **2002**, *59*, 794–804. [CrossRef]

37. Watkins, J.L.; Brierley, A.S. Verification of the acoustic techniques used to identify Antarctic krill. *ICES J. Mar. Sci.* **2002**, *59*, 1326–1336. [CrossRef]

38. Kang, M.H.; Fajaryanti, R.; Son, W.; Kim, J.-H.; La, H.S. Acoustic detection of krill scattering layer in the Terra Nova Bay Polynya, Antarctica. *Front. Mar. Sci.* **2020**, *7*, 1013. [CrossRef]

39. Korneliussen, R.J.; Ona, E. Synthetic echograms generated from the relative frequency response. *ICES J. Mar. Sci.* **2003**, *60*, 636–640. [CrossRef]

40. McGehee, D.E.; O'Driscoll, R.L.; Traykovski, L.M. Effects of orientation on acoustic scattering from Antarctic krill at 120 kHz. *Deep Sea Res. Part II Top. Stud. Oceanogr.* **1998**, *45*, 1273–1294. [CrossRef]

41. Demer, D.A.; Conti, S.G. Reconciling theoretical versus empirical target strengths of krill: Effects of phase variability on the distorted-wave Born approximation. *ICES J. Mar. Sci.* **2003**, *60*, 429–434. [CrossRef]

42. Bairstow, F.; Gastauer, S.; Finley, L.; Edwards, T.; Brown, C.T.; Kawauchi, S.; Cox, M.J. Improving the Accuracy of Krill Target Strength Using a Shape Catalog. *Front. Mar. Sci.* **2021**, *8*, 290. [CrossRef]

43. Hewitt, R.P.; Demer, D.A. Dispersion and abundance of Antarctic krill in the vicinity of Elephant Island in the 1992 austral summer. *Mar. Ecol. Prog. Ser.* **1993**, *99*, 29–39. [CrossRef]

44. Reiss, C.S.; Cossio, A.M.; Loeb, V.; Demer, D.A. Variations in the biomass of Antarctic krill (*Euphausia superba*) around the South Shetland Islands, 1996–2006. *ICES J. Mar. Sci.* **2008**, *65*, 497–508. [CrossRef]

45. Choi, S.G.; Chae, J.; Chung, S.; Oh, W.; Yoon, E.; Sung, G.; Lee, K. Density estimation of Antarctic krill in the south Shetland Island (subarea 48.1) using dB-difference method. *Sustainability* **2020**, *12*, 5701–5715. [CrossRef]

46. Iguchi, N.; Ikeda, T. Experimental study on brood size, egg hatchability, and early development of a euphausiid *Euphausia pacifica* from Toyama Bay, Southern Japan Sea. *Bull. Jpn. Sea Natl. Fish. Res. Inst.* **1994**, *44*, 49–57.

47. Choi, J.W.; Park, W.G. Variations of Marine Environments and Zooplankton Biomass in the Yellow Sea during the Past Four Decades. *J. Mar. Sci. Eng.* **2013**, *25*, 1046–1054. (In Korean)

Journal of
*Marine Science
and Engineering*

Article

Influences of Seasonal Variability and Potential Diets on Stable Isotopes and Fatty Acid Compositions in Dominant Zooplankton in the East Sea, Korea

Jieun Kim [1], Hee-Young Yun [1], Eun-Ji Won [1], Hyuntae Choi [1], Seok-Hyeon Youn [2] and Kyung-Hoon Shin [1],*

1   Department of Marine Science and Convergent Technology, Hanyang University, Ansan 15588, Republic of Korea
2   Ocean Climate & Ecology Research Division, National Institute of Fisheries Science, Busan 46083, Republic of Korea
*   Correspondence: shinkh@hanyang.ac.kr; Tel.: +82-31-400-5536; Fax: +82-31-416-6173

Abstract: Despite their crucial roles in transporting primary productions in marine food webs, the trophic dynamics of zooplankton throughout the seasons have rarely been studied. In this study, four dominant zooplankton taxa with phytoplankton size composition and productivity were collected over four seasons in the East Sea, which is known to change more rapidly than global trends. We then analyzed the $\delta^{13}C$ and $\delta^{15}N$ values and fatty acid composition of zooplankton. The heavy $\delta^{13}C$ values in February and August 2021 were observed with high concentrations of total chlorophyll-*a*, and the $\delta^{13}C$ differences among the four zooplankton taxa in the coastal region (site 105-05) were most pronounced in February 2021. The relative amounts of eicosapentaenoic acid (C20:5(n-3)) and docosahexaenoic acid (C22:6(n-3)), indicators of phytoplankton nutritional quality, were also highest in February 2021. Non-metric multivariate analyses showed dissimilarity among zooplankton taxa during the high productivity period based on chlorophyll-*a* concentrations (51.6%), which may be due to an increase in available foods during the highly productive season. In conclusion, the dietary intake of zooplankton can be reduced by the transition of phytoplankton, which has important implications for the impact of climate change on planktonic ecosystems in the East Sea.

Keywords: food web; trophic dynamics; primary production; chlorophyll-*a* size fraction; phytoplankton

Citation: Kim, J.; Yun, H.-Y.; Won, E.-J.; Choi, H.; Youn, S.-H.; Shin, K.-H. Influences of Seasonal Variability and Potential Diets on Stable Isotopes and Fatty Acid Compositions in Dominant Zooplankton in the East Sea, Korea. *J. Mar. Sci. Eng.* **2022**, *10*, 1768. https://doi.org/10.3390/jmse10111768

Academic Editor: Gerardo Gold Bouchot

Received: 11 October 2022
Accepted: 15 November 2022
Published: 17 November 2022

Publisher's Note: MDPI stays neutral with regard to jurisdictional claims in published maps and institutional affiliations.

1. Introduction

As zooplankton are crucial mediators of trophic transfer, zooplankton community shifts may reflect changes in primary production and affect consumers at higher trophic levels [1,2]. Alterations in the structures of diverse zooplankton populations have been implicated as the potential drivers of food web shifts in marine ecosystems [1,3], and such a trophic-based context is known to be affected by environmental variability [4,5]. For instance, water temperature can influence the distribution, physiology, and abundance of zooplankton [6,7], and stratification of seawater causes a decrease in the body size of zooplankton [8]. Moreover, different feeding strategies among diverse zooplankton taxa (e.g., herbivores, omnivores, and carnivores) could cause a variety of responses to changes in the ecosystem. The seasonal heterogeneity of primary production changes the diversity of edible food sources for each zooplankton taxon [5]. Thus, coastal and offshore ecosystem management involves tracking an actual change in ecological status that responds to environmental variability (spatial and temporal dynamics). However, the zooplankton population shift under naturally variable environmental conditions in local ecosystems is often poorly characterized, as do primary production, ecological parameters and community, and climate change.

Climate change causes high water temperature and ocean acidification, and triggers changes in physiology, production, and size composition in the phytoplankton community [9–11]. In particular, the ocean warming trend of the East Sea/Japan Sea (hereafter the East Sea) has been faster than the global trend over the last 50–60 years [12]. The East Sea is a semi-enclosed marginal sea of the Northwest Pacific Ocean, adjacent to the Korean Peninsula, Russia, and the Japanese islands. The East Sea has highly dynamic environmental conditions with seasonal upwelling, eddies, and mixing of water masses between the Tsushima and Oyashio currents [13]. Consequently, recent studies in the East Sea have reported that such a continuous ocean temperature rise induces a decreasing trend in primary productivity with an increasing proportion of small-sized phytoplankton in the basal food web [9,10]. A likely reason for such an increase in the proportion of small-sized phytoplankton could be because as phytoplankton cells decrease in size, their surface area-to-volume ratio increases, and the thickness of the diffusion boundary layer decreases. This may be advantageous over larger phytoplankton in nutrient-poor environments [14]. However, Kang et al. [10] reported that small-sized phytoplankton have lower calorific values per chlorophyll-*a* concentration than large phytoplankton, suggesting that small-sized phytoplankton could provide a more energy-inefficient food source for upper trophic level consumers. Nevertheless, previous studies of zooplankton conducted in the East Sea have focused on monitoring spatiotemporal changes in species abundance and richness [15–17] for water temperature and salinity. It is difficult to understand the ecological changes at the base of the pelagic ecosystem, which is composed of diverse and complex trophic relationships of zooplankton with phytoplankton, protozoa, detritus, and sinking particles in the water column.

Approaches to trophic relationships reflect in situ ecological changes through informing about the changes in niche breadth, interspecies dietary competition, the nutritional quality of organisms, and the trophic position of consumers in environments based on primary productivity [4,5,18–20]. In particular, stable isotope ratios and fatty acid (FA) concentrations have been used increasingly to provide dietary information in consumer tissues [5,19,20]. The carbon stable isotope (δ^{13}C) value of animal tissues indicates diet sources distributed in the habitat, as the value generally shows little isotopic enrichment during trophic transfer (<1‰). The nitrogen stable isotope (δ^{15}N) value can provide information on trophic positions in marine ecosystems due to a constant 3.4‰ increase per trophic level from prey to its direct consumer [21,22]. C and N stable isotope compositions have been used widely to investigate changes in the food diversity of zooplankton based on seasonal ocean currents [20,23], effects of eutrophication caused by anthropogenic activities on the trophic level of zooplankton [24], and differences in the ecological niche of zooplankton by size [19]. FAs can also be functional parameters, and information on plankton-specific FAs can provide more detailed C source to the δ^{13}C results of zooplankton [20,25]. In particular, some FAs are called essential FAs as these cannot be synthesized de novo in the consumer body, and thus the relative contributions of diatom and dinoflagellate in diet are directly reflected in consumers [5,20,23,26,27]. For instance, eicosapentaenoic acid (EPA, C20:5(n-3)) and docosahexaenoic acid (DHA, C22:6(n-3)) are specific markers for the ratio of diatom and dinoflagellate abundance. In addition, the total lipid content can identify the most nutritionally sufficient diet for consumers at the upper trophic level [5]. Overall, stable isotopes, FA biomarkers, and total lipid content are valuable tools for revealing the effects of environmental variability on trophic interactions between zooplankton consumers and primary producers at the base of the pelagic ecosystem.

It is still challenging to improve our understanding of the effects of temporal and spatial variability on the trophic structure of zooplankton despite several studies that have used stable isotope analysis and FA parameters to describe the trophic structure and to specify dietary resources successfully in pelagic ecosystems [24,27]. In this regard, we addressed how the four dominant zooplankton taxa in the East Sea are affected by environmental conditions and phytoplankton community changes over a one-year sampling period of August 2020 to August 2021. Stable isotope compositions and FA profiles in zooplankton from the coastal site 105-05, where vertical mixing and upwelling by wind occur actively and frequently, were compared with those from the offshore site 105-11, where vertical mixing and upwelling are less frequent. Identifying the ecological change in response to the fluctuation of primary producers coinciding with environmental changes helps us to broadly characterize the ecological and biochemical processes under current climate change.

2. Materials and Methods

2.1. Analysis of Environmental Parameter

Seasonal sampling was conducted in the East Sea using the ocean research vessel Tamgu3 (797 tons; National Institute of Fisheries Sciences, Busan, Republic of Korea) during four seasons from 2020 to 2021 (August, October, February, and April, representing summer, autumn, winter, and spring, respectively) (Figure 1 and Table 1). In 105-05 and 105-11, water samples were collected using Niskin bottles, which were attached to a conductivity-temperature-depth (CTD)/rosette sampler (SBE911 plus, Seabird Electronics Inc., Bellevue, WA, USA) to measure dissolved inorganic nutrients such as nitrate (NO_3^{2-}), nitrite (NO_2^-), ammonium (NH_4^+), silicate (SiO_3^{2-}), phosphate (PO_4^{3-}), and size-fractioned chlorophyll-*a* concentrations. Temperature, salinity, and dissolved oxygen were measured using a CTD/rosette sampler.

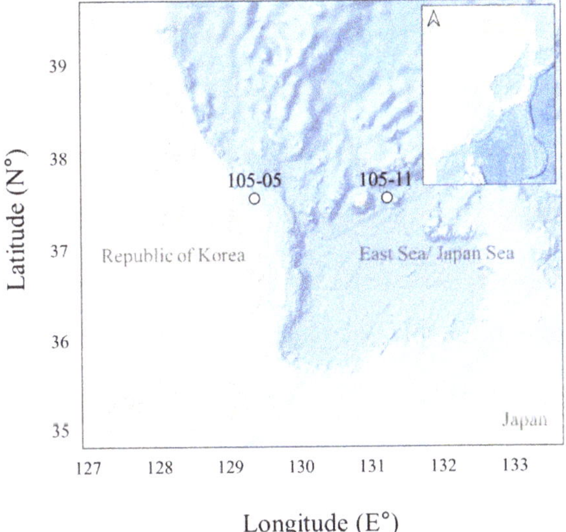

Figure 1. The sampling locations of the coastal (105-05) and offshore (105-11) regions during August 2020, October 2020, February 2021, April 2021, and August 2021.

Table 1. Information of sampling sites of this study.

Region	Station	Latitude	Longitude	Sample Collection Depth (m)	Bottom Depth (m)	Species
East Sea	105-05	37.55	129.38	100	280	Euchaetidae Chaetognatha Euphausiid Amphipod
	105-11	37.55	131.24	100	1140	Euchaetidae Chaetognatha Euphausiid Amphipod

0.1 L seawater was passed through a 0.45 µM disposable membrane filter unit and stored at −20 °C for analyzing dissolved inorganic nutrient concentrations. Nutrient concentrations were measured using an automatic analyzer (Quattro, Seal Analytical,

Norderstedt, Germany) at the National Institute of Fisheries Science (NIFS), Korea. The sum of NO_3^{2-}, NO_2^-, and NH_4^+ was calculated as dissolved inorganic N (DIN).

For total chlorophyll-*a*, 0.1–0.4 L of seawater was filtered through a GF/F (ø = 25 mm; Whatman). Size-fractioned chlorophyll-*a* was separated sequentially through membrane filters with pore sizes of 20 and 2 µM (ø = 47 mm; Whatman). The filtered sample was shaded from light using aluminum foil and stored at −20 °C. Before analysis, pigments were extracted with 90% acetone in the dark at 4 °C for 12–24 h. Chlorophyll-*a* concentrations were measured using a fluorometer, following Parson et al. [28] (Turner Designs, 10-AU, San Jose, CA, USA).

2.2. Zooplankton Sampling

Zooplankton were collected over four seasons, from August 2020 to August 2021. The plankton net (RN80, diameter 80 cm, mesh size 300 µM) was towed vertically to the surface after horizontal drawing for 10 min at a speed of 2 knots using the oblique tow method at site 105-05 in the coastal region and site 105-11 in the offshore of the East Sea (Figure 1, Table 1). Then, zooplankton were selectively sorted and isolated into four different groups (Euchaetidae, Chaetognatha, Euphausiid, and Amphipod) using an optical microscope and stored at −20 °C until further use. The zooplankton were freeze-dried for 48 h before sample pretreatment.

2.3. Stable Isotope Analysis of Zooplankton

The $\delta^{13}C$ and $\delta^{15}N$ values of the samples were analyzed in triplicate using an elemental analyzer (EA, Vario PYROcube, Elementar, Germany) equipped with an isotope ratio mass spectrometer (IRMS, Isoprime 100, Isoprime, UK), as described in a previous study. Briefly, inorganic C was extracted using 1 M HCl, whereas lipids were extracted using chloroform/methanol treatment. For analysis, approximately 0.1 and 1.0 mg of samples for carbon and nitrogen analysis, respectively, were wrapped in a tin capsule. Stable isotopes are stated in conventional δ notation as follows:

$$\delta X \ (\text{‰}) = \left[\left(\frac{R_{Sample}}{R_{Standard}} \right) - 1 \right] \times 1000 \tag{1}$$

where X and R indicate the isotope (C or N) and corresponding ratio of $^{13}C/^{12}C$ or $^{15}N/^{14}N$, respectively. $\delta^{13}C$ and $\delta^{15}N$ standards (IAEA CH-3 and N-1, respectively) with known isotopic ratios were measured every 10 sample runs to confirm the precision of the analysis instrument. The standard deviations of the samples in the entire analysis set were less than 0.3‰ for $\delta^{13}C$ and $\delta^{15}N$.

2.4. Fatty Acid Composition Analysis

The total lipids in approximately 2 mg of zooplankton samples were extracted using the method established by Folch et al. [29] and modified by Choi et al. [26]. Briefly, after total lipids were extracted using a dichloromethane/methanol solution, 100 µL of 20 ppm surrogate (nonadecanoic acid, C19:0) was added to the sample. Saponification was performed using KOH solution in MeOH, and methylation and derivatization were performed using Boron trifluoride-methanol solution and FA methyl esters (FAMEs). Methyl heneicosanoic acid (C21:0) was added at the same amount as the surrogate for use as an internal standard. FAME concentrations were determined in triplicate using gas chromatography (GC/FID, HP-7890A, Agilent Technologies, Santa Clara, CA, USA) equipped with a flame ionization detector on a DB5 column (60 m in length × 0.25 mm inner diameter × 0.25 µM film thickness, Agilent Technologies, Santa Clara, CA, USA). FAs were confirmed by comparing the retention time of FAME standards (Supelco, Bellefonte, PA, USA) and mass spectra of gas chromatography equipped with mass spectrometry (GC/MS, HP-7820A, Agilent Technologies, Santa Clara, CA, USA).

2.5. Statistical Analysis

All data are presented as the mean and standard deviation of three replicates. To confirm the effects of seasonality on the stable isotope ratio and FA composition in the zooplankton groups, a one-way analysis of variance (ANOVA) was performed using IBM SPSS statistics 27. Tukey's-b test was conducted for post hoc analysis. Non-metric multidimensional scaling (NMDS) was used for ordinations based on the Bray—Curtis similarity index using PAST software [30] to identify the differences between zooplankton taxa according to environmental differences. Feature scaling is a method of standardizing data for statistical processing to values within the desired range and can eliminate errors due to different values, signs, and variability. We selected seven FAs that generated up to 75% cumulative dissimilarity between zooplankton taxa using a similarity percentage (SIMPER) analysis (contributing at least 5% per FA). These seven FAs and carbon and nitrogen stable isotope ratios were used for NMDS analysis. In this study, NMDS was performed after normalizing the data to a range of 0 to 1 using the feature scaling method. We analyzed stress values from NMDS analysis and confirmed that all the values were less than 0.2. We performed SIMPER analysis to confirm dissimilarity between zooplankton taxa. Furthermore, the FAs that contributed the most to each taxon were identified using SIMPER analysis.

3. Results

3.1. Environmental Conditions

The surface water temperature ranged from 12.1–23.0 °C at 105-05 and 10.6–30.2 °C at 105-11 (Figures 2 and 3). Both sites showed low water temperatures in winter, i.e., February 2021, and the highest in summer, i.e., August 2020 and August 2021. However, salinity showed constant values in the range of 33.1–34.5 psu at 105-05, and 33.1–34.2 psu at 105-11, and a slight decline in August 2021. The water temperature and salinity data represent the stratification status of the study area according to the sampling seasons. At the 105-05 site, strong stratification occurred during and after the summer in August 2020, October 2020, and August 2021. During these seasons, a thermocline layer was formed at a depth of approximately 30 m, and in February and April 2021, the thermocline layer deepened to nearly 50 m.

The seasonal fluctuation in dissolved inorganic nutrient concentrations was slightly different between the sampling sites (Figures 2 and 3). In the section within 100 m, nutrient variability in the coastal region at site 105-05 was more pronounced than in the offshore at site 105-11. In October 2020, the section where the nutrient content decreased sharply from 12.8 to 5.3 μM was formed around 40 m, whereas in February, the section where the nutrient dropped from 17.4 to 9.3 μM was formed around 60 m. However, DIN concentration measured at site 105-11 maintained a relatively low concentration between 2.7–11.7 μM in October 2020 and 7.6–9.1 μM in February 2021.

During the sampling period, depth-integrated total chlorophyll-*a* concentrations within 50 m in 105-05 peaked in February and August 2021 (141.9 mg m^{-2} and 73.5 mg m^{-2}) (Figure 4). In particular, the chlorophyll-*a* concentrations of micro-size phytoplankton in February and August 2021 accounted for the dominant part of total chlorophyll-*a* concentrations, i.e., 114.3 and 56.8 mg m^{-2}, contributing to 80.6 and 77.2% of total 141.9 and 73.5 mg m^{-2}, respectively. Moreover, at 105-05, the chlorophyll-*a* concentrations of nano-size phytoplankton, 3–20 μM, were the highest in April 2021 (21.7 mg m^{-2}, 66.8%), and those of the pico-size sample were the highest in August 2020 and October 2020 (20.4 and 23.2 mg m^{-2} contributing to 60.6 and 75.0%, respectively). In contrast, the total chlorophyll-*a* concentrations at 105-11 peaked in April 2021 at 79.6 mg m^{-2} during the study period. Micro-sized chlorophyll-*a* was maintained at a low concentration of less than 18.0 mg m^{-2} throughout the year, except for April 2021. Nano-size chlorophyll-*a* showed the highest concentrations in April 2021 (51.1 mg m^{-2}, 64.2%), and pico-size chlorophyll-*a* showed the highest concentrations in October 2020 and February 2021 (21.6 and 23.8 mg m^{-2} contributing to 72.2 and 48.6%, respectively).

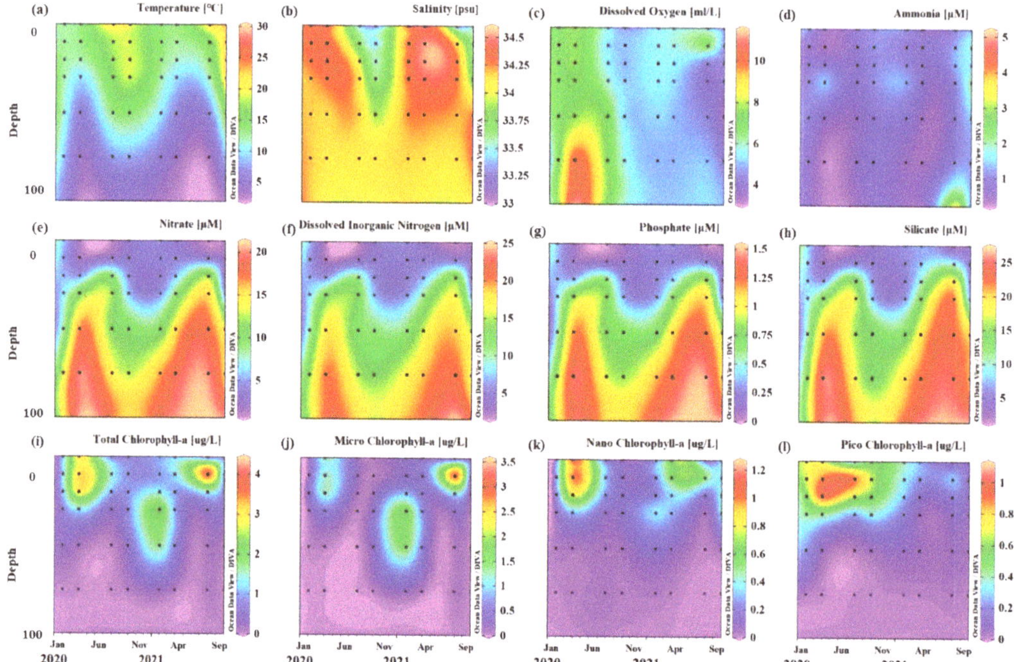

Figure 2. Monthly variation of temperature (**a**), salinity (**b**), dissolved oxygen (**c**), NH_4^+ concentration (**d**), NO_3^{2-} concentration (**e**), dissolved inorganic N concentration (**f**), PO_4^{3-} concentration (**g**), SiO_3^{2-} concentration (**h**), total chlorophyll-*a* concentration (**i**), micro chlorophyll-*a* concentration (**j**), nano chlorophyll-*a* concentration (**k**), and pico chlorophyll-*a* concentration (**l**) from 0 to 100 m in site 105-05. The white triangular mark indicates the month in which the zooplankton samples were collected.

3.2. Spatiotemporal Isotope Fluctuations in Zooplankton Taxa

Spatiotemporal variations in $\delta^{13}C$ and $\delta^{15}N$ values of zooplankton are shown in Figure 5 and Table S1. The $\delta^{13}C$ and $\delta^{15}N$ values of zooplankton collected at 105-05 ranged from −23.6 to −17.9‰ and 2.6 to 10.9‰, respectively. The average $\delta^{13}C$ values of zooplankton at 105-05 were significantly different, with the highest value of −19.2‰ in August 2021, followed by −20.3‰ in February 2021 (one-way ANOVA, $p < 0.01$). Furthermore, the average $\delta^{15}N$ of zooplankton showed significant differences among the seasons (one-way ANOVA, $p < 0.01$), with the highest value of 7.8‰ in February 2021. However, the $\delta^{13}C$ (−22.6 to −18.9‰) and $\delta^{15}N$ (4.7 to 9.8‰) values of zooplankton at 105-11 were not significantly different among seasons.

The absolute value of the isotope ratio and the difference in $\delta^{13}C$ values among the zooplankton taxa showed seasonal characteristics. At site 105-05, $\delta^{13}C$ values were considerably different among zooplankton taxa in February 2021, April 2021, and August 2021. $\delta^{13}C$ values of zooplankton taxa ranged from −22.6 to −20.8‰ and from −23.6 to −22.2‰ in August 2020 and October 2020, respectively, with little difference among taxa. However, a significant difference occurred in April 2021 (−22.9 to −20.3‰, one-way ANOVA, $p < 0.05$), and the largest difference occurred in February 2021 (−22.4 to −17.9‰, one-way ANOVA, $p < 0.01$). At 105-11, however, $\delta^{13}C$ values among zooplankton taxa were significantly different for all seasons except April 2021, when there were not enough comparable samples.

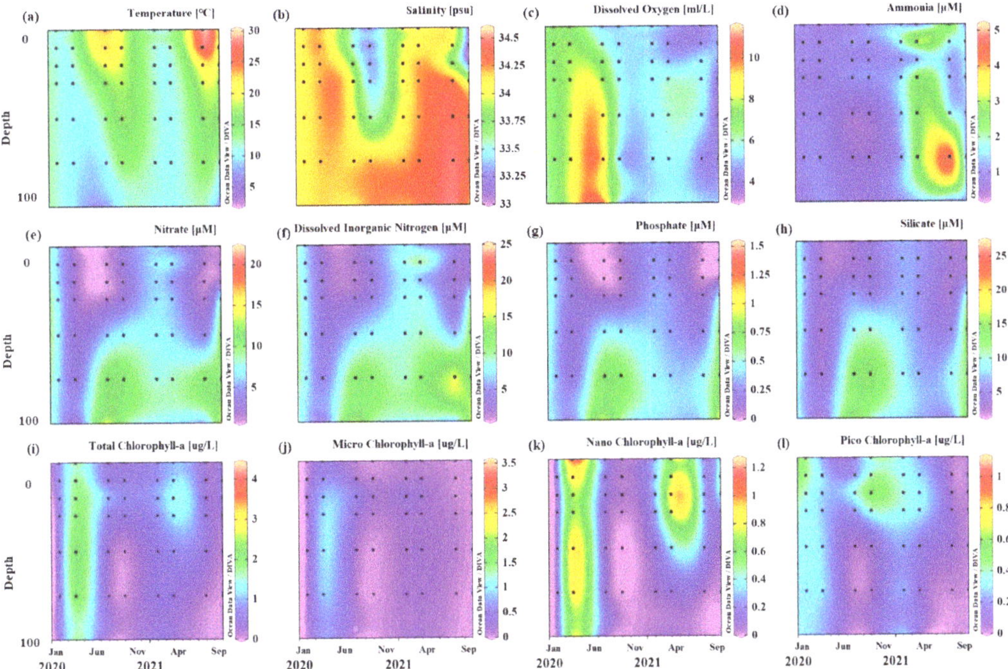

Figure 3. Monthly variation of temperature (**a**), salinity (**b**), dissolved oxygen (**c**), NH_4^+ concentration (**d**), NO_3^{2-} concentration (**e**), dissolved inorganic N concentration (**f**), PO_4^{3-} concentration (**g**), SiO_3^{2-} concentration (**h**), total chlorophyll-*a* concentration (**i**), micro chlorophyll-*a* concentration (**j**), nano chlorophyll-*a* concentration (**k**), and pico chlorophyll-*a* concentration (**l**) from 0 to 100 m at site 105-11. The white triangular mark indicates the month in which the zooplankton samples were collected.

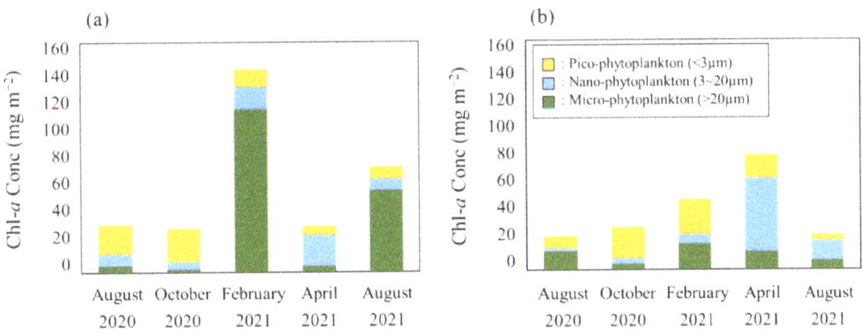

Figure 4. Seasonal variations in different size fractioned chlorophyll-*a* concentrations at 105-05 (**a**) and 105-11 (**b**) sites.

At 105-05, a significant difference in $\delta^{15}N$ values among zooplankton occurred from August 2020 to April 2021 (one-way ANOVA, $p < 0.05$). In addition, among the zooplankton taxa collected from both study sites, Chaetognatha showed the highest $\delta^{15}N$ values, followed by Euchaetidae. The $\delta^{15}N$ value of Chaetognatha showed the highest value of 10.2‰ in February 2021 and the lowest value of 7.3‰ in August 2020. Similarly, Euchaetidae and Euphausiid also showed the highest $\delta^{15}N$ values in February 2021 (8.4 and 7.9‰, respectively) and the lowest values in August 2020 (5.6 and 5.0‰, respectively). Amphipods showed the highest value of 6.8‰ in August 2021, which is different from other taxa. However, at 105-11, there was no difference in the $\delta^{15}N$ values of zooplankton based on season.

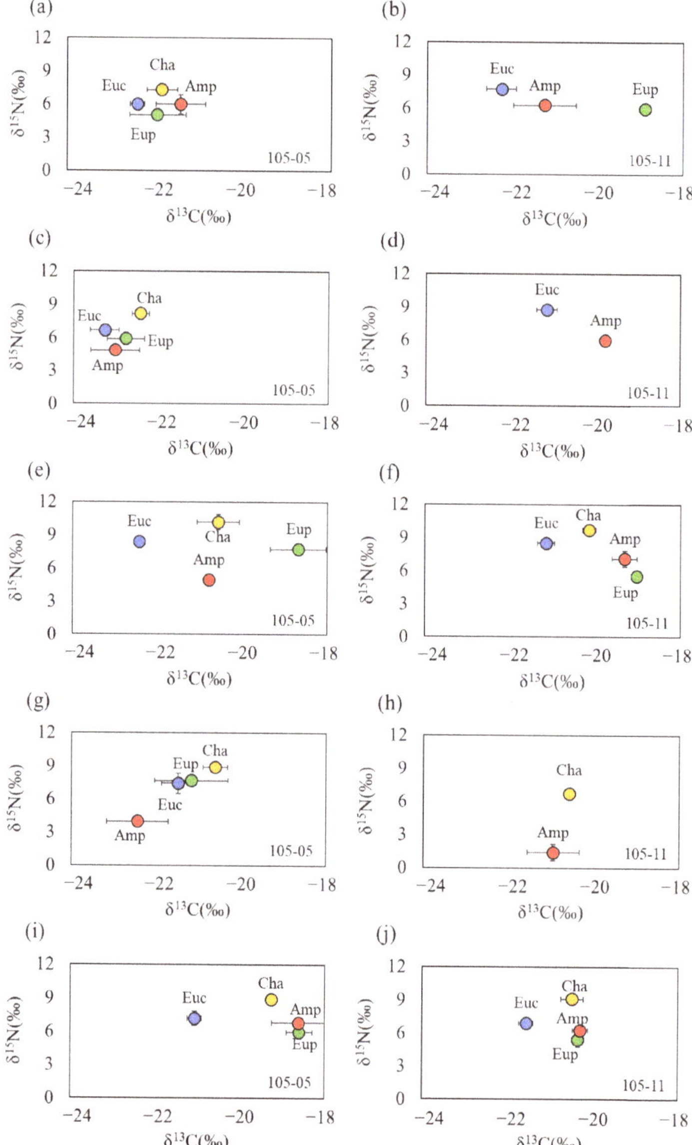

Figure 5. Spatiotemporal (August 2020: (**a**,**b**); October 2020: (**c**,**d**); February 2021: (**e**,**f**); April 2021: (**g**,**h**); August 2021: (**i**,**j**)) $\delta^{13}C$ and $\delta^{15}N$ biplots of four zooplankton taxa collected at 105-05 (**a**,**c**,**e**,**g**,**i**) and at 105-11 (**b**,**d**,**f**,**h**,**j**) sites. Each taxon is depicted with different color (Euchaetidae: blue; Chaetognatha: yellow; Euphausiid: green; Amphipod: red).

3.3. Changes in Total Fatty Acid Concentrations and the Proportion of Major Fatty Acids in Zooplankton Taxa

The total amount of FAs per unit dry weight of zooplankton showed seasonal variability (one-way ANOVA, $p < 0.01$) (Figure 6, Tabls S2–S5). The highest concentration of average total FA in both study sites showed in February 2021 (105-05: 179.5 mg g^{-1} dw; 105-11: 161.2 mg g^{-1} dw) compared to other sampling times.

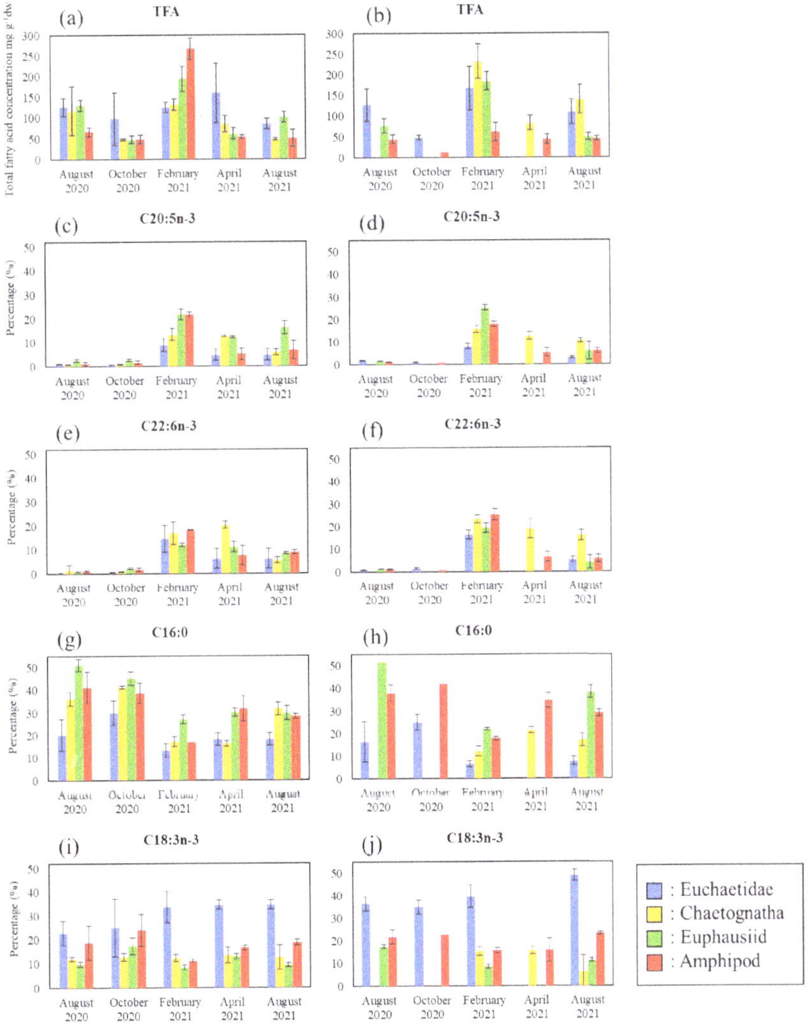

Figure 6. Changes in the total fatty acid (TFA) concentrations (**a,b**), proportions of C20:5(n-3) (**c,d**), C22:6(n-3) (**e,f**), C16:0 (**g,h**), C18:3(n-3) (**i,j**) best representing seasonal dissimilarity in 105-05 (**a,c,e,g,i**) and 105-11 (**b,d,f,h,j**) based on SIMPER analysis. Each color represents different zooplankton taxa (Blue: Euchaetidae, Yellow: Chaetognatha, Green: Euphausiid, Red: Amphipod).

Four FAs, C16:0, C18:3(n-3), C20:5(n-3), and C22:6(n-3), 63% of total FAs, were chosen based on SIMPER analysis, allowing the selection of FAs that contributed the most distinctive seasonal variation at both 105-05 and 105-11. Among these FAs, C16:0 accounted for the highest proportion in the zooplankton body, averaging 29.1 and 25.3% at 105-05 and 105-11, respectively. C18:3(n-3) was the second most abundant compound, with 18.0 and 22.2% at 105-05 and 105-11, respectively. C16:0 and C18:3(n-3) did not exhibit any significant seasonal variability. However, C20:5(n-3) and C22:6(n-3) showed seasonal variation. At both sampling sites, the proportion of C20:5 (n-3) and C22:6 (n-3) was the highest at 15.4 and 16.4%, respectively, in February 2021. In August and October 2020, these two FAs were less than 1.5 and 1.2%, respectively. Moreover, the seasonal trends in C20:5(n-3) and C22:6(n-3) were similar in all zooplankton taxa (one-way ANOVA, $p < 0.01$).

3.4. Estimating Changes in the Ecological Niche of Zooplankton

NMDS was performed using the variables of stable isotope ratios and FA compositions of the dominant zooplankton species to determine variations in the ecological niche of zooplankton taxa based on phytoplankton productivity (Figure 7). Productivity was categorized as high primary productivity (chlorophyll-*a* concentrations > 50 mg m^{-2}) in February 2021 and August 2021 at 105-05 and April 2021 at 105-11, and low (chlorophyll-*a* concentrations < 50 mg m^{-2}) in August 2020, October 2020, and April 2021 at 105-05 and August 2020, October 2020, February 2021, and August 2021 at 105-11. The dissimilarity of zooplankton taxa during low and high productivity periods was 37.8 and 51.6%, respectively, suggesting a high dissimilarity of zooplankton in the high production period rather than the low production period. The dissimilarity between Euchaetidae and Euphausiid was as high as 68.0% during the high-production period compared to 42.6% during the low-production period. In addition, the dissimilarity between Euphausiid and Amphipod during the high production period was 34.4%, which was large relative to 29.7% in the low production period.

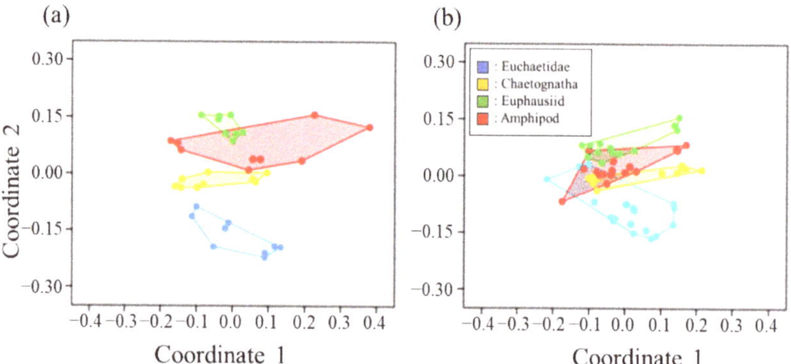

Figure 7. NMDS ordination of fatty acids and stable isotopes from four zooplankton taxa (Euchaetidae, Chaetognatha, Euphausiid, and Amphipod) during high productivity period (**a**) and low productivity period (**b**). Each color represents the different zooplankton taxa (Blue: Euchaetidae, Yellow: Chaetognatha, Green: Euphausiid, Red: Amphipod). The stress values of each graph were 0.15 and 0.14. Both values are less than 0.2, which indicates that the given data set is suitable for visualization within an acceptable level of distortion.

4. Discussion

In predicting changes in ecosystem structure according to environmental changes, understanding the organisms that have connectivity in feeding relationships is the most basic and important topic. In trophic relationships, changes in diet organisms affect consumers, and changes in primary production caused by physicochemical factors such as light, water temperature, and nutrients affect the trophic relationship that follows.

In this study, the increase in total chlorophyll-*a* and micro-size chlorophyll-*a* concentrations observed in February 2021 also signifies the proliferation of primary producers in February, possibly regarded as winter, rather than other times in the East Sea and can be explained by the reinforcement of vertical mixing in terms of salinity and temperature. Vertical mixing has been reported in the East Sea, and many studies have measured seasonal phytoplankton blooms in the study area [31,32]. The effects of vertical mixing on the stable isotope ratios of zooplankton have been reported frequently in other regions. For instance, the stable isotope ratio of zooplankton in the Red Sea is related to upwelling-induced diatom bloom [33]. In this study, total chlorophyll-*a* concentrations and the proportion of micro-sized chlorophyll-*a* at 105-05 increased with the depth of the mixing layer in February 2021. Simultaneously, the overall $\delta^{15}N$ of the zooplankton showed the highest value of 7.8‰. The isotopic values of nitrogen reflect these sources. For example, vertical mixing in the Red Sea supplies ^{15}N-enriched $NO_3{}^{2-}$ to the euphotic layer, and the actively growing diatom mainly uptakes ^{15}N-enriched $NO_3{}^{2-}$ rather than other N sources, e.g., $NH_4{}^+$, transferred to zooplankton through trophic transfer and thus shows a heavier $\delta^{15}N$ value in zooplankton [34]. In addition, in the Gulf of Riga and the Baltic Sea, the $\delta^{15}N$ value of suspended particulate matter (SPM) increased by 3.9‰ after phytoplankton bloom compared to pre-bloom under the N recycling of already assimilated N pool [35]. Likewise, seston isotope ratios in the Yellow Sea showed that N sources exported to bottom water during summer and autumn are transported to the euphotic layer through intense vertical mixing in winter and transferred to the local basal food web [36]. Thus, the results of chlorophyll-*a* concentrations and isotopic ratios of N indicate the supply of regenerated N sources through vertical mixing in February 2021. However, we did not measure bulk isotope ratios or the C:N ratio from particulate organic matter (POM) data. In addition, N sources supplied through vertical mixing imprint the enriched $\delta^{15}N$ signals in SPM and phytoplankton and are likely to be transferred into zooplankton within the local food web by recycling in the water column [37]. The $\delta^{13}C$ values of zooplankton at 105-05 increased sharply in February 2021 and August 2021. The ^{13}C-enrichment of zooplankton tissue in February 2021 could be due to high primary production caused by intense vertical mixing in winter, similar to the ^{15}N-enrichment. At 105-05, the $\delta^{13}C$ value of zooplankton rapidly increased as the proportion of micro-sized phytoplankton increased (February 2021, August 2021). The effect of phytoplankton community changes on isotope variables in consumers has been reported in other studies in which the $\delta^{13}C$ of POM under low temperature and high nutrition environments (i.e., upwelling regions) becomes higher when micro-sized phytoplankton (such as diatoms) are dominant over nano- and pico-sized phytoplankton [1,2]. In the Gulf of Riga and the Baltic Sea, ^{15}N and ^{13}C enrichment occurred as the populations of diatoms, dinoflagellates, and the ciliate *Mesodinium rubrum* increased [35]. Several studies have suggested that $\delta^{13}C$ values in rapidly growing diatoms can be increased through changes in the C source and metabolic processes [38–41]. In addition, there is a difference in $\delta^{13}C$ values of phytoplankton-derived POM in regions with and without seasonal upwelling on the western Indian shelf, showing that rapidly growing micro-sized phytoplankton were higher in upwelling regions, with relatively low temperatures and high nutrition conditions. In contrast, at other study sites where nano- and pico-sized phytoplankton having slow growth rates were dominant, the $\delta^{13}C$ value was approximately 3‰ lower than that of upwelling area with large-sized phytoplankton, i.e., diatom [42]. These results indicate that contrasting biogeochemical conditions influenced by phytoplankton communities and growth rates show that the basial diet source produced by vertical mixing is conveyed to the zooplankton community with the accumulation of the imprint of ^{13}C enrichment.

As expected, (e.g., consumer $\delta^{15}N$ increase per trophic level [22]), carnivorous Chaetognatha showed the highest $\delta^{15}N$ value (7.1–10.9‰) throughout the sampling periods over amphipods (complex feeder from detritivorous, carnivorous to herbivorous properties, varied on species specificity) (Figure 5). Moreover, the $\delta^{15}N$ variation pattern in all four zooplankton was dependent on seasonality, particularly at 105-05, but weakly at 105-11. In the context of $\delta^{13}C$ variability, in contrast to our expectation, consumer $\delta^{13}C$ variables become comparable if they shared basal resources [19,24,43], and the isotope variables were different among dominant zooplankton at specific times in February 2021, April 2021, and August 2021. Consistent with other studies [3–5], this study supports the view that the diversity of prey in local habitats is one of the main factors influencing temporally variable isotope values of zooplankton.

The results of the SIMPER analysis showed that the FAs best classified among the zooplankton taxa of 105-05 and 105-11 were C16:0, C18:3(n-3), C20:5(n-3), C22:6(n-3), C16:3(n-3), C18:1(n-9), and C14:0. In particular, the ratio of the biological indicators of diatoms (e.g.,

EPA) and dinoflagellates (e.g., DHA) measured in this study was relatively low during August 2020 and October 2020, called as the highly stratified period. Additionally, during this period, the total FA content of zooplankton and the contribution of phytoplankton were low. Phytoplankton synthesize essential FAs, i.e., EPA and DHA, that consumers cannot produce in vivo and can be delivered through trophic transfer by feeding. Thus, seasonal changes in the composition of FAs in zooplankton are closely related to dietary resources and certain primary producers, such as diatoms and dinoflagellates [44–46]. Several possibilities may have led to the determination of quantitative and qualitative changes in the main dietary resources of zooplanktonic FAs. First, as the productivity of phytoplankton decreases owing to the physicochemical conditions of the ocean, the diet consumption of zooplankton decreases, and accordingly, fewer FAs might be stored in their tissue [47]. In addition, several studies have reported that high temperatures sustained during August and October 2020 cause heat stress to zooplankton and weaken lipid storage efficiency [48,49]. Furthermore, high temperatures reduce the total FA concentration and change the relative abundance of each FA synthesized by phytoplankton [50–52]. As shown in Figures 2–4, an increase in pico-size phytoplankton ratio in August 2020 and October 2020 causes much less efficiency in the direct grazing of zooplankton in our study and a decrease in FA assimilation. These results suggest that the nutritional quality and ecological characteristics of intermediate species in the pelagic food web depend on the physicochemical aspects of the ocean, such as temperature and nutrient concentration, as reported in previous studies. As a result, the low total FA content and C20:5(n-3) and C22:6(n-3) abundances of zooplankton during the high-temperature period in this study suggested that zooplankton might take up little essential FAs and energy from diatoms and dinoflagellates under low productive conditions.

The typical seasonal isotopic variability of zooplankton taxa observed in this study supports the idea that they reflect information on taxonomic characteristics and environmental impacts in their tissue. In 105-05, δ^{13}C and δ^{15}N values of zooplankton taxa showed seasonal variability (Figure 5). During August and October 2020, the four zooplankton taxa showed relatively similar isotope ratios compared with other seasons, and after Feb 2021, the zooplankton showed a larger range of isotope ratios among taxa. In addition, the isotope ratios of zooplankton showed discrepancies according to the study sites. In 105-05, δ^{15}N was the lowest in Amphipods, but in 105-11, Euphausiid had the highest δ^{15}N value among species. Despite the spatiotemporal variation of δ^{15}N of zooplanktonic taxa, Chaetognatha had the heaviest δ^{15}N value for all sites and seasons.

The difference in δ^{13}C values between the zooplankton taxa was evident in February 2021 and August 2021. During this period, the δ^{13}C of Euphausiid was higher than that of other zooplankton. In contrast, Euchaetidae had relatively low δ^{13}C values. In 105-05, the seasonal variation in the carbon stable isotope ratio of zooplankton was significant, whereas, in 105-11, a relatively consistent isotope trend was observed. Although we found a spatiotemporal trend in zooplankton isotope ratios, we could not establish a complete spatiotemporal dataset because of the low zooplankton density during the study period. We confirmed that the environmental heterogeneity and characteristics of taxa were reflected in zooplankton tissues. Unfortunately, we were unable to collect some taxa, causing difficulty in understanding the overall trophic structure variation, but we inferred that more diverse results could be obtained if additional datasets were collected in the future.

NMDS analysis using the stable isotope ratio and FA proportion revealed a clear separation among zooplankton taxa in the productive season, and the stress value was less than 0.2 (Figure 7). The stress value refers to the degree to which a given dataset is suitable for visualization within an acceptable level of distortion. When the stress value is 0.2 or more, there is possibility to produce a plot that is not suitable for interpretation. In general, when consumers have more diet options with plenty of diet sources, differences in isotope ratios between consumers might be broad, as individuals opportunistically switch their diets [19,53,54]. However, in the less-productive period, zooplankton groups overlapped relative to the productive period. This might be due to limited diets during the less productive period, and thus, the range of δ^{13}C and δ^{15}N values for consumers might be narrowly distributed under harsh environmental conditions. Henschke et al. [18] and Kozak et al. [19] confirmed ecological niche separation by analyzing the δ^{13}C and δ^{15}N of zooplankton communities inhabiting water masses with high nutrient conditions. In addition, previous research on the west coast of Vancouver Island showed that NMDS clustering using the FA compositions of zooplankton indicated a distinct pattern in the functional groups of zooplankton during the post-bloom period from the pre-bloom period, i.e., spread vs. overlapped [5]. Ultimately, it can be applied to effectively describe the interspecific competition among zooplankton taxa according to the magnitude of primary production (e.g., Schoo et al. [20]). To understand the impact of climate change on micro-ecosystems in the East Sea, long-term monitoring of zooplankton trophic dynamics with stable isotope and FA analyses is necessary.

5. Conclusions

In this study, we confirmed that the stable isotopes and FA compositions of zooplankton directly reflect changes in the phytoplankton biomass in the East Sea. An increase in pico-size chlorophyll-*a* proportion relative to micro-size chlorophyll-*a* caused the communities of zooplankton taxa to overlap among zooplankton in August 2020 and October 2020, indicating relatively simplified options of diet choice in habitat environment for zooplankton that might suffer from diet limitations in the East Sea during the low-productivity season. On the other hand, during the highly productive seasons, each zooplankton group was well segregated because of the diversification of diet caused by the proliferation of primary producers during February 2021. This result indicates that the dietary intake of zooplankton can be reduced (restricted) owing to species succession to small-sized phytoplankton. This has important implications for predicting the impact of climate change on micro-ecosystems in the East Sea. In the future, long-term monitoring of the trophic dynamics of zooplankton is required to understand the disturbances in micro-ecosystems caused by climate change in the East Sea.

Supplementary Materials: The following supporting information can be downloaded at: https://www.mdpi.com/article/10.3390/jmse10111768/s1, Table S1, Carbon (δ^{13}C) and nitrogen (δ^{15}N) stable isotope ratios of zooplankton taxa among four seasons in the East Sea, Korea; Table S2, Spatiotemporal fatty acid percent composition of zooplankton taxa (Euc: Euchaetidae, Cha: Chaetognatha, Eup: Euphausiid, and Amp: Amphipod) in coastal region (105-05) of the East Sea; Table S3, Spatiotemporal fatty acid percent composition of zooplankton taxa (Euc: Euchaetidae, Cha: Chaetognatha, Eup: Euphausiid, and Amp: Amphipod) in offshore region (105-11) of the East Sea; Table S4, Seasonal total fatty acid concentrations (mg g^{-1} dw) in Zooplankton Taxa in coastal regions (105-05) in the East Sea; Table S5, Seasonal total fatty acid concentrations (mg g^{-1} dw) in zooplankton taxa in offshore regions (105-11) in the East Sea.

Author Contributions: Conceptualization, J.K., H.C. and K.-H.S.; methodology, J.K.; software, J.K.; validation, J.K., H.-Y.Y., E.-J.W. and H.C.; formal analysis, J.K.; investigation, J.K.; resources, S.-H.Y.; data curation, J.K. and H.C.; writing—original draft preparation, J.K. and H.C.; writing—review and editing, J.K., H.-Y.Y., E.-J.W., H.C. and K.-H.S.; visualization, J.K.; supervision, K.-H.S.; project administration; K.-H.S. and S.-H.Y.; funding acquisition; K.-H.S. All authors have read and agreed to the published version of the manuscript.

Funding: This research was supported by the National Institute of Fisheries Science (NIFS), Korea (Analysis of the species composition and structure of the marine food web, No. R2022073).

Institutional Review Board Statement: Not applicable.

Informed Consent Statement: Not applicable.

Data Availability Statement: All data supporting the results of this study are provided in this manuscript and Supplementary Information File (https://www.mdpi.com/article/10.3390/jmse10111768/s1).

Acknowledgments: This work was supported by the National Institute of Fisheries Science (NIFS) grant ('Analysis of the species composition and structure of the marine food web'; R2022073) funded by the ministry of oceans and fisheries, republic of Korea. Also, this research is supported by the National Research Foundation of Korea (NRF) grants funded by the Ministry of Science and ICT (MSIT) (2022M3I6A1085992). We would like to thank the NIFS researchers and the captain and crew for their assistance with the sampling.

Conflicts of Interest: The authors declare no conflict of interest.

References

1. Lomartire, S.; Marques, J.C.; Gonçalves, A.M. The key role of zooplankton in ecosystem services: A perspective of interaction between zooplankton and fish recruitment. *Ecol. Indic.* **2021**, *129*, 107867. [CrossRef]
2. Sterner, R. Role of zooplankton in aquatic ecosystems. In *Encyclopedia of Inland Waters*; Elsevier Inc.: Amsterdam, The Netherlands, 2009; pp. 678–688. [CrossRef]
3. Winder, M.; Jassby, A.D. Shifts in Zooplankton Community Structure: Implications for Food Web Processes in the Upper San Francisco Estuary. *Estuaries Coasts* **2011**, *34*, 675–690. [CrossRef]
4. Choi, H.; Ha, S.-Y.; Lee, S.; Kim, J.-H.; Shin, K.-H. Trophic dynamics of zooplankton before and after polar night in the Kongsfjorden (Svalbard): Evidence of trophic position estimated by δ^{15}N analysis of amino acids. *Front. Mar. Sci.* **2020**, *7*, 489. [CrossRef]
5. Stevens, C.; Sahota, R.; Galbraith, M.; Venello, T.; Bazinet, A.; Hennekes, M.; Yongblah, K.; Juniper, S. Total lipid and fatty acid composition of mesozooplankton functional group members in the NE Pacific over a range of productivity regimes. *Mar. Ecol. Prog. Ser.* **2022**, *687*, 43–64. [CrossRef]
6. Athira, T.R.; Nefla, A.; Shifa, C.T.; Shamna, H.; Aarif, K.M.; Almaarofi, S.S.; Rashiba, A.P.; Reshi, O.R.; Jobiraj, T.; Thejass, P.; et al. The impact of long-term environmental change on zooplankton along the southwestern coast of India. *Environ. Monit. Assess.* **2022**, *194*, 316. [CrossRef]
7. Richardson, A.J. In hot water: Zooplankton and climate change. *ICES J. Mar. Sci.* **2008**, *65*, 279–295. [CrossRef]
8. Gao, X.; Chen, H.; Govaert, L.; Wang, W.; Yang, J. Responses of zooplankton body size and community trophic structure to temperature change in a subtropical reservoir. *Ecol. Evol.* **2019**, *9*, 12544–12555. [CrossRef]
9. Joo, H.; Son, S.; Park, J.-W.; Kang, J.J.; Jeong, J.-Y.; Kwon, J.-I.; Kang, C.-K.; Lee, S.H. Small phytoplankton contribution to the total primary production in the highly productive Ulleung Basin in the East/Japan Sea. *Deep. Sea Res. Part II Top. Stud. Oceanogr.* **2017**, *143*, 54–61. [CrossRef]
10. Kang, J.J.; Jang, H.K.; Lim, J.-H.; Lee, D.; Lee, J.H.; Bae, H.; Lee, C.H.; Kang, C.-K.; Lee, S.H. Characteristics of Different Size Phytoplankton for Primary Production and Biochemical Compositions in the Western East/Japan Sea. *Front. Microbiol.* **2020**, *11*, 560102. [CrossRef]
11. Winder, M.; Sommer, U. Phytoplankton response to a changing climate. *Hydrobiologia* **2012**, *698*, 5–16. [CrossRef]
12. Han, I.-S.; Lee, J.-S. Change the annual amplitude of sea surface temperature due to climate change in a recent decade around the Korean Peninsula. *J. Korean Soc. Mar. Environ. Saf.* **2020**, *26*, 233–241. [CrossRef]
13. Chang, K.-I.; Zhang, C.-I.; Park, C.; Kang, D.-J.; Ju, S.-J.; Lee, S.-H.; Wimbush, M. (Eds.) *Oceanography of the East Sea (Japan Sea)*, 1st ed.; Springer International Publishing: Cham, Switzerland, 2016. [CrossRef]
14. Marañón, E.; Cermeño, P.; Latasa, M.; Tadonléké, R.D. Temperature, resources, and phytoplankton size structure in the ocean. *Limnol. Oceanogr.* **2012**, *57*, 1266–1278. [CrossRef]
15. Ashjian, C.J.; Davis, C.S.; Gallager, S.M.; Alatalo, P. Characterization of the zooplankton community, size composition, and distribution in relation to hydrography in the Japan/East Sea. *Deep. Sea Res. Part II Top. Stud. Oceanogr.* **2005**, *52*, 1363–1392. [CrossRef]
16. Lee, B.R.; Park, W.; Kang, H.K.; Lee, H.W.; Ji, H.S.; Choi, J.H. Comparison of zooplankton communities in the East Sea, East China Sea and Philippine Sea. *J. Environ. Biol.* **2019**, *40*, 861–870. [CrossRef]
17. Lee, H.; Choi, J.; Im, Y.; Oh, W.; Hwang, K.; Lee, K. Spatial–Temporal Distribution of the Euphausiid *Euphausia pacifica* and Fish Schools in the Coastal Southwestern East Sea. *Water* **2022**, *14*, 203. [CrossRef]
18. Henschke, N.; Everett, J.D.; Suthers, I.M.; Smith, J.A.; Hunt, B.P.V.; Doblin, M.A.; Taylor, M.D. Zooplankton trophic niches respond to different water types of the western Tasman Sea: A stable isotope analysis. *Deep. Sea Res. Part I Oceanogr. Res. Pap.* **2015**, *104*, 1–8. [CrossRef]
19. Kozak, E.R.; Franco-Gordo, C.; Godínez-Domínguez, E.; Suárez-Morales, E.; Ambriz-Arreola, I. Seasonal variability of stable isotope values and niche size in tropical calanoid copepods and zooplankton size fractions. *Mar. Biol.* **2020**, *167*, 37. [CrossRef]
20. Schoo, K.L.; Boersma, M.; Malzahn, A.M.; Löder, M.G.J.; Wiltshire, K.H.; Aberle, N. Dietary and seasonal variability in trophic relations at the base of the North Sea pelagic food web revealed by stable isotope and fatty acid analysis. *J. Sea Res.* **2018**, *141*, 61–70. [CrossRef]
21. Post, D.M. Using stable isotopes to estimate trophic position: Models, methods, and assumptions. *Ecology* **2002**, *83*, 703–718. [CrossRef]

22. Vander Zanden, M.J.; Rasmussen, J.B. Variation in δ^{15}N and δ^{13}C trophic fractionation: Implications for aquatic food web studies. *Limnol. Oceanogr.* **2001**, *46*, 2061–2066. [CrossRef]
23. Ahn, I.-Y.; Elias-Piera, F.; Ha, S.-Y.; Rossi, S.; Kim, D.-U. Seasonal Dietary Shifts of the Gammarid Amphipod *Gondogeneia antarctica* in a Rapidly Warming Fjord of the West Antarctic Peninsula. *J. Mar. Sci. Eng.* **2021**, *9*, 1447. [CrossRef]
24. Ke, Z.; Li, R.; Chen, D.; Zhao, C.; Tan, Y. Spatial and seasonal variations in the stable isotope values and trophic positions of dominant zooplankton groups in Jiaozhou Bay, China. *Front. Mar. Sci.* **2022**, *9*, 968. [CrossRef]
25. El-Sabaawi, R.; Dower, J.F.; Kainz, M.; Mazumder, A. Characterizing dietary variability and trophic positions of coastal calanoid copepods: Insight from stable isotopes and fatty acids. *Mar. Biol.* **2009**, *156*, 225–237. [CrossRef]
26. Choi, H.; Won, H.; Kim, J.H.; Yang, E.J.; Cho, K.H.; Lee, Y.; Kang, S.H.; Shin, K.H. Trophic Dynamics of *Calanus hyperboreus* in the Pacific Arctic Ocean. *J. Geophys. Res. Ocean.* **2021**, *126*, e2020JC017063. [CrossRef]
27. Hiltunen, M.; Strandberg, U.; Brett, M.T.; Winans, A.K.; Beauchamp, D.A.; Kotila, M.; Keister, J.E. Taxonomic, Temporal, and Spatial Variations in Zooplankton Fatty Acid Composition in Puget Sound, WA, USA. *Estuaries Coasts* **2022**, *45*, 567–581. [CrossRef]
28. Parson, T.R.; Maita, Y.; Lalli, C.M. *A Manual of Biological and Chemical Methods for Seawater Analysis*; Pergamon Press: Oxford, UK, 1984.
29. Folch, J.; Lees, M.; Sloane Stanley, G.H. A simple method for the isolation and purification of total lipids from animal tissues. *J. Biol. Chem.* **1957**, *226*, 497–509. [CrossRef]
30. Hammer, Ø.; Harper, D.A.T.; Ryan, P.D. PAST-palaeontological statistics, ver. 1.89. *Palaeontol. Electron* **2001**, *4*, 1–9.
31. Kim, H.-C.; Yoo, S.; Oh, I.S. Relationship between phytoplankton bloom and wind stress in the sub-polar frontal area of the Japan/East Sea. *J. Mar. Syst.* **2007**, *67*, 205–216. [CrossRef]
32. Lee, M.; Ro, H.; Kim, Y.-B.; Park, C.-H.; Baek, S.-H. Relationship of Spatial Phytoplankton Variability during Spring with Eutrophic Inshore and Oligotrophic Offshore Waters in the East Sea, Including Dokdo, Korea. *J. Mar. Sci. Eng.* **2021**, *9*, 1455. [CrossRef]
33. Kürten, B.; Al-Aidaroos, A.M.; Kürten, S.; El-Sherbiny, M.M.; Devassy, R.P.; Struck, U.; Zarokanellos, N.; Jones, B.H.; Hansen, T.; Bruss, G. Carbon and nitrogen stable isotope ratios of pelagic zooplankton elucidate ecohydrographic features in the oligotrophic Red Sea. *Prog. Oceanogr.* **2016**, *140*, 69–90. [CrossRef]
34. Domingues, R.B.; Barbosa, A.B.; Sommer, U.; Galvão, H.M. Ammonium, nitrate and phytoplankton interactions in a freshwater tidal estuarine zone: Potential effects of cultural eutrophication. *Aquat. Sci.* **2011**, *73*, 331–343. [CrossRef]
35. Tunēns, J.; Aigars, J.; Poikāne, R.; Jurgensone, I.; Labucis, A.; Labuce, A.; Liepiṇa-Leimane, I.; Buša, L.; Vīksna, A. Stable Carbon and Nitrogen Isotope Composition in Suspended Particulate Matter Reflects Seasonal Dynamics of Phytoplankton Assemblages in the Gulf of Riga, Baltic Sea. *Estuaries Coasts* **2022**, *45*, 2112–2123. [CrossRef]
36. Wu, Z.; Yu, Z.; Song, X.; Wang, W.; Zhou, P.; Cao, X.; Yuan, Y. Key nitrogen biogeochemical processes in the South Yellow Sea revealed by dual stable isotopes of nitrate. *Estuar. Coast. Shelf Sci.* **2019**, *225*, 106222. [CrossRef]
37. Bode, A.; Lamas, A.F.; Mompeán, C. Effects of Upwelling Intensity on Nitrogen and Carbon Fluxes through the Planktonic Food Web off A Coruña (Galicia, NW Spain) Assessed with Stable Isotopes. *Diversity* **2020**, *12*, 121. [CrossRef]
38. Fry, B.; Wainright, S.C. Diatom sources of ^{13}C-rich carbon in marine food webs. *Mar. Ecol. Prog. Ser.* **1991**, *76*, 149–157. [CrossRef]
39. Bardhan, P.; Karapurkar, S.G.; Shenoy, D.M.; Kurian, S.; Sarkar, A.; Maya, M.V.; Naik, H.; Varik, S.; Naqvi, S.W.A. Carbon and nitrogen isotopic composition of suspended particulate organic matter in Zuari Estuary, west coast of India. *J. Mar. Syst.* **2015**, *141*, 90–97. [CrossRef]
40. Perry, R.I.; Thompson, P.A.; Mackas, D.L.; Harrison, P.J.; Yelland, D.R. Stable carbon isotopes as pelagic food web tracers in adjacent shelf and slope regions off British Columbia, Canada. *Can. J. Fish. Aquat. Sci.* **1999**, *56*, 2477–2486. [CrossRef]
41. Rau, G.H.; Riebesell, U.; Wolf-Gladrow, D. A model of photosynthetic ^{13}C fractionation by marine phytoplankton based on diffusive molecular CO_2 uptake. *Mar. Ecol. Prog. Ser.* **1996**, *133*, 275–285. [CrossRef]
42. Silori, S.; Sharma, D.; Chowdhury, M.; Biswas, H.; Bandyopadhyay, D.; Shaik, A.U.R.; Cardinal, D.; Mandeng-Yogo, M.; Narvekar, J. Contrasting phytoplankton and biogeochemical functioning in the eastern Arabian Sea shelf waters recorded by carbon isotopes (SW monsoon). *Mar. Chem.* **2021**, *232*, 103962. [CrossRef]
43. Ying, R.; Cao, Y.; Yin, F.; Guo, J.; Huang, J.; Wang, Y.; Zheng, L.; Wang, J.; Liang, H.; Li, Z.; et al. Trophic structure and functional diversity reveal pelagic-benthic coupling dynamic in the coastal ecosystem of Daya Bay, China. *Ecol. Indic.* **2020**, *113*, 106241. [CrossRef]
44. Budge, S.M.; Parrish, C.C. Lipid biogeochemistry of plankton, settling matter and sediments in Trinity Bay, Newfoundland. II. Fatty acids. *Org. Geochem.* **1998**, *29*, 1547–1559. [CrossRef]
45. Dalsgaard, J.; John, M.S.; Kattner, G.; Muller-Navarra, D.; Hagen, W. Fatty acid trophic markers in the pelagic marine food environment. *Adv. Mar. Biol.* **2003**, *46*, 226–340. [CrossRef]
46. El-Sabaawi, R.W.; Sastri, A.R.; Dower, J.F.; Mazumder, A. Deciphering the seasonal cycle of copepod trophic dynamics in the Strait of Georgia, Canada, using stable isotopes and fatty acids. *Estuaries Coasts* **2010**, *33*, 738–752. [CrossRef]
47. Lee, R.F.; Hagen, W.; Kattner, G. Lipid storage in marine zooplankton. *Mar. Ecol. Prog. Ser.* **2006**, *307*, 273–306. [CrossRef]
48. Sokolova, I.M.; Lannig, G. Interactive effects of metal pollution and temperature on metabolism in aquatic ectotherms: Implications of global climate change. *Clim. Res.* **2008**, *37*, 181–201. [CrossRef]
49. Werbrouck, E.; Van Gansbeke, D.; Vanreusel, A.; De Troch, M. Temperature affects the use of storage fatty acids as energy source in a benthic copepod (*Platychelipus littoralis*, Harpacticoida). *PLoS ONE* **2016**, *11*, e0151779. [CrossRef]
50. Jin, P.; Gonzàlez, G.; Agustí, S. Long-term exposure to increasing temperature can offset predicted losses in marine food quality (fatty acids) caused by ocean warming. *Evol. Appl.* **2020**, *13*, 2497–2506. [CrossRef]

51. Shin, K.-H.; Hama, T.; Yoshie, N.; Noriki, S.; Tsunogai, S. Dynamics of fatty acids in newly biosynthesized phytoplankton cells and seston during a spring bloom off the west coast of Hokkaido Island, Japan. *Mar. Chem.* **2000**, *70*, 243–256. [CrossRef]
52. Shin, K.-H.; Hama, T.; Handa, N. Effect of nutrient conditions on the composition of photosynthetic products in the East China Sea and surrounding waters. *Deep. Sea Res. Part II Top. Stud. Oceanogr.* **2003**, *50*, 389–401. [CrossRef]
53. Landry, M.R. Switching between herbivory and carnivory by the planktonic marine copepod *Calanus pacificus*. *Mar. Biol.* **1981**, *65*, 77–82. [CrossRef]
54. López-Ibarra, G.A.; Bode, A.; Hernández-Trujillo, S.; Zetina-Rejón, M.J.; Arreguín-Sánchez, F. Trophic position of twelve dominant pelagic copepods in the eastern tropical Pacific Ocean. *J. Mar. Syst.* **2018**, *187*, 13–22. [CrossRef]

Journal of
*Marine Science
and Engineering*

Article

Discovery of Pelagic Eggs of Two Species from the Rare Mesopelagic Fish Genus *Trachipterus* (Lampriformes: Trachipteridae)

Hae-young Choi [1], Hee-chan Choi [2], Sung Kim [3,*], Hyun-ju Oh [1] and Seok-hyun Youn [1,*]

[1] Ocean Climate & Ecology Research Division, National Institute of Fisheries Science, Busan 46083, Korea; chlgodud1030@gmail.com (H.-y.C.); hyunjuoh@korea.kr (H.-j.O.)

[2] Fisheries Resources and Environment Division, East Sea Fisheries Research Institute, Gangneung 25435, Korea; gmlckschl82@korea.kr

[3] Marine Ecosystem Research Center, Korea Institute of Ocean Science & Technology, Busan 49111, Korea

* Correspondence: skim@kiost.ac.kr (S.K.); younsh@korea.kr (S.-h.Y.); Tel.: +82-51-664-3278 (S.K.); +82-52-720-2233 (S.-h.Y.)

Abstract: The ecology of the mesopelagic fish genus *Trachipterus*, which is rarely found in oceans, remains unclear. In this study, we found 22 eggs of *T. trachypterus* and *T. jacksonensis* around the Ulleung Basin of the East/Japan Sea during ichthyoplankton surveys from 2019 to 2021. The eggs were identified through genetic relationships with the genus *Trachipterus* based on partial sequences (COI and 16S) or concatenated sequences of 13 protein-coding genes and 2 rRNA genes of mitochondrial DNA. *T. trachypterus* eggs were discovered in all seasons, but more frequently during the winter. One *T. jacksonensis* egg that appeared during the autumn was the first in the northwestern Pacific Ocean. Identifying *Trachipterus* pelagic eggs would provide insight into their spawning ecology and biogeography.

Keywords: mesopelagic fish; mitochondrial DNA sequence; pelagic fish eggs; spawning; *Trachipterus jacksonensis*; *Trachipterus trachypterus*; Ulleung Basin

Citation: Choi, H.-y.; Choi, H.-c.; Kim, S.; Oh, H.-j.; Youn, S.-h. Discovery of Pelagic Eggs of Two Species from the Rare Mesopelagic Fish Genus *Trachipterus* (Lampriformes: Trachipteridae). *J. Mar. Sci. Eng.* **2022**, *10*, 637. https://doi.org/10.3390/jmse10050637

Academic Editor: Roberto Carlucci

Received: 15 March 2022
Accepted: 4 May 2022
Published: 7 May 2022

Publisher's Note: MDPI stays neutral with regard to jurisdictional claims in published maps and institutional affiliations.

1. Introduction

Members of the genus *Trachipterus* (Lampriformes, Trachipteridae), which comprise six species (*T. altivelis*, *T. jacksonensis*, *T. arcticus*, *T. fukuzakii*, *T. trachypterus*, and *T. ishikawae*) [1,2], have elongated and compressed bodies with large eyes and are distributed in several oceans [3,4]. They are rarely caught by deep fishing gear or found inshore [5,6]. The species of *Trachipterus* have been reported based on a few specimens [7–10].

The first nominal species of *Trachipterus* is *T. trachypterus* [11]. *Trachipterus* has been described using juvenile specimens, which reside in shallower waters than adults. Due to its rarity and the morphological changes at early developmental stages, *T. trachypterus* larvae of different sizes are often incorrectly identified [12]. *T. trachypterus* eggs were first reported as three specimens in the Gulf of Napoli [12]. The Mediterranean Sea, where *T. trachypterus* eggs, larvae, and adults have been found several times, is a known spawning ground [13].

Fish eggs are a key indicator of spawning and evidence of species intrusion [14,15]. Because most marine teleost fish release large numbers of pelagic eggs, their egg distribution density is higher than that of spawners [16]. Even in rare species, the probability of finding for eggs is higher than that for adult fish during the spawning season [15]. In addition, eggs with a shorter pelagic duration than larvae are in or close to spawning grounds [17]. The spawning grounds of the Japanese eel, *Anguilla japonica*, were revealed through DNA barcoding-based identification of their eggs after prolonged research [18].

DNA barcoding enables species-level identification of fish eggs [19]. Pelagic marine fish eggs are typically transparent and round [20]. Eggs of the same species may have morphologically different embryos depending on developmental stage [21]. Unlike dramatic changes in morphology, DNA remains constant throughout life history [22,23]. Intra- and inter-specific genetic distances, typically based on mitochondrial DNA (mtDNA) sequences (COI, 12S, 16S, etc.), are analyzed to identify fish eggs and larvae species [24–26].

We conducted COI barcoding of pelagic eggs collected from the East/Japan Sea from 2019 to 2021 and surveyed the literature that applied DNA barcoding to fish eggs near the study area. The eggs in this study were finally determined to belong to *T. trachypterus* and *T. jacksonensis* based on their COI, 16S rRNA, and mitogenome sequences. Here, we report *T. trachypterus* spawning and the first finding of *T. jacksonensis* based on eggs in the East/Japan Sea.

2. Materials and Methods

2.1. Pelagic Fish Egg Collection

Pelagic fish eggs were collected from 13 stations in the East/Japan Sea (Figure 1) using a zooplankton net (mouth diameter: 80 cm; mesh size: 300 μm) during research cruises on R/V Tamgu 3 in four seasons (winter: February and March; spring: April; summer: June and August; autumn: October and November) from 2019 to 2021. The water depths at the stations are 126–2203 m. The net was towed obliquely from 10 m above the bottom to the surface at the stations with a water depth of less than 300 m (st. 103-05, 209-05, and 209-07). At the other stations, the net was lowered to a depth of 300 m and was towed as above. Samples were preserved in 95% ethanol. Temperature and salinity were measured using CTD (SBE 911plus, Sea-Bird Scientific Inc., Bellevue, WA, USA).

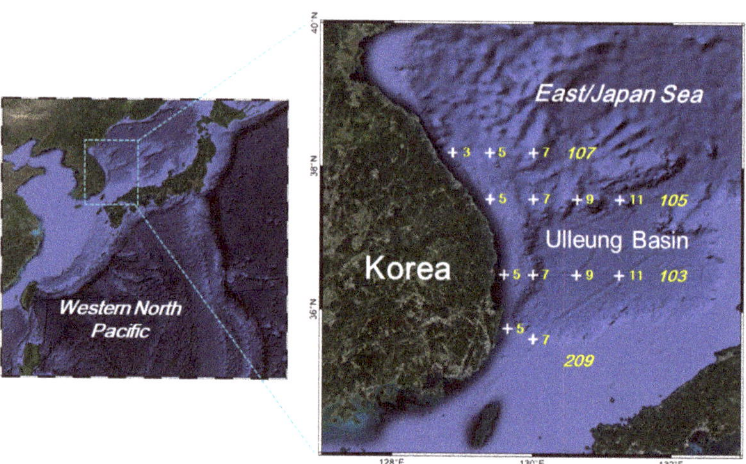

Figure 1. Pelagic fish egg sampling stations in the East/Japan Sea, 2019–2021. Numbers with plus marks, station name; italic number, station lines. The base map is a Google Satellite map drawn using QGIS [27] through the XYZ Tiles tool (https://mt1.google.com/vt/lyrs=s&x=\{x\}&y=\{y\}&z=\{z\}, accessed on 10 March 2022).

Fish eggs were sorted from the samples using a stereomicroscope (M125C, Leica, Wetzlar, Germany). Among them, the 22 largest eggs were selected and photographed using a camera mounted on a stereomicroscope (SMZ18, Nikon, Tokyo, Japan).

2.2. Genomic DNA Extraction, PCR, and Sequencing and Sequence Analysis

Genomic DNA was extracted from 22 eggs according to the protocol of the DNeasy Blood & Tissue Kit (Qiagen, Hilden, Germany). The COI gene of mtDNA was amplified using the primers VF2_t1 (5′-CAACCAACCACAAAGACATTGGCAC-3′), FishF2_t1 (5′-TCGACTAATCATAAAGATATCGGCAC-3′), FishR2_t1 (5′-ACTTCAGGGTGACCGAA GAATCAGAA-3′), and FR1d_t1 (5′-ACCTCAGGGTGTCCGAARAAYCARAA-3′) [28,29]. The 20 µL PCR mixture was composed of 10 µL of 2X DNA free-Taq Master Mix including PCR buffer, dNTPs mixture, and Taq DNA polymerase (CellSafe, Gyeonggi, Korea), 0.2 µL of each of the four primers, 2 µL of genomic DNA, and 7.2 µL of distilled water. The PCR program consisted of initial denaturation at 94 °C for 3 min; 35 cycles of denaturation at 94 °C for 30 s, annealing at 52 °C for 40 s, extension at 72 °C for 1 min; and a final extension at 72 °C for 7 min. The PCR products were sequenced on a 3730xl DNA Analyzer (Applied Biosystems, Foster City, CA, USA).

Other mtDNA regions of the 7 eggs among the 22 eggs were also amplified for comparison with those of related taxa. The 16S rRNA gene of one egg was amplified using the 16Sar (5-CGCCTGTTTATCAAAAACAT-3) and 16Sbr (5-CCGGTCTGAACTCAGATCACGT-3) primers [30]. PCR and sequencing methods were used for COI analysis, except that the annealing temperature was 56 °C. Six eggs were selected to analyze mitogenome sequences in the consideration of sample conditions; three eggs were collected in 2019 and 2020 each. To obtain sufficient DNA to analyze the complete mitogenome of the six eggs, the whole genome was amplified following the REPLI-g Mini Kit (Qiagen) protocol. The amplified products were sequenced using a NovaSeq 6000 (Illumina, San Diego, CA, USA). A total of 385,684,576–474,309,190 raw reads (length: 150 bp) were obtained from the six eggs. The reads were mapped to a reference sequence (GenBank accession number: NC_003166, [31]) using Geneious R11 [32] mapper. The resulting consensus sequences were annotated using MitoFish [33] and Geneious R11. The three mitogenome sequences were constructed from three eggs (2002E3–E5), except for the d-loop region (mean coverage, 112 ± 119×–7866 ± 11,917×; 16,159–16,163 bp), but COI (1467–1551 bp) and 16S (1595–1683 bp) sequences were obtained from the other three eggs (1902E1, 1902E2, and 1904E1).

The COI, 16S, and mitogenome sequences from the 22 eggs were searched using BLAST [34] and BOLD systems [35] to identify related taxa. The sequences of the eggs, related taxa, and outgroups were aligned using Clustal Omega [36]. The sequences were used to analyze the maximum likelihood (ML) tree based on the best-fit substitution model [37–39] and Kimura 2-parameter distances in MEGA X (ver. 11.0.10) [40]. The egg sequences were submitted to NCBI GenBank under accession numbers OM527130–OM527151, OM527153, OM574770–OM574772, and ON231742–ON231747 (Table 1). We also investigated the literature using DNA barcoding for the species identification of eggs around the study area, and sequences from the eggs of Shin et al. [41] were used in this study.

Table 1. GenBank accession numbers of eggs of this study and literature.

No.	Specimen ID	COI	16S	Mitogenome Except for d-Loop	Remark
1	1902E1	OM527131 ON231745	ON231742		this study
2	1902E2	OM527132 ON231746	ON231743		this study
3	1902E3	OM527133			this study
4	1902E4	OM527134			this study
5	1902E5	OM527135			this study
6	1902E6	OM527136			this study
7	1904E1	OM527137 ON231747	ON231744		this study
8	1904E2	OM527138			this study
9	1904E3	OM527139			this study
10	2002E1	OM527140			this study

Table 1. *Cont.*

No.	Specimen ID	COI	16S	Mitogenome Except for d-Loop	Remark
11	2002E2	OM527141			this study
12	2002E3	OM527142 OM574770	OM574770	OM574770	this study
13	2002E4	OM527143 OM574771	OM574771	OM574771	this study
14	2002E5	OM527144 OM574772	OM574772	OM574772	this study
15	2002E6	OM527145			this study
16	2002E7	OM527146			this study
17	2002E8	OM527147			this study
18	2004E1	OM527148			this study
19	2004E2	OM527149			this study
20	2102E1	OM527150			this study
21	2108E1	OM527151			this study
22	2110E1	OM527153	OM527130		this study
23	DFRCI 367	MZ596220.1			[41]
24	DFRCI 368	MZ596221.1			[41]
25	DI_4	MH144581.1			[41]
26	DI_7		MH144584.1		personal communication

Underlined sequences were analyzed from shotgun sequencing.

3. Results

3.1. Genetic Identification of Eggs

Twenty-two egg specimens were identified as *T. trachypterus* and *T. jacksonensis* according to genetic relationships based on COI, 16S, and mitogenome sequences.

The COI sequences of the 22 eggs and 3 eggs from the reference [41] and genus *Trachipterus* consisted of three distinct clades (between genetic distances: average $\pm$ standard deviation, 0.158 ± 0.057; min, 0.083; max, 0.227; Table S1), with either two or three species in the maximum likelihood (ML) tree (Figure 2). Of the three clades, COI_Clade 1 had sequences from 24 eggs, *T. altivelis* and *T. trachypterus*, and COI_Clade 2 had sequences from one egg of this study, *T. arcticus*, *T. jacksonensis*, and *Trachipterus* sp. Although each clade consisted of sequences from different species, the genetic distances within the clades (COI_Clade 1, 0.009 ± 0.006; COI_Clade 2, 0.010 ± 0.007; and COI_Clade 3, 0.009 ± 0.005) were less than between the clades (0.158 ± 0.057), indicating that each clade represented the species.

Species of the two clades (COI_ Clades 1 and 2; Figure 2a), including the eggs, were reanalyzed based on 16S rRNA and mitogenome sequences. Among the eggs of COI_Clade 1, COI and 16S rRNA sequences for three eggs (1902E1, 1902E2, and 1904E1) and mitogenome sequences excluding the d-loop for three eggs (2002E3–E5) were obtained from mitogenome analysis (Table 1). The COI sequences of the six eggs were also located in COI_Clade 1. Concatenated sequences (13 protein-coding genes and two rRNAs) of the three eggs (2002E3–E5) formed a clade with those of *T. trachypterus* (NC_003166.1) (genetic distance, 0.007 ± 0.003; Table S2), and they were distinct from those of *Desmodema polystictum* and *Zu cristatus* of Trachipteridae (0.258 ± 0.012) (Figure 2b). The 16S rRNA sequences of the six eggs from the same samples of COI_Clade 1 and one egg (MH144584.1) from [41] formed a clade with *T. trachypterus* and *T. altivelis* with very small genetic distances (0.002 ± 0.001; min, 0.000; max, 0.004; Table S3) (Figure 2c). Interestingly, the *T. trachypterus* 16S sequence (DQ027909.1, [42]) was distinct from 16S_Clade 1, although it was analyzed from the same specimen with the *T. trachypterus* COI sequence (DQ027978.1) of COI_Clade 1.

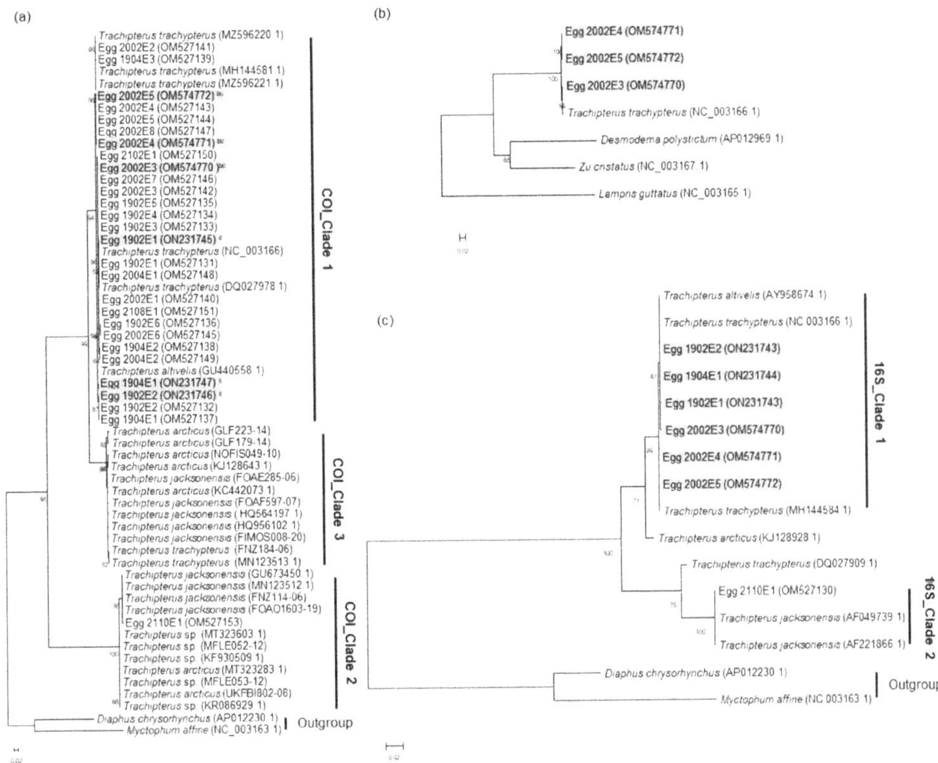

Figure 2. Maximum likelihood (ML) tree constructed using mitochondrial DNA sequences of pelagic fish eggs, *Trachipterus* species, and outgroups. (**a**) COI ML tree (based on the HKY + G + I model) including COI sequences from 25 eggs. Sequences of eggs with superscripts (bc, c) were also analyzed in the trees in (**b,c**). (**b**) Thirteen protein-coding genes and two rRNA genes ML tree (based on the GTR + G + I model), including concatenated sequences from the three eggs. (**c**) 16S rRNA ML tree (based on the K2 + G model), including 16S rRNA sequences from seven eggs. Bootstrap values (1000 replicates) over 50% are shown on the branches. Sequences shaded in gray were obtained using shotgun sequencing.

One egg (2110E1) formed COI_Clade 2 with *T. arcticus*, *T. jacksonensis*, *T. altivelis*, and *Trachipterus* sp. comprised a clade with only *T. jacksonensis* (genetic distance, 0.000 ± 0.000) in the 16S ML tree (16S_Clade 2; Figure 2c). The 16S sequence of *T. arcticus* (KJ128928.1) diverged sharply (0.107) from the clade of *T. jacksonensis* with the egg (2110E1; OM527130). The *T. arcticus* 16S rRNA sequence (KJ128928.1) was obtained from the same specimen as the *T. arcticus* COI sequence (KJ128643.1) in COI_Clade 3.

3.2. Pelagic Eggs of Two Trachipterus Species

The average diameter of *T. trachypterus* eggs was 3.2 ± 0.1 mm (min, 2.7 mm; max, 3.6 mm) (Figure 3; Table S4). The diameter of *T. jacksonensis* egg was 2.2 mm, smaller than that of *T. trachypterus* eggs. The internal morphology of eggs preserved in 95% ethanol was difficult to investigate. The common characteristics of *T. trachypterus* eggs that could be confirmed were a narrow perivitelline space and a lack of oil globules. The developed embryos had melanophores on the head, dorsal side, and on the yolk sac around the middle of the body.

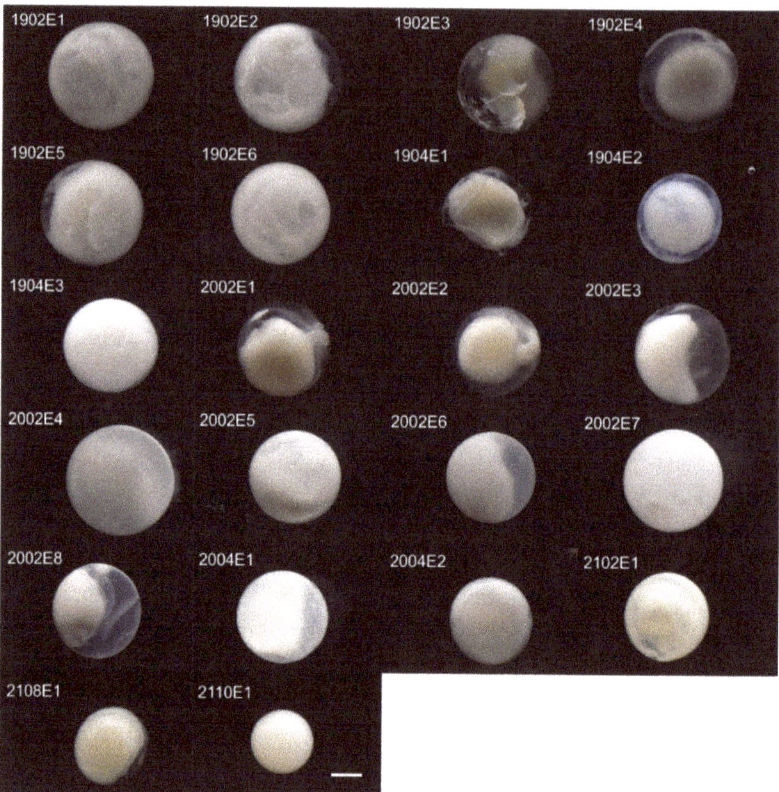

Figure 3. Pelagic eggs of two *Trachipterus* species preserved in 95% ethanol. 1902E1–2108E1, *T. trachypterus*; 2110E1, *T. jacksonensis*. Scale bar, 1.0 mm.

3.3. Distribution of Trachipterus Eggs

T. trachypterus eggs appeared in all seasons around the Ulleung Basin and Dokdo, and one *T. jacksonensis* egg appeared once in autumn (Figure 4). The occurrence frequency of *T. trachypterus* eggs was highest during the winter and lowest during the summer and autumn. The surface temperature of the stations where eggs were detected ranged from 10.0 °C to 25.9 °C (Table S4). The mean surface temperature of the study area was the lowest during the winter and peaked in the summer (Table S5). Unlike the surface, the bottom temperature was constant at approximately 1 °C. Thermocline was generated by the large difference between the temperatures of the surface and bottom. The depth and strength of the thermocline varied according to season and station. The center of the thermocline was usually at a depth of 100–200 m (Figure S1). Salinity was approximately 33–34 psu, and there was no significant difference depending on season and depth, unlike the temperature.

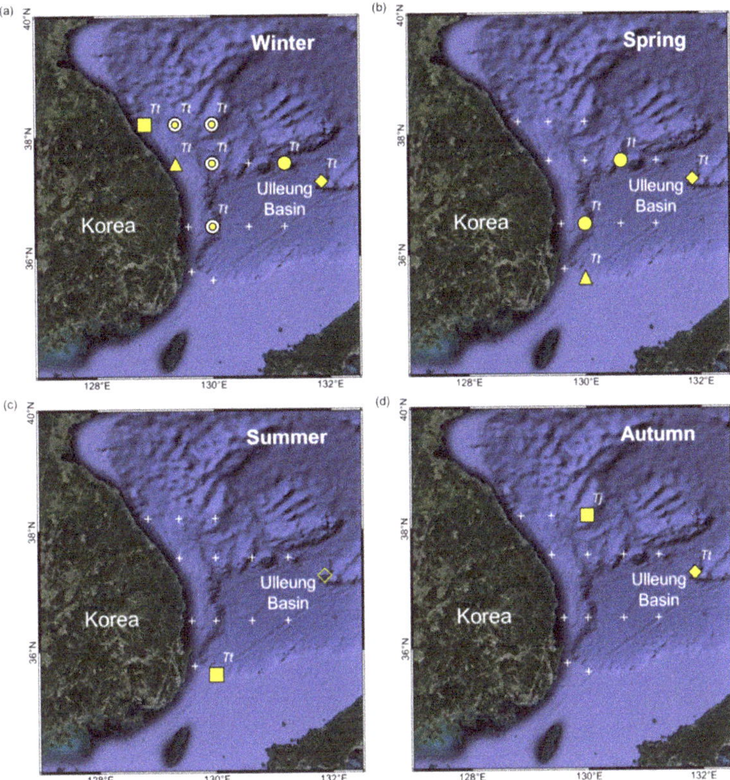

Figure 4. Distributions of pelagic eggs of two *Trachipterus* species for the four seasons: (**a**) winter, (**b**) spring, (**c**) summer, and (**d**) autumn. *Tt, T. trachypterus; Tj, T. jacksonensis;* Sampling year of *Trachipterus* eggs: filled circle, 2019; fisheye, 2019 and 2020; filled triangle, 2020; filled square, 2021 of this study; filled diamond, 2017–2019 of Shin et al. [41]; no egg detection: cross, this study; empty diamond, Shin et al. [41].

4. Discussion

DNA barcoding of fish eggs enables the investigation of the spawning ecology of various species [43,44]. Studies on fish eggs have been limited to a few species due to the difficulty in morphology-based egg identification [45]. Egg morphological characteristics were described based on larval fish hatching from eggs or eggs obtained from adult fish [46,47]. Recently, larval fish, which have more morphological features available for identification than eggs, have been analyzed using DNA barcoding to improve the accuracy of species identification [48]. However, DNA barcoding has limitations in determining species due to the lack of comparable DNA sequences, unsuitable DNA regions for barcoding, or sequences from misidentified specimens.

4.1. Genetic Identification of Eggs

The first DNA barcode for species identification of 22 eggs in this study was the COI sequence, which contains a lot of information available to identify fish. The maximum likelihood (ML) tree based on the COI sequences of the 25 eggs, including those of Shin et al. [41] and *Trachipterus*, showed three distinct clades of *Trachipterus* (Figure 2a). Each clade was regarded as a species by considering the genetic distances of the intra- and inter-clades of the COI sequences of fishes [23,28]. However, the coexistence of sequences from two or three species within each clade was problematic: COI_Clade 1, *T. altivelis*, and

T. trachypterus with 24 eggs; COI_Clade 2, *T. arcticus*, *T. jacksonensis*, *Trachipterus* sp., and one egg (2110E1); and COI_Clade 3, *T. arcticus*, *T. jacksonensis*, and *T. trachypterus*. This could be because COI sequences were analyzed from misidentified samples or were not appropriate for distinguishing the species.

The *T. trachypterus* mitogenome sequence was the key to determining the species of COI_Clade 1. The mitogenome sequences of three eggs (2002E3–E5) located in COI_Clade 1 formed a clade with that of *T. trachypterus* (NC_003166.1) (Figure 2b). In addition, genetic distances between the 16S rRNA sequences of the seven eggs (1902E1, 1902E2, 1904E1, 2002E3–E5, and DI_7 (MH144584.1)) and *T. trachypterus* (NC_003166, MH144581.1) were also very small (0.000–0.004; Table S3). These genetic relationships were a criterion for identifying 24 eggs, including those of Shin et al. [41] in COI_Clade 1 as *T. trachypterus*.

COI_Clade 2, including one egg (2110E1; OM527153), was determined to be *T. jacksonensis* based on the distinct 16S_Clade 2 with sequences of *T. jacksonensis* and the one egg (2110E1; OM527130) in the 16S ML tree (Figure 2c). In addition, COI_Clade 3 was identified as *T. arcticus* based on the *T. arcticus* sequences (COI, KJ128643.1; 16S rRNA, KJ128928.1) showing the same position in the COI and 16S rRNA ML tree (Figure 2a,c).

Interestingly, the *T. trachypterus* 16S sequence (DQ027909.1, [42]) from the same sample with COI sequence (DQ027978.1, [42]) in the COI_Clade 1 was independently clustered in the 16S ML tree (Figure 2c). This indicates that the 16S rRNA sequence (DQ027909.1) was mishandled during the experiment or sequence analysis. In addition, *T. altivelis* (GU440558.1) of COI_Clade 1 and *T. altivelis* (AY958674.1) of 16S_Clade 1 would be derived from misidentification.

4.2. Trachipterus Eggs

T. trachypterus is widely distributed in the Mediterranean Sea and Atlantic Indo-Pacific Oceans [5,6,9,49–60]. Its spawning area is the Mediterranean Sea, where eggs of *Trachypterus taenia* (synonym for *T. trachypterus*) have been found [12]. Lo Bianco [12] collected three live eggs of *T. taenia* from a sampling gear lowered to a depth of 100–150 m in the Gulf of Napoli in February, May, and October 1905–1907. Spawning was assumed to occur year-round. Three eggs (diameter, 2.90–2.95 mm) were in late developmental stages. In this study, we confirmed that *T. trachypterus* spawned in all seasons and peaked in winter based on eggs continuously collected during the survey periods. If the *T. trachypterus* eggs in this study had similar ecological characteristics to the eggs from the Gulf of Napoli, the spawning depth could be estimated as the thermocline layer.

T. jacksonensis is distributed in the Southern Hemisphere (Australia, New Zealand, Africa, Brazil, and others) [61–70]. Adults have not been reported in the Northern Hemisphere. The eggs detected once in autumn in this study were the first in the Northern Hemisphere. Similarly to *T. jacksonensis*, *Peristedion liorhynchus* has not been recorded in Korean waters, but its eggs and larvae have been recorded [15]. This suggests that eggs may be useful for the detection of rare species.

5. Conclusions

In this study, we discovered *T. trachypterus* and *T. jacksonensis* eggs in the Ulleung Basin of the East/Japan Sea of which their sequences could be used for post verification. The Ulleung Basin is a spawning area for *T. trachypterus* along with the Mediterranean Sea; this is the first report of *T. jacksonensis* egg in the northwestern Pacific. Finding *Trachipterus* species eggs will elucidate their spawning ecology and geographic distribution.

Supplementary Materials: The following supporting information can be downloaded at: https://www.mdpi.com/article/10.3390/jmse10050637/s1, Figure S1: Vertical distribution of temperature at stations where *Trachipterus* eggs appeared in this study; Table S1: Pairwise genetic distances of the COI genes of pelagic eggs of this study, *Trachipterus* species, and outgroups. Outlined values belong to each clade of the maximum likelihood tree in Figure 2a; Table S2: Pairwise genetic distances of the concatenated mtDNA sequences of 13 protein-coding genes and two rRNA genes of pelagic eggs of this study, *Trachipterus* species, and outgroups; Table S3: Pairwise genetic distances of the 16S rRNA genes of pelagic egg of this study, *Trachipterus* species, and outgroups. Outlined values belong to each clade of the maximum likelihood tree in Figure 2c; Table S4: Information on *Trachipterus* eggs of this study and literature; Table S5: Mean and standard deviation values of temperature and salinity from 2019 to 2021 in the East/Japan Sea.

Author Contributions: Conceptualization, H.-y.C.; methodology, H.-y.C. and S.K.; validation, H.-y.C. and S.K.; formal analysis, H.-y.C. and S.K.; investigation, H.-c.C.; data curation, H.-y.C.; writing—original draft preparation, H.-y.C.; writing—review and editing, H.-y.C., H.-c.C., S.K. and S.-h.Y.; visualization, H.-y.C.; supervision, S.-h.Y.; project administration, H.-j.O.; funding acquisition, H.-j.O. All authors have read and agreed to the published version of the manuscript.

Funding: This research was funded by the National Institute of Fisheries Science (NIFS) grant (Study on the ecophysiology and occurrence prediction of jellyfish; R2022052).

Institutional Review Board Statement: Not applicable.

Informed Consent Statement: Not applicable.

Data Availability Statement: The Sequence data that support the findings of this study are deposited in NCBI/GenBank (https://www.ncbi.nlm.nih.gov/genbank/) under accession numbers OM527130-OM527151, OM527153, OM574770-OM574772, and ON231742-ON231747.

Acknowledgments: We appreciate the captain and crews of the Tamgu3 research vessel for the survey. We are also grateful to the researchers in the NIFS for their assistance with sampling.

Conflicts of Interest: The authors declare no conflict of interest.

References

1. Integrated Taxonomic Information System (ITIS). Available online: http://www.itis.gov (accessed on 25 January 2022).
2. Eschmeyer's Catalog of Fishes. Available online: http://researcharchive.calacademy.org/research/ichthyology/catalog/fishcatmain.asp (accessed on 25 January 2022).
3. Palmer, G. Trachipteridae. In *Fishes of the North-Eastern Atlantic and the Mediterranean*; Whitehead, P.J.P., Bauchot, M.L., Hureau, J.C., Nielsen, J., Tortonese, E., Eds.; United Nations Educational, Scientific and Cultural Organization: Paris, France, 1986; Volume 2, pp. 729–732.
4. Olney, J.E.; Richards, W.J. Trachipteridae: Dealfishes and ribbonfishes. In *Early Stages of Atlantic Fishes an Identification Guide for the Western Central North Atlantic*; Richard, W.J., Ed.; Taylor and Francis: Boca Raton, FL, USA, 2006; pp. 1019–1026.
5. Borme, D.; Voltolina, F. On the occurrence of ribbon fish *Trachipterus trachypterus* (Gmelin, 1789) in the Gulf of Trieste (northern Adriatic Sea). *Ann. Ser. Hist. Nat.* **2006**, *16*, 181.
6. Tiralongo, F.; Lillo, A.O.; Tibullo, D.; Tondo, E.; Martire, C.L.; D'Agnese, R.; Macali, A.; Mancini, E.; Giovos, I.; Coco, S.; et al. Monitoring uncommon and non-indigenous fishes in Italian waters: One year of results for the AlienFish project. *Reg. Stud. Mar. Sci.* **2019**, *28*, 100606. [CrossRef]
7. MoriTz, T.; STüMer, D.; JaKobSen, K.; JaKobSen, J. Observations on two live specimens of *Trachipterus arcticus* (Lampriformes: Trachipteridae) from the Azores. *Cybium* **2015**, *39*, 78–80.
8. Angulo, A.; Lopez-Sanchez, M.I. New records of lampriform fishes (Teleostei: Lampriformes) from the Pacific coast of lower Central America, with comments on the diversity, taxonomy and distribution of the Lampriformes in the eastern Pacific Ocean. *Zootaxa* **2017**, *4236*, 573–591. [CrossRef] [PubMed]
9. Lipej, L.; Trkov, D.; Mavrič, B. Occurrence of ribbon fish (*Trachipterus trachypterus*) in Slovenian waters (northern Adriatic Sea). *Ann. Ser. Hist. Nat.* **2018**, *28*, 129–134.
10. Han, K.H.; Lee, S.H.; Kim, C.C.; Yu, T.S. Description of Morphology and Osteology of the Slender Ribbonfish, *Trachipterus ishikawae* Jordan & Snyder, 1901. *Korean J. Ichthyol.* **2020**, *32*, 136–142.
11. Martin, J.; Hilton, E.J. A taxonomic review of the family Trachipteridae (Lampridiformes), with an emphasis on taxa distributed in the Western Pacific Ocean. *Zootaxa* **2021**, *5039*, 301–351. [CrossRef]
12. Lo Bianco, S. Uova e larve di *Trachypterus taenia* Bl. *Mitt. Zool. Stn. Neapel.* **1908**, *19*, 1–17.
13. Palmer, G. The dealfishes (Trachipteridae) of the Mediterranean and north-east Atlantic. *Bull. Br. Mus. Nat. Hist. Zool.* **1961**, *7*, 335–351. [CrossRef]
14. Lewis, L.A.; Richardson, D.E.; Zakharov, E.V.; Hanner, R. Integrating DNA barcoding of fish eggs into ichthyoplankton monitoring programs. *Fish. Bull.* **2016**, *114*, 153–165. [CrossRef]
15. Choi, H.Y.; Chin, B.S.; Park, G.S.; Kim, S. Evidence of intrusion of a rare Species, *Peristedion liorhynchus*, into Korean waters based on high-throughput sequencing of the mixed fish eggs. *Korean J. Icthyol.* **2022**, *1*, 8–15. [CrossRef]
16. Houde, E.D. Fish early life dynamics and recruitment variability. *Am. Fish. Soc. Symp.* **2018**, *2*, 17–29.
17. Choi, H.Y.; Oh, J.; Kim, S. Genetic identification of eggs from four species of Ophichthidae and Congridae (Anguilliformes) in the northern East China Sea. *PLoS ONE* **2018**, *13*, e0195382. [CrossRef] [PubMed]

18. Tsukamoto, K.; Chow, S.; Otake, T.; Kurogi, H.; Mochioka, N.; Miller, M.J.; Aoyama, J.; Kimura, S.; Watanabe, S.; Yoshinaga, T.; et al. Oceanic spawning ecology of freshwater eels in the western North Pacific. *Nat. Commun.* **2011**, *2*, 179. [CrossRef]
19. Shao, K.T.; Chen, K.C.; Wu, J.H. Identification of marine fish eggs in Taiwan using light microscopy, scanning electric microscopy and mtDNA sequencing. *Mar. Freshw. Res.* **2002**, *53*, 355–365. [CrossRef]
20. Ahlstrom, E.H.; Moser, H.G. Characters useful in identification of pelagic marine fish eggs. *CalCOFI Rep.* **1980**, *21*, 121–131.
21. Kendall, A.W.; Ahlstrom, E.H., Jr.; Moser, H.G. Early Life History Stages of Fishes and Their Characters. In *Ontogeny and Systematics of Fishes*; Moser, H.G., Richards, W.J., Cohen, D.M., Fahay, M.P., Kendall, A.W., Richardson, S.L., Eds.; American Society of Ichthyologists and Herpetologists: Kansas, MO, USA, 1984; pp. 31–33.
22. Avise, J.C. *Molecular Markers, Natural History and Evolution*; Springer: Boston, MA, USA, 1994.
23. Hebert, P.D.; Cywinska, A.; Ball, S.L.; DeWaard, J.R. Biological identifications through DNA barcodes. *Proc. R. Soc. B Biol. Sci.* **2003**, *270*, 313–321. [CrossRef]
24. Choi, H.Y.; Jang, Y.S.; Oh, J.N.; Kim, S. Morphology of a Larval Hammerjaw *Omosudis lowii* Gunther 1887 (Aulopiformes, Omosudidae) Identified by Partial Mitochondrial 12S rRNA Gene Analysis. *Korean J. Ichthyol.* **2020**, *32*, 239–244. [CrossRef]
25. Hou, G.; Xu, Y.; Chen, Z.; Zhang, K.; Huang, W.; Wang, J.; Zhou, J. Identification of Eggs and Spawning Zones of Hairtail Fishes *Trichiurus* (Pisces: Trichiuridae) in Northern South China Sea, Using DNA Barcoding. *Front. Environ. Sci.* **2021**, *9*, 703029. [CrossRef]
26. Habib, K.A.; Neogi, A.K.; Rahman, M.; Oh, J.; Lee, Y.H.; Kim, C.G. DNA barcoding of brackish and marine water fishes and shellfishes of Sundarbans, the world's largest mangrove ecosystem. *PLoS ONE* **2021**, *16*, e0255110. [CrossRef]
27. QGIS Geographic Information System. Available online: http://www.qgis.org (accessed on 25 January 2022).
28. Ward, R.D.; Zemlak, T.S.; Innes, B.H.; Last, P.R.; Hebert, P.D. DNA barcoding Australia's fish species. *Philos. Trans. R. Soc. B Biol. Sci.* **2005**, *360*, 1847–1857. [CrossRef] [PubMed]
29. Ivanova, N.V.; Zemlak, T.S.; Hanner, R.H.; Hebert, P.D. Universal primer cocktails for fish DNA barcoding. *Mol. Ecol. Notes* **2007**, *7*, 544–548. [CrossRef]
30. Palumbi, S. Nucleic acids II: The polymerase chain reaction. In *Molecular Systematics*; Hillis, D.M., Moritz, C., Mable, B.K., Eds.; Sinauer Associates: Sunderland, MA, USA, 1996; pp. 205–247.
31. Miya, M.; Kawaguchi, A.; Nishida, M. Mitogenomic exploration of higher teleostean phylogenies: A case study for moderate-scale evolutionary genomics with 38 newly determined complete mitochondrial DNA sequences. *Mol. Biol. Evol.* **2001**, *18*, 1993–2009. [CrossRef] [PubMed]
32. Kearse, M.; Moir, R.; Wilson, A.; Stones-Havas, S.; Cheung, M.; Sturrock, S.; Buxton, S.; Cooper, A.; Markowitz, S.; Duran, C.; et al. Geneious Basic: An integrated and extendable desktop software platform for the organization and analysis of sequence data. *Bioinformatics* **2012**, *28*, 1647–1649. [CrossRef]
33. Iwasaki, W.; Fukunaga, T.; Isagozawa, R.; Yamada, K.; Maeda, Y.; Satoh, T.P.; Sado, T.; Mabuchi, K.; Takeshima, H.; Miya, M.; et al. MitoFish and MitoAnnotator: A mitochondrial genome database of fish with an accurate and automatic annotation pipeline. *Mol. Biol. Evol.* **2013**, *30*, 2531–2540. [CrossRef]
34. Altschul, S.F.; Gish, W.; Miller, W.; Myers, E.W.; Lipman, D.J. Basic local alignment search tool. *J. Mol. Biol.* **1990**, *215*, 403–410. [CrossRef]
35. Ratnasingham, S.; Hebert, P.D. BOLD: The Barcode of Life Data System (http://www.barcodinglife.org). *Mol. Ecol. Notes* **2007**, *7*, 355–364. [CrossRef]
36. Sievers, F.; Wilm, A.; Dineen, D.; Gibson, T.J.; Karplus, K.; Li, W.; Li, W.; Lopez, R.; McWilliam, H.; Remmert, M.; et al. Fast, scalable generation of high-quality protein multiple sequence alignments using Clustal Omega. *Mol. Syst. Biol.* **2011**, *7*, 539. [CrossRef]
37. Hasegawa, M.; Kishino, H.; Yano, T. Dating the human-ape split by a molecular clock of mitochondrial DNA. *J. Mol. Evol.* **1985**, *22*, 160–174. [CrossRef]
38. Nei, M.; Kumar, S. *Molecular Evolution and Phylogenetics*; Oxford University Press: New York, NY, USA, 2000.
39. Kimura, M. A simple method for estimating evolutionary rate of base substitutions through comparative studies of nucleotide sequences. *J. Mol. Evol.* **1980**, *16*, 111–120. [CrossRef]
40. Kumar, S.; Stecher, G.; Li, M.; Knyaz, C.; Tamura, K. MEGA X: Molecular evolutionary genetics analysis across computing platforms. *Mol. Biol. Evol.* **2018**, *35*, 1547. [CrossRef]
41. Shin, U.C.; Yoon, S.; Kim, J.K.; Choi, G. Species composition of Ichthyoplankton off Dokdo in the East Sea. *Korean J. Fish. Aquat. Sci.* **2021**, *54*, 498–507.
42. Sparks, J.S.; Dunlap, P.V.; Smith, W.L. Evolution and diversification of a sexually dimorphic luminescent system in ponyfishes (Teleostei: Leiognathidae), including diagnoses for two new genera. *Cladistics* **2005**, *21*, 305–327. [CrossRef]
43. Harada, A.E.; Lindgren, E.A.; Hermsmeier, M.C.; Rogowski, P.A.; Terrill, E.; Burton, R.S. Monitoring spawning activity in a southern California marine protected area using molecular identification of fish eggs. *PLoS ONE* **2015**, *10*, e0134647. [CrossRef]
44. Lin, H.Y.; Chiu, M.Y.; Shih, Y.M.; Chen, I.S.; Lee, M.A.; Shao, K.T. Species composition and assemblages of ichthyoplankton during summer in the East China Sea. *Cont. Shelf Res.* **2016**, *126*, 64–78. [CrossRef]
45. Hernandez-Vazquez, S. Distribution of eggs and larvae from sardine and anchovy off California and Baja California, 1951–1989. *CalCOFI Rep.* **1994**, *35*, 94–107.

46. Leis, J.M.; Moyer, J.T. Development of eggs, larvae and pelagic juveniles of three Indo-Pacific ostraciid fishes (Tetraodontiformes): *Ostracion meleagris*, *Lactoria fornasini* and *L. diaphana*. *Jpn. J. Icthyol.* **1985**, *32*, 189–202.
47. Kim, K.M.; Kim, S.Y.; Song, M.Y.; Song, H.Y. Morphological development of egg and larvae of *Hemiculter leucisculus*. *Korean J. Ichthyol.* **2020**, *32*, 222–231. [CrossRef]
48. Ko, H.L.; Wang, Y.T.; Chiu, T.S.; Lee, M.A.; Leu, M.Y.; Chang, K.Z.; Chen, W.Y.; Shao, K.T. Evaluating the accuracy of morphological identification of larval fishes by applying DNA barcoding. *PLoS ONE* **2013**, *8*, e53451. [CrossRef]
49. Amaoka, K.; Nakaya, K.; Yabe, M. Fishes of Usujiri and adjacent waters in southern Hokkaido, Japan. *Bull. Fac. Fish. Hokkaido Univ.* **1989**, *40*, 254–277.
50. Meléndez, R.; Clément, A. *Trachipterus trachypterus* (Gmelin, 1789) en el sur de Chile (Pisces: Lampridiformes: Trachipteridae). *Cienc. Tecnol. Mar.* **1992**, *15*, 43–47.
51. Cortes, N.; Arriaza, M.; Oyarzún, C. Nuevos registros de *Trachipterus trachypterus* (Gmelin, 1789) para el Pacífico suroriental, con una revisión de ejemplares congenéricos de Chile (Osteichthyes, Trachipteridae). *Rev. Biol. Mar.* **1995**, *30*, 265–273.
52. Dulcic, J. First record of ribbon fish, Trachipterus trachypterus, from the eastern Adriatic. *Cybium* **1996**, *20*, 101–102.
53. Farias, I.; Moura, T.; Figueiredo, I.; Vieira, A.R.; Serra-Pereira, B.; Serrano Gordo, L. Northernmost occurrence of the ribbonfish *Trachipterus trachypterus* (Gmelin, 1789) in the NE Atlantic: The Portuguese continental shelf. *J. Appl. Ichthyol.* **2010**, *26*, 143–144. [CrossRef]
54. Guerriero, G.; Di Finizio, A.; Ciarcia, G. Biological pollution: Molecular identification of non-native species in the Central Tyrrhenian Sea. *Catrina* **2010**, *5*, 41–47.
55. Shinohara, G.; Shirai, S.M.; Nazarkin, M.V.; Yabe, M. Preliminary list of the deep-sea fishes of the Sea of Japan. *Bull. Natl. Mus. Nat. Sci. Ser. A* **2011**, *37*, 35–62.
56. Mytilineou, C.; Anastasopoulou, A.; Christides, G.; Bekas, P.; Smith, C.J.; Papadopoulou, K.N.; Lefkaditou, E.; Kavadas, S. New records of rare deep-water fish species in the Eastern Ionian Sea (Mediterranean Sea). *J. Nat. Hist.* **2013**, *47*, 1645–1662. [CrossRef]
57. Garibaldi, F. By-catch in the mesopelagic swordfish longline fishery in the Ligurian Sea (Western Mediterranean). *Collect. Vol. Sci. Pap. ICCAT* **2015**, *71*, 1495–1498.
58. Yapici, S. New and additional records of rare fish species from the Anatolian coasts of Turkey. *Mugla J. Sci. Technol.* **2019**, *5*, 13–16. [CrossRef]
59. Macali, A.; Semenov, A.; Paladini de Mendoza, F.; Dinoi, A.; Bergami, E.; Tiralongo, F. Relative Influence of Environmental Factors on Biodiversity and Behavioural Traits of a Rare Mesopelagic Fish, *Trachipterus trachypterus* (Gmelin, 1789), in a Continental Shelf Front of the Mediterranean Sea. *J. Mar. Sci. Eng.* **2020**, *8*, 581. [CrossRef]
60. Gökoğlu, M.; Özen, M.R. First Record of *Trachipterus trachypterus* (Gmelin, 1789) in the Gulf of Antalya (Turkey). *Acta Aquat. Turc.* **2021**, *17*, 505–507. [CrossRef]
61. Mancini, P.L.; de Amorim, A.F.; Arfelli, C.A. Observações em *Trachipterus jacksonensis* capturados no Brasil. In Proceedings of the III Congresso Brasileiro de Pesquisas Ambientais e Saúde (CBPAS' 2003), Santos, Brazil, 21–23 July 2003; pp. 56–58.
62. Scott, E.O.G. Observations on some Tasmanian fishes: Part XXX. *Pap. Proc. R. Soc. Tasm.* **1984**, *118*, 187–222. [CrossRef]
63. Macpherson, E.; Roel, B.A. Trophic relationships in the demersal fish community off Namibia. *S. Afr. J. Mar. Sci.* **1987**, *5*, 585–596. [CrossRef]
64. Roberts, C.D. Fishes of the Chatham Islands, New Zealand: A trawl survey and summary of the ichthyofauna. *N. Z. J. Mar. Freshw. Res.* **1991**, *25*, 1–19. [CrossRef]
65. Mincarone, M.M.; Lima, A.T.; Soto, J.M. Sobre a ocorrência do peixe-fita *Trachipterus jacksonensis* (Ramsay, 1881) (Lampridiformes, Trachipteridae) na costa brasileira. *Mare Magnum* **2001**, *1*, 121–124.
66. Letourneur, Y.; Chabanet, P.; DurviLLe, P.; Taquet, M.; Teissier, E.; Parmentier, M.; QUÉRO, J.C.; Pothin, K. An updated checklist of the marine fish fauna of Reunion Island, south-western Indian Ocean. *Cybium* **2004**, *28*, 199–216.
67. Smith, P.J.; Steinke, D.; McMillan, P.; McVeagh, S.M.; Struthers, C.D. DNA database for commercial marine fish. *N. Z. Aquat. Environ. Biodivers.* **2008**, *22*, 1–62.
68. Fricke, R.; Mulochau, T.; Durville, P.; Chabanet, P.; Tessier, E.; Letourneur, Y. Annotated checklist of the fish species (Pisces) of La Réunion, including a red list of threatened and declining species. *Stuttg. Beitr. Naturkunde A* **2009**, *2*, 1–168.
69. Satapoomin, U. The fishes of southwestern Thailand, the Andaman Sea-a review of research and a provisional checklist of species. *Phuket Mar. Biol. Cent. Res.* **2011**, *70*, 29–77.
70. Melo, M.R.S.D.; Caires, R.A.; Sutton, T.T. The scientific explorations for deep-sea fishes in Brazil: The known knowns, the known unknowns, and the unknown unknowns. In *Brazilian Deep-Sea Biodiversity*; Springer: Cham, Switzerland, 2020; pp. 153–216.

Journal of
Marine Science and Engineering

MDPI

Article

Feeding Strategy of the Wild Korean Seahorse (*Hippocampus haema*)

Myung-Joon Kim [1], Hyun-Woo Kim [2], Soo-Rin Lee [3], Na-Yeong Kim [1], Yoon-Ji Lee [1], Hui-Tae Joo [4], Seok-Nam Kwak [5] and Sang-Heon Lee [1,*]

[1] Department of Oceanography, Pusan National University, Geumjeong-gu, Busan 46241, Korea; mjune@pusan.ac.kr (M.-J.K.); knayo@pusan.ac.kr (N.-Y.K.); yoonji051@pusan.ac.kr (Y.-J.L.)

[2] Department of Marine Biology, Pukyong National University, Nam-gu, Busan 48513, Korea; kimhw@pknu.ac.kr

[3] Industry 4.0 Convergence Bionics Engineering, Pukyong National University, Nam-gu, Busan 48513, Korea; srlee090@pukyong.ac.kr

[4] Oceanic Climate & Ecology Research Division, National Institute of Fisheries Science, Gijang-gun, Busan 46083, Korea; huitae@korea.kr

[5] Environ-Ecological Engineering Institute Company Limited, Haeundae-gu, Busan 48058, Korea; seoknam@eeei.kr

* Correspondence: sanglee@pusan.ac.kr

Abstract: The feeding and spawning grounds for seahorses have been lost due to nationwide coastal developments in South Korea. However, little information on the feeding ecology of the Korean seahorse (*Hippocampus haema*) is currently available. The main objective in this study was to understand the feeding strategy of *H. haema* on the basis of DNA analysis of the contents of the guts. This is the first study on the feeding ecology of *H. haema*. Crustaceans were found to be major prey for *H. haema* in this study. Among the 12 identified species, arthropods were predominantly observed as potential prey of *H. haema* in this study. The *Caprella* sp. Was detected in all summer specimens followed by the *Ianiropsis* sp., whereas isopods were dominant, and amphipods accounted for a small proportion in winter specimens. According to the results in this study, there appears to be a seasonal shift in the major prey of *H. haema*. Moreover, a potential change in the habitats for adults was further discussed. Since this is a pilot study, further studies should be conducted for a better understanding of the feeding ecology of *H. haema*.

Keywords: wild seahorse; *H. haema*; feeding habits; NGS analysis

Citation: Kim, M.-J.; Kim, H.-W.; Lee, S.-R.; Kim, N.-Y.; Lee, Y.-J.; Joo, H.-T.; Kwak, S.-N.; Lee, S.-H. Feeding Strategy of the Wild Korean Seahorse (*Hippocampus haema*). *J. Mar. Sci. Eng.* **2022**, *10*, 357. https://doi.org/10.3390/jmse10030357

Academic Editor: Ka Hou Chu

Received: 8 January 2022
Accepted: 28 February 2022
Published: 3 March 2022

Publisher's Note: MDPI stays neutral with regard to jurisdictional claims in published maps and institutional affiliations.

1. Introduction

Seahorses are fascinating creatures for many people around the world due to their unique appearance and life-history characteristics that are different to those of common fish [1,2]. The overfishing of seahorses is occurring in some regions [3,4], and reckless coastal developments causing their habitat loss are also threatening their survival [2,5,6]. For this reason, the scientific community is making various efforts to reduce the loss of seahorse populations and preserve the *Hippocampus* species by adding them to lists or conventions, such as the red list of the IUCN (International Union for Conservation of Nature) and CITES (Convention on International Trade in Endangered Species of Wild Fauna and Flora). Until now, there is no record of seahorse overfishing along the Republic of Korea coastal area; however, the Korean coastal ecosystem, due to ongoing nationwide coastal development, has been losing its natural shelter abilities as a safe habitat and spawning ground for various fishes, including seahorses [2,7]. Despite these crises, very little ecological information on seahorses is currently available in the Republic of Korea [8,9]; thus, we do not know what ecological roles they play in coastal ecosystems.

Seahorses lead a sedentary life in coastal areas where the environment can be changed dynamically compared to the open ocean [10–13]. Five seahorse species have been reported in South Korean waters, mainly in the seaweed and seagrass beds throughout the southern coast [8,9,14–17]. Recently, *H. haema*, previously known as *H. coronatus*, was newly identified as a Korean species [18]. The first study on the morphometric characteristics and basic ecological data for the newly recorded Korean seahorse (*H. haema*) populations was conducted in Geoje-Hansan Bay [8]. However, no study has been conducted on their feeding ecology to date.

The study of the feeding habits of an organism is a cornerstone that can be used for efficient management, preservation, and artificial reproduction of specific organisms, and it is widely conducted for various fish [19–24]. Although several previous studies were conducted on the feeding habits of some seahorse species under laboratory conditions (*H. abdominalis* [25], *H. barbouri* [26], *H. erectus* [27], *H. guttulatus* [28], *H. hippocampus* [29], *H. kuda* [30], and *H. reidi* [31,32]), the information on wild seahorses is still insufficient [24,33–42].

A visual analysis (naked eye or microscope) of gut contents, often used in feeding ecology, can show what quantity of prey is present in the target fish's gut; however, this approach may overestimate some preys that have just been recently eaten or are difficult to digest. Often times, it is difficult to identify what kinds of prey they have eaten due to their easily digestible characteristics or their small size [39,42]. This difference in the digestibility of the prey items can cause some bias when evaluating their diet compositions [39,42]. Stable isotope analysis has the advantage of being able to perform nonlethal analysis of the target organism, but it is difficult to detect the recently eaten prey due to the time gaps in the turnover rate of isotopes [43,44]. On the other hand, metabarcoding has the advantage of being able to detect short-term feeding habits with a very small amount of sample, as well as to detect soft and highly digested items, which are not recognizable through morphological identification [42,45]. Moreover, metabarcoding is also a nonlethal method, depending on the sampling method (i.e., fecal sample [42] or flushing method [34,40]). Therefore, in this study, a DNA analysis method was applied to understand the feeding habits of small *Hippocampus* individuals.

In this study, the main objective was to understand the feeding strategy of the Korean seahorse species (*H. haema*) in the coastal environment on the basis of the contents of the guts in summer and winter, using molecular biological approaches. This is the first study on the feeding ecology of the Korean seahorse species (*H. haema*).

2. Materials and Methods

2.1. Study Area and Samplings

Among the samples collected by the Environ-Ecological Engineering Institute (EEEI) for the investigation of fish biota in the sargassum bed (*Sargassum piluliferum*), seven specimens were received frozen in summer (July 2017) and winter (January 2019) for a comparison of the feeding strategies of *H. haema*. The sampling area was Geoje-Hansan Bay, Republic of Korea (34°48′40″ N; 128°31′03″ E; Figure 1). The mean water depth of the sampling site was approximately 3 m. The two specimens taken from the samples reported in [8] were analyzed to confirm the possibility of feeding differences between each sampling group even in the summer season (Table 1). The body length and wet weight were measured at the home laboratory using a caliper (Mitutoyo, 0.01 mm, Kawasaki, Japan) and a scale (Mettler Toledo, 1 mg, Columbus, OH, USA) according to [46]. After those measurements, all specimens were stored at −80 °C in a freezer and brought to the Marine Ecological Laboratory in Pusan National University for further analysis. For all sampled specimens (except the two specimens from 2016), the length–weight relationship was calculated using the following equation:

$$Wt = a \times L^b,$$

where Wt is wet weight (g), a is the intercept, b is the slope, and L is the standard length (SL = straight line between snout and tail, mm). A *t*-test was conducted to verify the statistical difference between the measured values (SL and Wt) of the January and July groups (SPSS, version 12.0, Chicago, IL, USA). The two specimens in July 2016 were deliberately selected as relatively large fish; hence, they were excluded from the average and statistical analysis (for a simple DNA comparison).

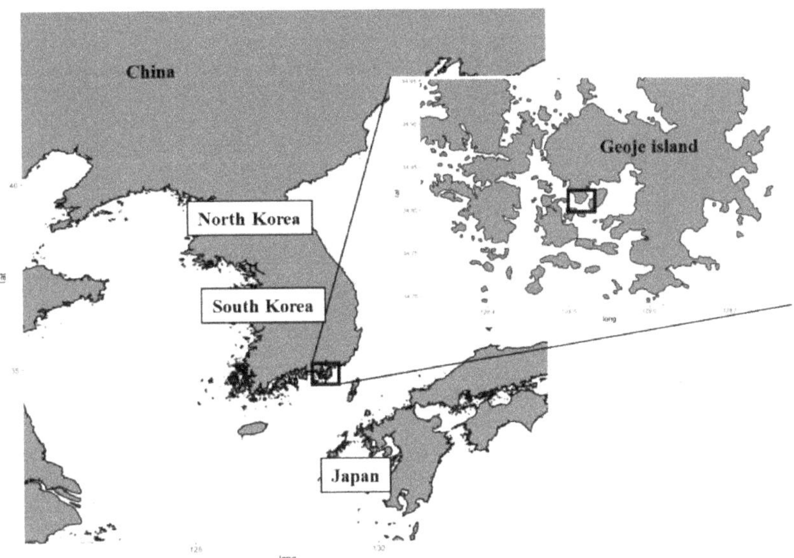

Figure 1. Map of the study area in Geoje-Hansan Bay, Republic of Korea (South Korea).

Table 1. Measurements of seahorses at each collection period and the *p*-value of the *t*-test.

Sampling Date		July 2017	January 2019	July 2016
	n	7	7	2
Range	*SL*	35.5–89.4	52.7–71.1	71.4–83.6
	Wt	0.078–1.500	0.300–0.721	0.681–0.814
Average	*SL*	58.43	60.08	77.50
	Wt	0.50	0.42	0.75
SD	*SL*	17.27	5.96	8.57
	Wt	0.49	0.15	0.09
t	*SL*	0.815		
	Wt	0.679		

2.2. Genomic DNA Extraction and NGS Library Construction

The genomic DNA of *H. haema* was extracted after complete homogenization with Tissue Lyser™ II (Qiagen, Hilden, Germany), by adding tissue lysis buffer to the gut of each of the 16 specimens six times, according to the instructions of the AccuPrep® Genomic DNA Extraction Kit (Bioneer, Daedeok-gu, Korea). ND-1000 (Thermo Scientific, Waltham, MA, USA) was used for the assay and quantification.

The NGS library of the gut contents of *H. haema* was constructed using primers (COIMISQ and NEXCOIMISQ, Table 2) targeting COI in the mitochondrial DNA region, which was previously used for the diet study of *Dissostichus mawsoni* [47]. The blocking primer (Table 2) was prepared the region of the universal primer COIMISQF1 in the nucleotide sequences of *Amphipoda*, *Copepoda*, *Mysidacea*, and *Isopoda*, known as prey organisms, and modified with a C3 spacer at the 3' end to suppress annealing of the *H. haema* sequence [48]. For the first PCR reaction, a final volume of 25 µL mixed solution consisting of 10 ng of genomic DNA, 100 µM of COIMISQ primers, 5 µL of 100 µM Blocking primer, 2 µL of 10 mM dNTPs (Takara Bio Inc., Kusatsu, Japan), 0.2 µL of Ex Taq Hot Start Version (Takara Bio Inc., Kusatsu, Japan), 2 µL of 10× Ex Taq buffer (Takara Bio Inc., Kusatsu, Japan), 3% DMSO, and distilled water was used. The reaction conditions and process of the first PCR were as follows: initial denaturation at 94 °C for 5 min, followed by 94 °C, 48 °C, and 72 °C for 30 s, respectively, repeated for 15 cycles. The final extension was conducted at 72 °C for 5 min. Prior to the second PCR reaction, the amplicons were purified using an AccuPrep® PCR Purification Kit (Bioneer, Republic of Korea). For the second PCR reaction, a final volume of 20 µL mixed solution consisting of 5 µL of the purified amplicons, 100 µM of NEXCOIMISQ primers, 2 µL of 100 µM Blocking primer, 2 µL of 10 mM dNTPs (Takara Bio Inc., Kusatsu, Japan), 0.2 µL of Ex Taq Hot Start Version (Takara Bio Inc., Kusatsu, Japan), 2 µL of 10× Ex Taq buffer (Takara Bio Inc., Kusatsu, Japan), and distilled water was used. The reacting conditions were repeated for 18 cycles under the same conditions as that of the first PCR. The amplicons of the second PCR were identified by 1.5% agarose gel electrophoresis (630 bp) and purified by the AccuPrep® PCR Purification Kit. The libraries constructed by the Nextera XT index Kit (Illumina, San Diego, CA, USA) were quantified with a QuantiFluor® Fluorometer (Promega, Madison, WI, USA), before using the MiSeq platform (Illunima, San Diego, CA, USA) for NGS analysis.

Table 2. Primers used in this study.

Name	Direction	Sequence (5′ to 3′)	Reference
COIMISQ	Forward	ATNGGNGGNTTYGGNAA	[47]
	Reverse	TANACYTCNGGRTGNCC	
NEXCOIMISQ	Forward	TCGTCGGCAGCGTCAGATGTGTATAAGAGACAGGGNGGNTTYGGNAAYTG	
	Reverse	GTCTCGTGGGCTCGGAGATGTGTATAAGAGACAGGGRTGNCCRAARAAYCA	
Blocking Primer		GCTTTGGTAATTGACTTG-C3 spacer	This study

2.3. Bioinformatic Analysis

Some parts of the sequence data of the NGS analysis, which featured a short length (less than 100 bp) and low quality (QV < 20), were disregarded. According to [49], the merged reads (more than 470 bp) constructed with the Mothur software package v1.41.3 were used. After the primer sequences were elided, the OTU (Operational Taxonomic Unit) clustering was conducted with Usearch v8.1.1861 [50] at 97%, and the chimeras were discarded. The OTUs were classified as species ($\geq$99%), genus (<99%, $\geq$90%), and unknown (<90%) on the NCBI GenBank database. Three specimens (No. 2 of 2017, Nos. 4 and 5 of 2019) were excluded from further analysis because no potential prey organisms were identified. The MEGA X Maximum Likelihood method was used for phylogenetic analysis [51].

3. Results

3.1. Morphometric Measurements

Summer and winter mean surface water temperatures are 24.9 °C and 9.3 °C, respectively, in the study area. A total of 16 seahorses were captured for this study. Table 1 shows the mean *SL* (mm) and *Wt* (g), as well as the results of the *t*-test for each collection period. Although no significant difference in mean *SL* and *Wt* for each collection period was observed, the July specimens had broader ranges in *SL* and *Wt* than the January specimens in this study. The length–weight relationship of each period's specimens is shown in Figure 2. The *b* values (slopes in Figure 2) of the July and January specimens were 3.237 and 3.034, respectively, which were greater than 3, indicating positive allometric growth.

3.2. NGS Analysis

As a result of the DNA analysis from the guts of 16 individuals, all specimens were identified as *H. haema*. Of the 2,263,757 reads (mean of 174,135 reads per sample) identified by sequencing, 41.87% were identified as seahorse and bacteria. These reads were excluded from further analysis.

Thus, 47.4% of the nucleotide sequences of the analyzed prey organisms were identified to the lowest taxonomic level (species). Among the 12 identified taxa, 10 were arthropods, and the remainder were *Bryozoa* and fish (Table 3). Among the 10 arthropods, four orders (*Harpacticoida, Caprella, Ianiropsis,* and *Mysida*) showed a relatively high ratio of over 80% in gut contents of some specimens. In July 2016 and 2017, *Caprella* sp. (amphipods) was detected in all specimens, especially for No. 1 (84.35%) and No. 7 (67.37%), whereas *Mysida* was mostly found in two of the total specimens. The *Ianiropsis* sp. (isopods) was the second most common species. On the other hand, in the January 2019 specimens, isopods tended to be dominant in their prey items, with amphipods accounting for a small proportion (Table 3).

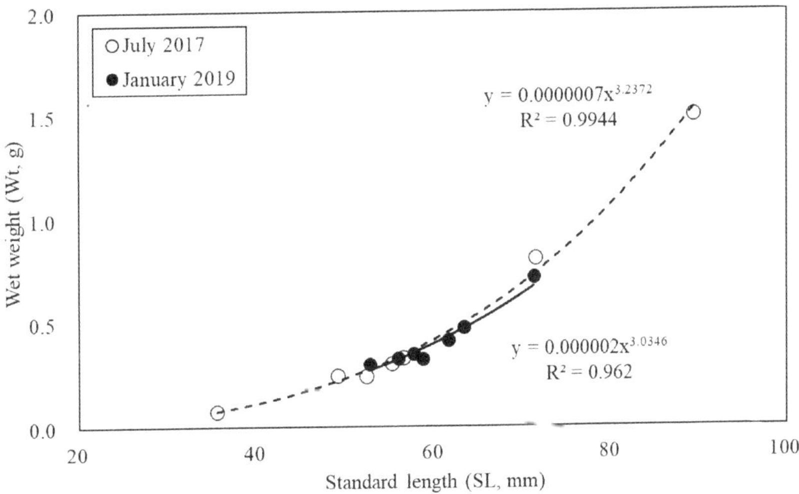

Figure 2. Length–weight relationship between specimens collected for each period. The white circles were taken in July 2017, and the black circles were taken in January 2019.

Table 3. The results of genetic analysis of gut (up to the species level). Bold font indicates the results of further analysis up to the order level for the results showing "unknown" in the analysis up to the species level.

Phylum	Class	Order	Species	Proportion (%)												
				July 2016		July 2017						January 2019				
				No. 11	No. 12	No. 1	No. 3	No. 4	No. 5	No. 6	No. 7	No. 1	No. 2	No. 3	No. 6	No. 7
Arthropoda	*Hexanauplia*	*Calanoida*	*Pseudodiaptomus* sp.		0.02						0.00			0.00		
		Harpacticoida		**43.67**	**97.76**		**6.67**									
		Harpacticoida		0.13	0.12											
			Amonardia sp.	0.07												

Table 3. *Cont.*

Phylum	Class	Order	Species	Proportion (%) July 2016 No. 11	July 2016 No. 12	July 2017 No. 1	July 2017 No. 3	July 2017 No. 4	July 2017 No. 5	July 2017 No. 6	July 2017 No. 7	January 2019 No. 1	January 2019 No. 2	January 2019 No. 3	January 2019 No. 6	January 2019 No. 7
	Malacostraca	**Amphipoda**					57.47	58.53	96.20	10.80			0.14			0.40
			Caprella sp.	2.12		84.35	9.33	21.58	3.37		67.37		3.88	0.15		0.13
			Crassicorophium crassicorne					0.13								
		Amphipoda	*Gammaridae* sp.	22.03									2.90			
			Leucothoe nagatai										0.01			
			Monocorophium acherusicum													0.00
		Isopoda						7.89							13.41	
			Ianiropsis epilittoralis										92.03	97.74		
		Isopoda	*Ianiropsis* sp.	9.44				11.87				88.67				99.39
			Munna japonica				15.08	0.00			0.00			0.63		0.01
		Mysida								89.02					86.59	
Bryozoa	*Gymnolaemata*	Ctenostomatida	*Amathia verticillate*	0.17		0.26	0.04		0.06	0.00	0.08					
Chordata	*Actinopterygii*	Perciformes	*Pictichromis paccagnellae*						0.23							
		Unknown		22.36	2.10	15.39	11.42		0.15	0.18	32.55	11.33	1.04	1.48		0.07
		Total		100	100	100	100	100	100	100	100	100	100	100	100	100

Since each specimen could be different in their energy budget requirements according to their body size (growth), the seahorse's diet would be different by size in the summer with various sizes of seahorses. However, among the specimens collected in July 2017, the largest seahorse (No. 1, *SL*: 89.4 mm) had a higher proportion of *Caprella* sp. (84.35%) at the species level, while the smallest one (No. 5, *SL*: 35.5 mm) had mostly Amphipoda (96.2%) at the order level (Table 3). In both cases, the Amphipoda order showed a tendency to occupy the highest ratio, although it was not possible to compare them due to the difference in the lowest taxonomic level of identification.

As a result of additional analysis of the 52.6% classified as "Unknown", clustering was confirmed at the order level (Figure 3). Because Figure 3 does not indicate the proportion of gut contents but just all items identified up to the order level, some items with very low percentages (e.g., Algae, Calanoida, Ctenostomatida, and Perciformes) are not shown in Table 3 or Figure 4a. The difference in the gut composition in each collection period was clear in comparison at the order level rather than the species level (Figure 4a). The proportions of Harpacticoida and Amphipoda had high ratios in July 2016 and 2017, respectively, while Isopoda was dominant in January 2019. This result was consistent with the similarity analysis in this study (Figure 4b).

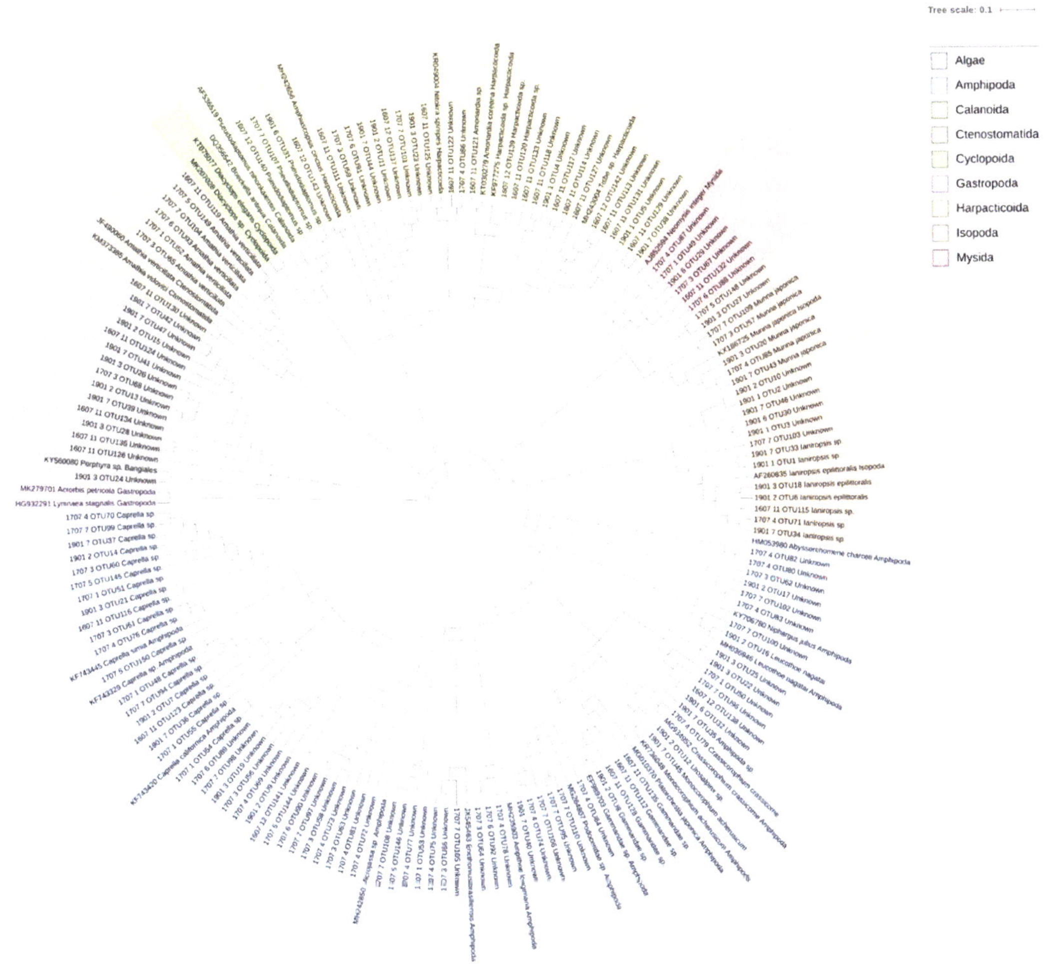

Figure 3. Phylogenetic tree of gut contents.

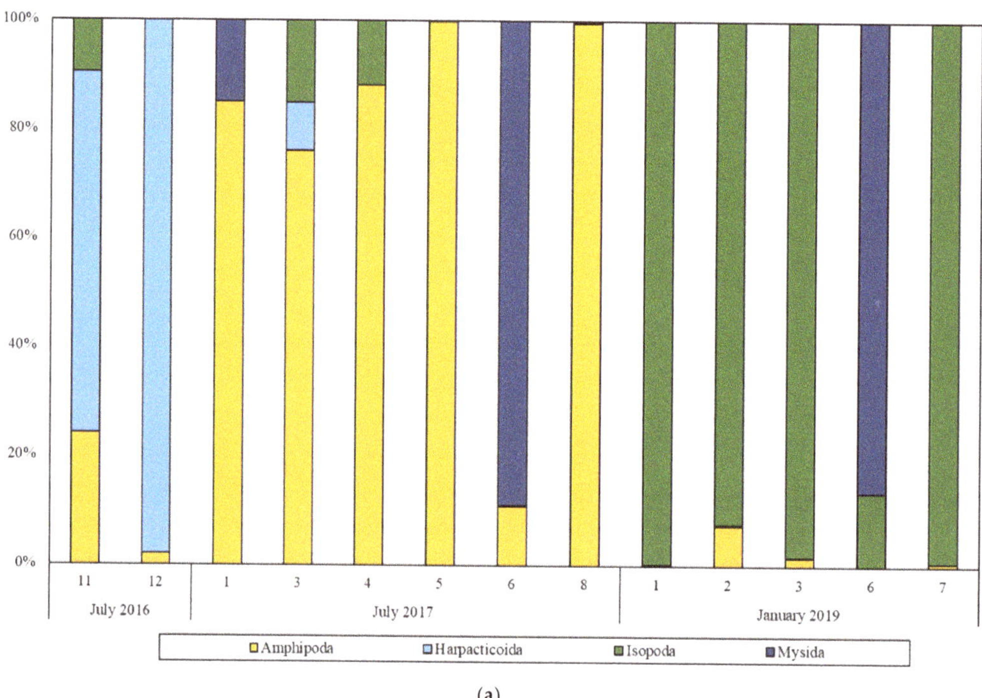

(**a**)

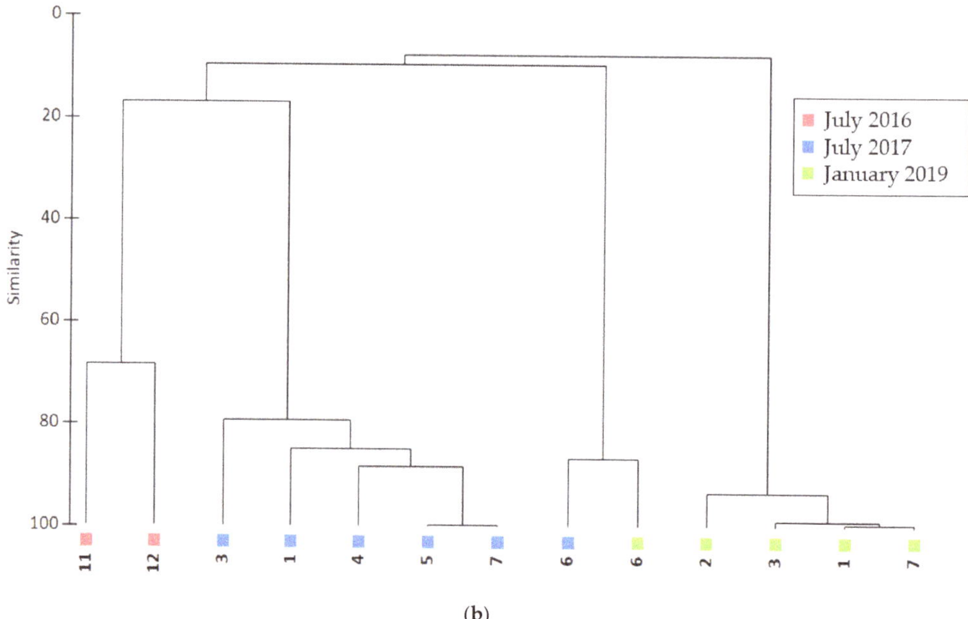

(**b**)

Figure 4. (**a**) The order-level taxonomic ratio of gut contents analyzed with COI primer. (**b**) Similarity analysis of gut contents by collecting period.

4. Discussion

To compare the seasonal feeding habits of *H. haema* during summer and winter, 14 specimens were collected from Geoje Hansan Bay in July (seven) and January (seven), when the seasonal characteristics were distinct. All the specimens were found in *Sargassum piluliferum* [8] and identified as the same species by DNA analysis [18]. There was no significant difference in mean *SL* and *Wt* by the two different seasons, but there was a relatively wider range in the mean *SL* and *Wt* in the July specimens. In other words, more varied sizes of seahorses were captured in July compared to January. According to previous studies, the breeding season of seahorse is from late spring to autumn in Korea [8,9], and there is a positive correlation between the growth of the genus *Hippocampus* (*H. whitei*, *H. guttulatus*, and *H. zosterae*) and water temperature within a certain range [13,52,53]. Thus, the appearance of seahorses of various sizes in July could be due to the recruitment of newborns and the relatively fast growth rates during the summertime.

NGS Analysis

A total of 12 taxa were identified in all gut contents except for three specimens without genetic information for any prey items. In comparison with other previous studies on the seahorse diet [24,34–42], Bryozoa was detected in a small proportion in this study, which might indicate that bryozoans were fed non-selectively with other prey.

Fish was detected only in one individual in this study. We presumed that other fish or eggs were misidentified due to the lack of NCBI data. *Pictichromis paccagnellae* is a tropical species which does not live in Korean waters. Although it is also assumed to be misidentified, it is likely that the genetic information could be other fish larvae or eggs, as actual feeding larvae have been reported in several different kinds of seahorses such as *H. abdominalis*, *H. reidi*, and *H. trimaculatus* [24,34,38].

Some "unknown" genetic information, which could not be confirmed even at the order level, suggests that there are still insufficient coastal zooplankton DNA library data for a variety of zooplankton in coastal areas.

Like other seahorse species reported previously [3,23,24,26,33–42,54–56], *H. haema* appears to feed mainly on crustaceans, according to the results in this study. However, the gut contents of *H. haema* were clearly differentiated by different sampling periods at the order level as shown in Figure 4a,b. *Caprella* was detected in most specimens in this study regardless of season, but a large proportion was recorded mainly in summer. In contrast, *Ianiropsis* accounted for a large proportion of gut contents in winter, showing a pattern opposite to that of *Caprella*. In fact, although *Caprella* appears in abundance on the coastal areas all year round [57–59], it is most often found in summer due to its strong tolerance to high water temperatures [57]. In comparison, large numbers of *Ianiropsis* are normally observed in winter [60].

Mysida species are generally observed throughout the year in coastal waters, which have an extreme seasonal temperature difference of more than 20 °C [61,62]. They stay and/or migrate in several swarms on the sandy bottom, near vegetation and rocks [61]. In general, seahorses forage accessible prey (slow and smaller than their mouths) within vegetation, but they occasionally hunt fast prey such as Mysida and Caridea on sandy bottoms [24,34]. The fact that Mysida was found only in some seahorses in this study indicates that several active individuals actually attempted to hunt them when the Mysida swarm reached their vicinity, but it was not easy for the seahorses to catch them due to the rapid reaction rate of Mysida. This can be supported by the low hunting success rates of *H. haema* toward a Mysida swarm observed in a breeding water tank (unpublished data).

Although Harpacticoida accounted for a large percentage of the specimens in July 2016, they occupied a very small portion for the prey items for *H. haema* in July 2017. Normally, Harpacticoids are a major item for some *Hippocampus* species (*H. zosterae*, [23]; *H. reidi*, [34]; and *H. subelongatus*, [54]). The seahorses can change their feeding capacity as they grow [3,23,24,63]. Indeed, in a laboratory environment, copepods (<1 mm) were eaten by all *H. haema* regardless of their size, whereas Artemia and Mysida (>1 mm) were mainly eaten by large seahorses (unpublished data). Like *Caprella*, harpacticoids appear frequently in the late spring–summer period [64,65], which coincides with the breeding season of *H. haema*. Therefore, with a maximum size of approximately 1 mm, harpacticoids could be an important food source for growing seahorses with their small mouth sizes.

The coincidence between the gut contents of *H. haema* and the seasonally thriving zooplankton in our study indicates that the main diet of *H. haema* can be modified according to the availability of prey. The authors of [24] found that wild *H. abdominalis* collected from Wellington Harbor did not have a seasonal shift in their habitats, but there was a major seasonal change in their prey. According to a previous study that analyzed feces from *H. guttulatus* using a genetic method, the seahorse diet differs depending on the habitat [42]. Although some differences among species are expected, this ability of *Hippocampus* species to flexibly change their diets in response to spatiotemporal changes in their surroundings could be one of the survival strategies for adapting to dynamic coastal environments.

5. Conclusions

Although specimens were collected once each in summer and winter, differences in feeding habits of *H. haema* could be clearly distinguished using DNA tools. However, due to the lack of genetic information on coastal zooplankton, it was difficult to identify up to the species level what seahorses mainly feed on. If coastal zooplankton can be identified up to the species level and their ecological characteristics understood, it could make a great contribution to understanding the role of the seahorse from an ecological perspective. In conclusion, this study is very important to provide the first information on the feeding behaviors of the Korean seahorse species, which is very valuable for their sustainable management and conservation in Korea.

Author Contributions: M.-J.K. and S.-H.L. conceived and designed the experiments; H.-W.K. and S.-R.L. performed the experiments; M.-J.K. and S.-R.L. analyzed the data; S.-H.L. and H.-W.K. validated the results; M.-J.K., N.-Y.K., Y.-J.L. and S.-N.K. investigated; M.-J.K., N.-Y.K. and Y.-J.L. curated the data; M.-J.K. wrote the original draft; M.-J.K. and S.-H.L. reviewed and edited the manuscript; S.-N.K. and H.-T.J. visualized the data; S.-H.L. supervised this research; S.-H.L. and H.-T.J. funded acquisition. All authors have read and agreed to the published version of the manuscript.

Funding: This work was supported by a 2 year research grant from Pusan National University.

Institutional Review Board Statement: Ethical review and approval were waived for this study due to all fish samples were dead when we received them.

Informed Consent Statement: Not applicable.

Data Availability Statement: Not applicable.

Acknowledgments: The authors would like to thank the anonymous reviewers and the handling editors who dedicated their time to providing the authors with constructive and valuable recommendations.

Conflicts of Interest: The authors declare no conflict of interest.

References

1. Gasparini, J.L.; Floeter, S.R.; Ferreira, C.E.L.; Sazima, I. Marine ornamental trade in Brazil. *Biodivers. Conserv.* **2005**, *14*, 2883–2899. [CrossRef]
2. Vincent, A.C.J.; Foster, S.J.; Koldewey, H.J. Conservation and management of seahorses and other Syngnathidae. *J. Fish Biol.* **2011**, *78*, 1681–1724. [CrossRef] [PubMed]
3. Foster, S.J.; Vincent, A.C.J. Life history and ecology of seahorses: Implications for conservation and management. *J. Fish Biol.* **2004**, *65*, 1–61. [CrossRef]
4. Martin-Smith, K.M.; Samoilys, M.A.; Meeuwig, J.J.; Vincent, A.C.J. Collaborative development of management options for an artisanal fishery for seahorses in the central Philippines. *Ocean Coast. Manag.* **2004**, *47*, 165–193. [CrossRef]
5. Curtis, J.M.R.; Ribeiro, J.; Erzini, K.; Vincent, A.C.J. A conservation trade-off? Interspecific differences in seahorse responses to experimental changes in fishing effort. *Aquat. Conserv. Mar. Freshw. Ecosyst.* **2007**, *17*, 468–484. [CrossRef]
6. Lourie, S.A.; Pritchard, J.C.; Casey, S.P.; Truong, S.K.; Hall, H.J.; Vincent, A.C.J. The taxonomy of Vietnam's exploited seahorses (family Syngnathidae). *Biol. J. Linn. Soc.* **1999**, *66*, 231–256. [CrossRef]

7.   Caldwell, I.R.; Vincent, A.C.J. A sedentary fish on the move: Effects of displacement on long-snouted seahorse (*Hippocampus guttulatus* Cuvier) movement and habitat use. *Environ. Biol. Fishes* **2012**, *96*, 67–75. [CrossRef]

8.   Kim, M.J.; Kim, H.C.; Lee, W.C.; Park, J.M.; Kwak, S.N.; Oh, Y.; Kang, M.G.; Lee, S.H. Ecological Characteristics of the New Recorded Seahorse (*Hippocampus haema*) in Geoje-Hansan Bay, Korea. *J. Coast. Res.* **2018**, *85*, 351–355. [CrossRef]

9.   Park, J.M.; Kwak, S.N. Length–weight relationships and reproductive characteristics of the crowned seahorse (*Hippocampus coronatus*) in eelgrass beds (*Zostera marina*) of Dongdae Bay, Korea. In *Marine Biology Research*; Taylor and Francis Ltd.: Abingdon, UK, 2014; Volume 11. [CrossRef]

10.   Bell, E.M.; Lockyear, J.F.; Mcpherson, J.M.; Marsden, A.D.; Vincent, A.C.J. First field studies of an Endangered South African seahorse, *Hippocampus capensis*. *Environ. Biol. Fishes* **2003**, *67*, 35–46. [CrossRef]

11.   Caldwell, I.R.; Vincent, A.C.J. Revisiting two sympatric European seahorse species: Apparent decline in the absence of exploitation. *Aquat. Conserv. Mar. Freshw. Ecosyst.* **2012**, *22*, 427–435. [CrossRef]

12.   Vincent, A.C.J.; Evans, K.L.; Marsden, A.D. Home range behaviour of the monogamous Australian seahorse, *Hippocampus whitei*. *Environ. Biol. Fishes* **2005**, *72*, 1–12. [CrossRef]

13.   Vincent, A.C.J.; Sadler, L.M. Faithful pair bonds in wild seahorses, *Hippocampus whitei*. *Anim. Behav.* **1995**, *50*, 1557–1569. [CrossRef]

14.   Choi, Y.-U.; Rho, S.; Jung, M.-M.; Lee, Y.-D.; Noh, G.-A. Parturition and Early Growth of Crowned Seahorse, *Hippocampus coronatus* in Korea. *J. Aquac.* **2006**, *19*, 109–118.

15.   Jung, M.-M.; Choi, Y.-U.; Lee, J.-E.; Kim, J.-W.; Kim, S.-C.; Lee, Y.-H.; Rho, S. Coexisting Fish Fauna in the Seahorse Habitats. *J. Aquac.* **2007**, *20*, 41–46.

16.   Kim, I.-S.; Lee, W.-O. First Record of the Seahorse Fish, *Hippocampus trimaculatus* (Pisces: Syngnathidae) from Korea. *Korean J. Zool.* **1995**, *38*, 74–77.

17.   Kim, S.-Y.; Kweon, S.-M.; Choi, S.-H. First Record of *Hippocampus sindonis* (Syngnathiformes: Syngnathidae) from Korea. *Korean J. Ichthyol.* **2013**, *25*, 42–45.

18.   Han, S.-Y.; Kim, J.-K.; Kai, Y.; Senou, H. Seahorses of the *Hippocampus coronatus* complex: Taxonomic revision, and description of *Hippocampus haema*, a new species from Korea and Japan (Teleostei, Syngnathidae). *ZooKeys* **2017**, *712*, 113–139. [CrossRef] [PubMed]

19.   Braga, R.R.; Bornatowski, H.; Vitule, J.R.S. Feeding ecology of fishes: An overview of worldwide publications. *Rev. Fish Biol. Fish.* **2012**, *22*, 915–929. [CrossRef]

20.   Huh, S.-H.; Kwak, S.N. Feeding habits of Favonigobius gymnauchen in the eelgrass (*Zostera marina*) bed in Kwangyang Bay. *J. Korean Fish. Soc.* **1998**, *31*, 372–379.

21.   Rosas-Luis, R.; Navarro, J.; Loor-Andrade, P.; Forero, M.G. Feeding ecology and trophic relationships of pelagic sharks and billfishes coexisting in the central eastern Pacific Ocean. *Mar. Ecol. Prog. Ser.* **2017**, *573*, 191–201. [CrossRef]

22.   Shreeve, R.S.; Collins, M.A.; Tarling, G.A.; Main, C.E.; Ward, P.; Johnston, N.M. Feeding ecology of myctophid fishes in the northern Scotia Sea. *Mar. Ecol. Prog. Ser.* **2009**, *386*, 221–236. [CrossRef]

23.   Tipton, K.; Bell, S.S. Foraging patterns of two syngnathid fishes: Importance of harpacticoid copepods. *Mar. Ecol. Prog. Ser.* **1988**, *47*, 31–43. [CrossRef]

24.   Woods, C.M.C. Natural diet of the seahorse *Hippocampus abdominalis*. *N. Z. J. Mar. Freshw. Res.* **2002**, *36*, 655–660. [CrossRef]

25.   Woods, C.M.C. Growth and survival of juvenile seahorse *Hippocampus abdominalis* reared on live, frozen and artificial foods. *Aquaculture* **2003**, *220*, 287–298. [CrossRef]

26.   Garcia, L.M.B.; Hilomen-Garcia, G.V.; Celino, F.T.; Gonzales, T.T.; Maliao, R.J. Diet composition and feeding periodicity of the seahorse *Hippocampus barbouri* reared in illuminated sea cages. *Aquaculture* **2012**, *358–359*, 1–5. [CrossRef]

27.   Vargas-Abúndez, J.A.; Simões, N.; Mascaró, M. Feeding the lined seahorse *Hippocampus erectus* with frozen amphipods. *Aquaculture* **2018**, *491*, 82–85. [CrossRef]

28.   Palma, J.; Stockdale, J.; Correia, M.; Andrade, J.P. Growth and survival of adult long snout seahorse (*Hippocampus guttulatus*) using frozen diets. *Aquaculture* **2008**, *278*, 55–59. [CrossRef]

29.   Segade, A.; Robaina, L.; Otero-Ferrer, F.; Romero, J.G.; Dominguez, L.M. Effects of the diet on seahorse (*Hippocampus hippocampus*) growth, body colour and biochemical composition. *Aquac. Nutr.* **2015**, *21*, 807–813. [CrossRef]

30.   Celino, F.T.; Hilomen-Garcia, G.V.; del Norte-Campos, A.G.C. Feeding selectivity of the seahorse, *Hippocampus kuda* (Bleeker), juveniles under laboratory conditions. *Aquac. Res.* **2012**, *43*, 1804–1815. [CrossRef]

31.   Felício, A.K.C.; Rosa, I.L.; Souto, A.; Freitas, R.H.A. Feeding behavior of the longsnout seahorse *Hippocampus reidi* Ginsburg, 1933. *J. Ethol.* **2006**, *24*, 219–225. [CrossRef]

32.   Olivotto, I.; Avella, M.A.; Sampaolesi, G.; Piccinetti, C.C.; Navarro Ruiz, P.; Carnevali, O. Breeding and rearing the longsnout seahorse *Hippocampus reidi*: Rearing and feeding studies. *Aquaculture* **2008**, *283*, 92–96. [CrossRef]

33.   Balasubramanian, R. Food and feeding habits of seahorse, *Hippocampus kelloggi* (Jordan and Snyder, 1902) in Cuddalore coastal water, Southeast coast of India. *Indian J. Educ. Inf. Manag.* **2017**, *6*, 1–8.

34.   da Costa Castro, A.L.; de Farias Diniz, A.; Martins, I.Z.; Vendel, A.L.; de Oliveira, T.P.R.; de Lucena Rosa, I.M. Assessing diet composition of seahorses in the wild using a non-destructive method: *Hippocampus reidi* (Teleostei: Syngnathidae) as a study-case. *Neotrop. Ichthyol.* **2008**, *6*, 637–644. [CrossRef]

35. Gurkan, S.; Taskavak, E.; Murat Sever, T.; Akalin, S. Gut Contents of Two European Seahorses *Hippocampus hippocampus* and *Hippocampus guttulatus* in the Aegean Sea, Coasts of Turkey. *Pak. J. Zool.* **2011**, *43*, 1197–1201.
36. Huh, S.-H.; Park, J.M.; Kwak, S.N.; Seong, B.J. Abundances and feeding habits of *Hippocampus coronatus* in an eelgrass (*Zostera marina*) bed of Dongdae Bay, Korea. *J. Korean Soc. Fish. Technol.* **2014**, *50*, 115–123. [CrossRef]
37. Storero, L.P.; Gonzélez, R.A. Feeding habits of the seahorse *Hippocampus patagonicus* in San Antonio Bay (Patagonia, Argentina). *J. Mar. Biol. Assoc. UK* **2008**, *88*, 1503–1508. [CrossRef]
38. Yip, M.Y.; Lim, A.C.O.; Chong, V.C.; Lawson, J.M.; Foster, S.J. Food and feeding habits of the seahorses *Hippocampus spinosissimus* and *Hippocampus trimaculatus* (Malaysia). *J. Mar. Biol. Assoc. UK* **2015**, *95*, 1033–1040. [CrossRef]
39. Valladares, S.; Soto, D.X.; Planas, M. Dietary composition of endangered seahorses determined by stable isotope analysis. *Mar. Freshw. Res.* **2017**, *68*, 831–839. [CrossRef]
40. Ape, F.; Corriero, G.; Mirto, S.; Pierri, C.; Lazic, T.; Gristina, M. Trophic flexibility and prey selection of the wild long-snouted seahorse *Hippocampus guttulatus* Cuvier, 1829 in three coastal habitats. *Estuar. Coast. Shelf Sci.* **2019**, *224*, 1–10. [CrossRef]
41. Manning, C.G.; Foster, S.J.; Vincent, A.C.J. A review of the diets and feeding behaviours of a family of biologically diverse marine fishes (Family Syngnathidae). In *Reviews in Fish Biology and Fisheries*; Springer International Publishing: Cham, Switzerland, 2019; pp. 1–25. [CrossRef]
42. Lazic, T.; Corriero, G.; Balech, B.; Cardone, F.; Deflorio, M.; Fosso, B.; Gissi, C.; Marzano, M.; Pesole, G.; Santamaria, M.; et al. Evaluating DNA metabarcoding to analyze diet composition of wild long-snouted seahorse *Hippocampus guttulatus*. In *2021 International Workshop on Metrology for the Sea; Learning to Measure Sea Health Parameters (MetroSea)*; IEEE: Reggio Calabria, Italy, 2021; pp. 257–261. [CrossRef]
43. Young, T.; Pincin, J.; Neubauer, P.; Ortega-García, S.; Jensen, O.P. Investigating diet patterns of highly mobile marine predators using stomach contents, stable isotope, and fatty acid analyses. *ICES J. Mar. Sci.* **2018**, *75*, 1583–1590. [CrossRef]
44. Zorica, B.; Ezgeta-Balić, D.; Vidjak, O.; Vuletin, V.; Šestanović, M.; Isajlović, I.; Čikeš Keč, V.; Vrgoč, N.; Harrod, C. Diet Composition and Isotopic Analysis of Nine Important Fisheries Resources in the Eastern Adriatic Sea (Mediterranean). *Front. Mar. Sci.* **2021**, *8*, 183. [CrossRef]
45. Piñol, J.; San Andrés, V.; Clare, E.L.; Mir, G.; Symondson, W.O.C. A pragmatic approach to the analysis of diets of generalist predators: The use of next-generation sequencing with no blocking probes. *Mol. Ecol. Resour.* **2014**, *14*, 18–26. [CrossRef] [PubMed]
46. Lourie, S.A.; Foster, S.J.; Cooper, E.W.T.; Vincent, A.C.J. A guide to the identification of seahorses. In *Project Seahorse and TRAFFIC North America*; Project Seahorse and TRAFFIC North America: Vancouver, Canada, 2004.
47. Yoon, T.H.; Kang, H.E.; Lee, S.R.; Lee, J.B.; Baeck, G.W.; Park, H.; Kim, H.W. Metabarcoding analysis of the stomach contents of the Antarctic Toothfish (*Dissostichus mawsoni*) collected in the Antarctic Ocean. *PeerJ* **2017**, *5*, e3977. [CrossRef] [PubMed]
48. Vestheim, H.; Jarman, S.N. Blocking primers to enhance PCR amplification of rare sequences in mixed samples–a case study on prey DNA in Antarctic krill stomachs. *Front. Zool.* **2008**, *5*, 12. [CrossRef] [PubMed]
49. Schloss, P.D.; Westcott, S.L.; Ryabin, T.; Hall, J.R.; Hartmann, M.; Hollister, E.B.; Lesniewski, R.A.; Oakley, B.B.; Parks, D.H.; Robinson, C.J.; et al. Introducing mothur: Open-source, platform-independent, community-supported software for describing and comparing microbial communities. *Appl. Environ. Microbiol.* **2009**, *75*, 7537–7541. [CrossRef] [PubMed]
50. Edgar, R.C. UPARSE: Highly accurate OTU sequences from microbial amplicon reads. *Nat. Methods* **2013**, *10*, 996–1000. [CrossRef] [PubMed]
51. Kumar, S.; Stecher, G.; Li, M.; Knyaz, C.; Tamura, K. MEGA X: Molecular evolutionary genetics analysis across computing platforms. *Mol. Biol. Evol.* **2018**, *35*, 1547–1549. [CrossRef] [PubMed]
52. Planas, M.; Blanco, A.; Chamorro, A.; Valladares, S.; Pintado, J. Temperature-induced changes of growth and survival in the early development of the seahorse *Hippocampus guttulatus*. *J. Exp. Mar. Biol. Ecol.* **2012**, *438*, 154–162. [CrossRef]
53. Strawn, K. Life History of the Pigmy Seahorse, *Hippocampus zosterae* Jordan and Gilbert, at Cedar Key, Florida. *Copeia* **1958**, *1958*, 16–22. Available online: https://about.jstor.org/terms (accessed on 7 February 2022). [CrossRef]
54. Kendrick, A.J.; Hyndes, G.A. Variations in the dietary compositions of morphologically diverse syngnathid fishes. *Environ. Biol. Fishes* **2005**, *72*, 415–427. [CrossRef]
55. Reid, G.K. An ecological study of the Gulf of the Mexico fishes, in the vicinity of Cedar Key, Florida. *Bull. Mar. Sci. Gulf Caribb.* **1954**, *4*, 1–12.
56. Teixeira, R.L.; Musick, J.A. Reproduction and food habits of the lined seahorse, *Hippocampus erectus* (Teleostei: Syngnathidae) of Chesapeake Bay, Virginia. *Rev. Brasil. Biol.* **2001**, *61*, 79–90. [CrossRef] [PubMed]
57. Caine, E.A. Ecology of Two Littoral Species of Caprellid Amphipods (Crustacea) from Washington, USA. *Mar. Biol.* **1980**, *56*, 327–335. [CrossRef]
58. Lolas, A.; Vafidis, D. Population dynamics of two caprellid species (Crustacea: Amphipoda: Caprellidae) from shallow hard bottom assemblages. *Mar. Biodivers.* **2013**, *43*, 227–236. [CrossRef]
59. Yun, S.G.; Byun, S.H.; Kwak, S.N.; Huh, S.-H. Seasonal Variation of Caprellids (Crustacea: Amphipoda) on Blades of *Zostera marina* in Kwangyang Bay, Korea. *J. Korean Fish.* **2002**, *35*, 105–109.
60. Edgar, G.J. Population regulation, population dynamics and competition amongst mobile epifauna associated with seagrass. *Mar. Biol. Ecol.* **1990**, *144*, 205–234. [CrossRef]

61. Bae, J.-Y.; Noh, G.E.; Park, W.-G. Population structure and life history of Neomysis nigra Nakazawa, 1910 (Mysida) on Jeju Island, South Korea. *Crustaceana* **2016**, *89*, 1–17. [CrossRef]
62. Ohtsuka, S.; Inagaki, H.; Onbe, T.; Gushima, K.; Yoon, Y.H. Direct observations of groups of mysids in shallow coastal waters of western Japan and southern Korea. *Mar. Ecol. Prog. Ser.* **1995**, *123*, 33–44. [CrossRef]
63. Ryer, C.H.; Orth, R.J. Feeding Ecology of the Northern Pipefish, *Syngnathus fuscus*, in a Seagrass Community of the Lower Chesapeake Bay. *Estuaries* **1987**, *10*, 330–336. [CrossRef]
64. Shin, A.; Kim, D.; Kang, T.; Oh, J.H. Seasonal fluctuation of the meiobenthic fauna community in the intertidal zone sediments of coastal areas in Jeju Island, Korea. *Environ. Biol. Res.* **2019**, *37*, 406–425. [CrossRef]
65. Song, S.J.; Ryu, J.; Khim, J.S.; Kim, W.; Yun, S.G. Seasonal variability of community structure and breeding activity in marine phytal harpacticoid copepods on *Ulva pertusa* from Pohang, east coast of Korea. *J. Sea Res.* **2010**, *63*, 1–10. [CrossRef]

Journal of
*Marine Science
and Engineering*

Article

The First Population Simulation for the *Zalophus japonicus* (Otariidae: Sea Lions) on Dokdo, Korea

Yoon-Ji Lee [1], Giphil Cho [2], Sangil Kim [3], Inseo Hwang [4], Seong-Oh Im [4], Hye-Min Park [4], Na-Yeong Kim [1], Myung-Joon Kim [1], Dasom Lee [1], Seok-Nam Kwak [5] and Sang-Heon Lee [1,*]

1   Department of Oceanography, Pusan National University, Geumjeong-gu, Busan 46241, Korea; yoonji051@pusan.ac.kr (Y.-J.L.); knayo@pusan.ac.kr (N.-Y.K.); mjune@pusan.ac.kr (M.-J.K.); ldasom91@naver.com (D.L.)
2   Finance·Fishery·Manufacture Industrial Mathematics Center on Big Data, Pusan National University, Geumjeong-gu, Busan 46241, Korea; giphil@pusan.ac.kr
3   Department of Mathematics, Pusan National University, Geumjeong-gu, Busan 46241, Korea; sangil.kim@pusan.ac.kr
4   Marine Ecosystem Management Team, Korea Marine Environment Management Corporation, Yeongdo-gu, Busan 49111, Korea; ishwang@koem.or.kr (I.H.); san94@koem.or.kr (S.-O.I.); shrimppark@koem.or.kr (H.-M.P.)
5   Environ-Ecological Engineering Institute Company Limited, Haeundae-gu, Busan 48058, Korea; seoknam@eeei.kr
*   Correspondence: sanglee@pusan.ac.kr; Tel.: +82-51-510-2256

Abstract: The Japanese sea lion (*Z. japonicus*) has been regarded as an extinct species since the last report on Dokdo in 1951. Not much ecological information on the *Z. japonicus* on Dokdo (hereafter Dokdo sea lion) is currently available. Using a discrete time stage-structured population model, we reconstructed the Dokdo sea lion population to explore the effect of human hunting pressure on them. This study provides the first estimate for the Dokdo sea lion population from 1900 to 1951. The reconstructed capture numbers of the Dokdo sea lion and the parameters estimated in this study were well matched with the recorded numbers and ecological parameters reported previously for the Californian sea lion. Based on the reconstructed population, their number rapidly declined after hunting started and it took less than 10 years for a 70% decline of the initial population, which would be considered to be an extinction risk. Since some caveats exist in this study, some caution about our results is necessary. However, this study demonstrates how rapidly human over-hunting can cause the extermination of a large local population. This study will be helpful to raise people's awareness about endangered marine animals such as local finless porpoises in Korea.

Keywords: sea lions; Dokdo; marine mammals; pinnipeds

Citation: Lee, Y.-J.; Cho, G.; Kim, S.; Hwang, I.; Im, S.-O.; Park, H.-M.; Kim, N.-Y.; Kim, M.-J.; Lee, D.; Kwak, S.-N.; et al. The First Population Simulation for the *Zalophus japonicus* (Otariidae: Sea Lions) on Dokdo, Korea. *J. Mar. Sci. Eng.* **2022**, *10*, 271. https://doi.org/10.3390/jmse10020271

Academic Editor: Roberto Carlucci

Received: 10 January 2022
Accepted: 11 February 2022
Published: 15 February 2022

Publisher's Note: MDPI stays neutral with regard to jurisdictional claims in published maps and institutional affiliations.

1. Introduction

The genus *Zalophus* consists of the Japanese (*Z. japonicus*), California (*Z. californianus*), and Galapagos (*Z. wollebaeki*) sea lions which are closely related species [1,2]. The three species were typically considered to be isolated geographically since their geographic distributions are widely separated from the temperate western (*Z. japonicus*) and eastern (*Z. californianus*) North Pacific to the tropical Galapagos Archipelago (*Z. wollebaeki*) [2]. Based on the results from ancient DNA analysis extracted from skeletal remains of Japanese sea lions, Sakahira and Niimi [3] reported that *Z. japonicus* diverged from *Z. californianus* approximately 2.2 million years ago in the late Pliocene epoch. Recently, Kim et al. [4] found that *Z. japonicus* was closely related to *Z. californianus* with 98.61% sequence identity based on the first completed mitochondrial genome sequence of *Z. japonicus* recently excavated from an archaeological site on Ulleungdo, Korea (Figure 1).

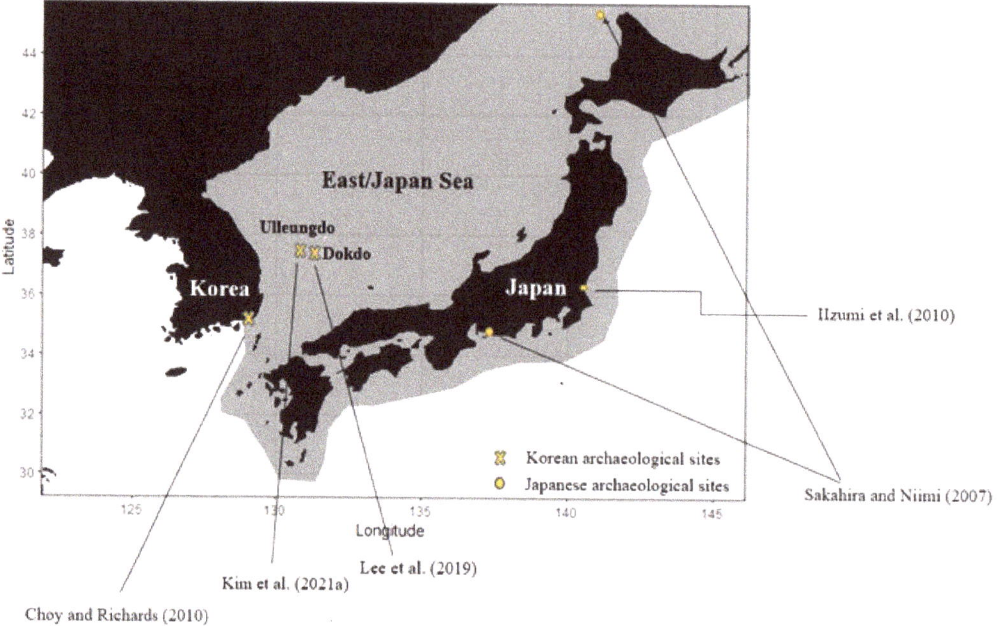

Figure 1. The possible historical distribution range from Melin et al. [5] (grey color) and archaeological sites of the *Zalophus japonicus* in the East/Japan Sea from [3,4,6–8].

Among the three genera of *Zalophus*, Japanese sea lions have been regarded as an extinct species since the last report of 50 to 60 sea lions on Dokdo in 1951 [1]. Only one last *Z. japonicus* was reported in the northern coastal area in Hokkaido in 1974, but specific identification was not conducted so that misidentification could have occurred [9]. Reynolds et al. [10] and Kovacs et al. [11] reported that the Japanese sea lion has been extinct for the past 60 years. However, several international research collaborations have been conducted to detect whether this species are still alive along the northwest Pacific region including Ulleungdo and Dokdo in Korea [12–14]. Although some unofficial reports of observations for the species were reported, it is considered that there may have been confusion with Steller's sea lion or the Northern fur seal [2] which have been occasionally observed along the eastern coast in Korea [15]. Recently, the International Union for Conservation of Nature (IUCN) listed *Z. japonicus* as an extinct species [16].

Previously, *Z. japonicus* had been distributed along the northwest Pacific coastline ranging from Russia (Kamchatka and Sakhalin) to coastal waters of Japan and Korea and especially Ulleungdo and Dokdo, which had been regarded as one of the main Korean habitats in the 1900s [1]. Based on numerous historical local Japanese names based on the sea lion, a large number of sea lions have observed in the Japanese coastal regions since prehistoric times ([17]; Figure 1). In Korea, several historical caves named after sea lion have been continuously used by local people on Ulleungdo and Dokdo [15,18]. Subsistence sea lion hunting by native coastal communities since prehistoric times was conducted in Korea [7] and Japan [3] as evidenced by various archaeological evidence. Choy and Richards [7] found that main food sources for the prehistoric coastal communities in the southeastern coast of the Korean Peninsula were marine animals and large marine mammals including sea lions, based on isotopic compositions of bone collagen excavated from an archaeological site in Yeongdo-gu, Busan city, South Korea. This prehistoric subsistence hunting probably did not have a large effect on the population but relatively modern commercial harvesting hunting with modern sophisticated techniques caused a substantial decline in the population of sea lions in southern California and Mexico [2]. The population of the Japanese sea lion had nearly disappeared in the coastal waters of Japan by the early 1900s mainly due to over-catching by Japanese fishermen for their meat and oil [17]. This commercial hunting pressure appears to have reduced the geographical ranges of *Z. japonicus* to remote islands in the East/Japan Sea, namely Ulleungdo and Dokdo which were previously pristine and unpopulated [6,18]. The population of *Z. japonicus* on Dokdo drastically declined during the early decades of the 1900s since the Japanese commercial harvesting started in 1904 and only 50 to 60 animals were reported by the 1950s (Figure 2; [2,6,18]). However, the major potential reasons for the drastic decline in their population have not confirmed whether there were some environmental changes in their major habitats during El Niño events or human over-catching [18]. The environmental changes could cause a reduction in major prey and a change in their behavioral responses (e.g., migration, breeding, etc.) during El Niño events [2].

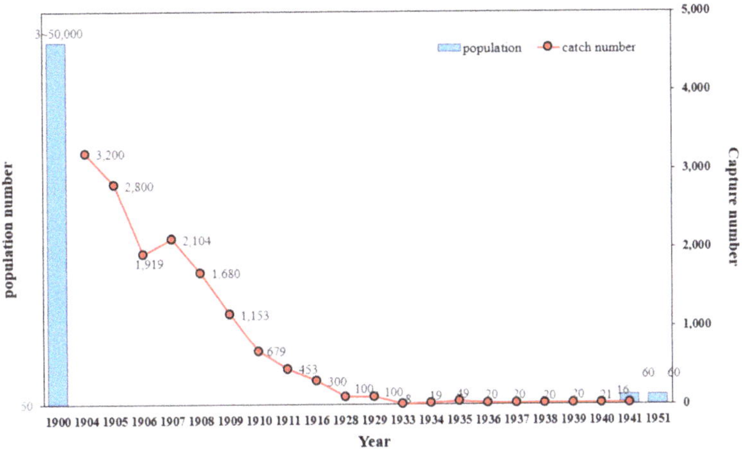

Figure 2. The recorded populations and annual capture numbers of the Dokdo sea lion from 1900 to 1951. The population numbers were from [1,2,6,18] and the annual capture numbers from 1904 to 1941 were based on [13,19].

Not much information on the ecology and specific population of *Z. japonicus* is currently available, since they had been extinct before various scientific research works were conducted. Based on the fragmented historical and official records from Korea and Japan, we reconstructed the population of *Z. japonicus* on Dokdo (hereafter the Dokdo sea lion) using population models with ecological parameters of the California sea lion (*Z. californianus*) which is genetically closely related to the Dokdo sea lion [3,4]. The main objectives of this study was to reconstruct the population size of the Dokdo sea lion from a simple fitting curve with the existing historical records and to explore the effect of human hunting pressure on the population. This study will be helpful to raise people's awareness for critically endangered marine mammals.

2. Materials and Methods

2.1. Historical Hunting Records

At the current stage, limited quantitative data on the Dokdo sea lion and their capture numbers by Japanese fisherman are available. The reported populations of the Dokdo sea lion in 1900, 1941, and 1951 and the annual capture numbers from 1904 to 1941 were obtained from the historical records and unpublished documents from Korea and Japan ([2,13,14,18]; Figure 1).

2.2. Stage-Structured Model

A discrete time stage-structured population model was applied for the Dokdo sea lion in this study. The discrete time stage-structured population model has been described in various articles and textbooks of fishery mathematical modelling [20–22]. The model is divided into two groups of sea lion compartments, immature (x) and mature (y) groups, and the difference equations for the two groups are as follows:

$$x_{t+1} = be^{-(M_y+F_y)}y_t - (1-a)e^{-(M_x+F_x)}x_t$$
$$y_{t+1} = ae^{-(M_x+F_x)}x_t - e^{-(M_y+F_y)}y_t \tag{1}$$

In model (1), b is the birth rate of the immature at time t, M_x and M_y represent the natural mortality rates of immature and mature groups, respectively. F_x and F_y represent the catch mortality rates of the two groups, respectively.

Specifically, $e^{-(M_x+F_x)}$ and $e^{-(M_y+F_y)}$ describe the survival probabilities for immature and mature individuals, respectively. $be^{-(M_y+F_y)}y_t$ represents the birth population into the immature by surviving mature individuals. $ae^{-(M_x+F_x)}x_t$ is the number of individuals which become mature from the surviving immature population where a a is the proportion of immature individuals becoming mature individuals. In our model, no large immigration and emigration were considered since the Ulleungdo and Dokdo as major habitats for the Dokdo sea lion are remote islands in the East/Japan Sea.

2.3. Parameter Estimation

We estimated five parameters, the birth rate (b), natural mortalities of immature and mature M_x, M_y and natural mortalities of immature and mature F_x, F_y, by fitting the actual catch data for the Dokdo sea lion by using a MATLAB-embedded function, lsqcurvefit. This is a least square estimation method for the five parameters that minimizes the gap of catch data between the actual data and the simulated results from Equation (1). The simulated results of the catch at time t, denoted by F_t for year were calculated as:

$$F_t = \frac{F_x}{F_x + M_x}(1 - e^{-(M_x+F_x)})x_t + \frac{F_y}{F_y + M_y}(1 - e^{-(M_y+F_y)})y_t \tag{2}$$

The Dokdo sea lions were distinguished between immature and mature based on an age of five years of old when reproduction becomes possible, and the value of a is expressed as 1/5 [2].

3. Results and Discussion

The historical geographical distribution of *Z. japonicus* were the eastern coasts of Korea, coastal waters of Japan, and the Russian territories such as southwestern Sakhalin, Kuril's islands, and Kamchatka ([1]; Figure 1). In the mid-1800s, the population of *Z. japonicus* was estimated in a range of 30,000–50,000 around Korea and Japan [2]. They hunted for prey sometimes a long distance away but did not seasonally migrate [16]. Although prehistoric subsistence hunting by native coastal communities had been conducted in Korea and Japan [3,17], a substantial decline of the Dokdo sea lion population in the early 1900s could have been caused by the over-catching by Japanese fishermen for their meat and oil [17], which had happened in southern California and Mexico [2,5]. Based on the historical and official bibliographic records [2,6,7,19], the populations and annual capture numbers of the Dokdo sea lion were plotted from 1900 to 1951 (Figure 2). Japanese fisherman commercially started to hunt a large number of Dokdo sea lions (3200 individuals) in 1904 and their hunting number declined to a couple of hundred by the early 1910s after 10 years of hunting (Figure 2). A major hunting of sea lions was conducted by the Japanese Jukdo Fishing Joint Stock Company during the 1904–1925 period and then the hunting was irregularly done by the Okido people [19]. They hunted mostly females and sub-adults which annually increased up to 86% in 1908 since the large male sea lions were hard to catch [19]. Even after the large decline in the population (based on their capture number), they still caught a small number live for circus performances until 1941 rather than hunting for their products [18]. A total of approximately 15,000 Dokdo sea lions were hunted for the 40-year Japanese fishing period from 1904 to 1941. After hunting stopped, 50–60 individuals, which was approximately 0.2% of the total population at the beginning of the commercial hunting in 1904, were reported in 1951 [1].

The reconstructed capture numbers of the Dokdo sea lion are matched well with the recorded numbers (Figure 3). The parameters estimated in this study (Table 1) were within the ranges reported previously for the California sea lion [23–25]. For example, the birth rate for the female California sea lion ranged from 38% to 54% at Los Islotes, Baja California Sur, Mexico, from the 1978 to 1982 study period [24]. Actually, the birth rate can vary with age, being significantly higher rates in young and middle-age groups (0.52–0.80) than the older age-groups (0.06) [25]. For the natural mortality, 50% and 10% of immature and mature sea lions, respectively, were estimated in this study. Normally, the natural mortality of the California sea lion is substantially higher in immature pup group during their pupping season [23]. The annual pup mortality of the California sea lion averaged from 3 years observation is approximately 50% at Los Islotes, Baja California Sur, Mexico [24]. The fishing mortality was estimated substantially higher in mature groups than immature groups. Given these parameters, the reconstructed population estimates matched well with several data points for the numbers of the Dokdo sea lion (Figure 4). Based on the reconstructed population of the Dokdo sea lion, their number rapidly declined to approximately 50% of the initial population in 1908 after 5 years' fishing activity and to approximately 6% of the initial population by the end of 20 years fishing in 1923. Since there is no quantitative time-series data on the actual population size of the Dokdo sea lion, the precision of the results for their historical population reconstruction might be low. Several caveats are associated with the results in this study. Firstly, the records on the historical population estimates of the Dokdo sea lion are not quantitative. Secondly, our simple fitting curve population models have several assumptions for ecological parameters estimated for the Dokdo sea lion since not much information is currently available for a robust model. Moreover, no other environmental conditions were considered in this study for their population dynamics. Therefore, our results in this study should be considered with caution since our approach to reconstructing the historical population estimates of the Dokdo sea lion is a simple mathematical technique

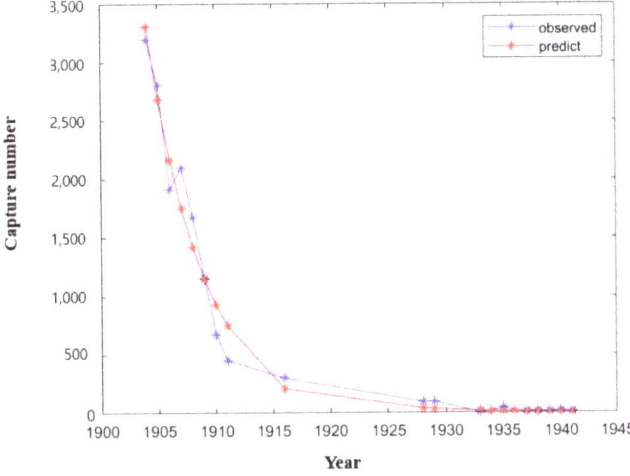

Figure 3. The comparison between observed and estimated annual capture numbers of the Dokdo sea lion.

Table 1. Description of discrete time stage-structured model parameters.

Parameter	Description	Value	Reference
b	Birth rate	0.5	estimated
a	The proportion of immature to mature	1/5	Heath and Perrin (2009)
M_x	The natural mortality of the immature individuals (year^{-1})	0.5	estimated
M_y	The natural mortality of the mature individuals (year^{-1})	0.1	estimated
F_x	The fishing mortality of the immature individuals (year^{-1})	0.0138	estimated
F_y	The fishing mortality of the mature individuals (year^{-1})	0.2428	estimated

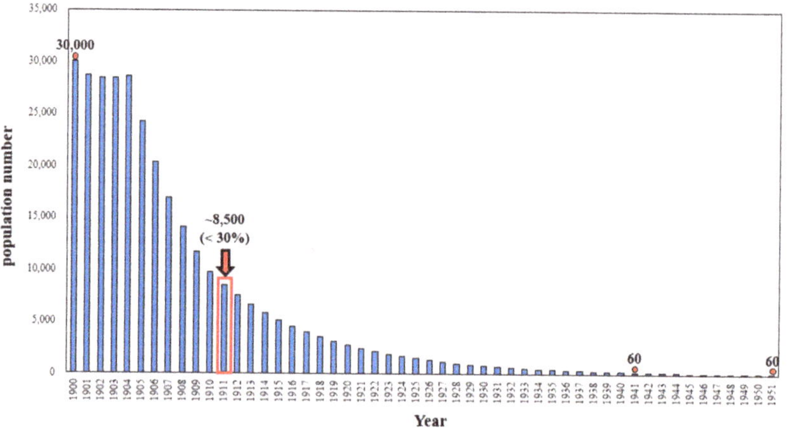

Figure 4. The reconstructed population numbers of the Dokdo sea lion from 1900 to 1951. The red circles are population numbers reported previously.

A sudden major decline in a population to extinction generally results from short-term natural catastrophic or anthropogenic events or chronic, gradual decline of a population [26]. Mangel and Tier [27] concluded that catastrophic events could cause a sudden major decline in population for a relatively short time period and thus be a more important factor for determining their persistence over time than any other factor. Catastrophic events usually control the birth and death processes which affects the dynamics and extinction of a population [26]. A population with a 70% decline for years would be considered to be an extinction risk [26]. Based on our simulation result, the population of the Dokdo sea lion was less than 30% in 1911 within the 10-year fishing period (Figure 4). It took only less than 10 years for the 70% decline of the Dokdo population by Japanese commercial hunting. Four possible mechanisms for human-caused extinction are over-hunting, which is the most obvious extinction mechanism, introduction of new species, habitat destruction which is currently a major human-caused extinction risk, and secondary cascade extinction [28]. The result in this study obviously showed that the extinction of a large local population of the Dokdo sea lion could have been caused by primary human-caused extermination, over-hunting. However, any environmental change was not considered for our discrete time stage-structured population model. The dramatic decline of the Dokdo sea lion population could have been caused by the increase in sea surface temperature. For the California sea lion, the annual population growth rate can be largely influenced by the increasing sea surface temperature which is strongly associated with major prey for the sea lions in the highly productive California coastal waters [29,30]. The latitudinal difference in resource availability can influence foraging behavior and thus survival rate of local resident sea lion species associated with a changing environment [29]. Indeed, the local productivity in Ulleungdo and Dokdo are lower than what is observed in California or Baja California [31,32]. Moreover, the Galapagos sea lion had a smaller population than the California sea lion [5] and one that was probably more vulnerable to extinction than the

California sea lion. The Dokdo sea lion in the mid-1800s once had a similar population size (30,000–50,000) [2] as the Galapagos sea lion (approximately 40,000) in 1978. Therefore, the Dokdo sea lion would be even more vulnerable to extinction with a changing environment.

We examined the population dynamics of the Dokdo sea lion to determine the major population decline and the recovery population simulations with several fishing ban scenarios during the Japanese fishing period for 40 years (Figure 5). If we assumed that they stopped their hunting in 1910 after 6 years of hunting, then it would take approximately 120 years to fully recover the initial population in 1900 (Figure 5). If they stopped after 14 years hunting in 1920, it would take approximately 220 years for the population recovery (Figure 5). Based on the results from the simple simulations, it is very surprising to realize how much heavy commercial fishing impacts the natural population recovery of the specific target species. Recently, critically important ecological roles of marine mammals were reported in various marine ecosystems [33–35]. Their ecological roles could vary from bottom-up stimulation of phytoplankton production and further climate regulation through a whale pumping mechanism to top-down predators affecting the abundance and spatial distribution of their prey and, consequently, hierarchical ecosystem structure [34,35]. Unfortunately, we cannot know the ecological roles of the Dokdo sea lion on the East/Japan Sea ecosystem since they were extinct before scientific research on them was undertaken.

Figure 5. The population recovery scenarios for the Dokdo sea lion. (**a**) hunting cessation after 6 years in 1910, (**b**) hunting cessation after 14 years in 1920.

In this study, we demonstrated the huge impact of human hunting on the local population of the Dokdo sea lion which is consequently extinct, using a simple fitting curve method based on the fragmented historical and official records on the Dokdo sea lion population. This study provides the first estimate for the annual population size of the Dokdo sea lion from 1900 to 1951. This simple approach in this study could be extended to other pinnipeds and marine mammals. The risk of extinction is a globally important concern for various marine mammals, in particular otariids [26]. It is essential to understand major extinct causes and potential subsequent consequences of the extinct species for conservation biology [36]. Given the importance of over-catching for the extinction of the Dokdo sea lion, the analysis conducted in this study should be considered for other local marine mammals including pinnipeds. For example, a large number of local finless porpoises (*Neophocaena asiaeorientalis*) are distributed along coastal areas in Korea. Approximately 36,000 finless porpoises were estimated in the west coast of Korea in 2004 based on models of the detection function in a line transect [37]. However, the population density of the species rapidly declined by approximately 70% in less than one decade from 2004 to 2011 mainly due to bycatching related to stranding and fishing gear [38]. In the Bohai Sea, China, approximately 8000 spotted seals in the 1940s declined to 2000 in the 1980s due to human hunting and then less than 1500 were estimated to remain in recent years [39,40]. Because of their habitat loss and human impacts, the spotted seal is categorized as a critically endangered species in China and Korea [41]. Approximately

400 spotted seals are annually observed along the coast of Baengnyeong islands and some small population group (around 10) of the seals are found in Galolim Bay, South Korea [15]. Therefore, it is important to protect marine mammals such as finless porpoises and spotted seals which has been declining in numbers in coastal areas in Korea and China. In the case of marine mammals, the increase in the population does not appear in a short period of time due to the characteristics of their life history by which one female gives birth to one pup and it takes a long time to nurse the young pup. Indeed, it is quite surprising to realize how long it would take for their population recovery based on the simple simulations in this study (Figure 5). Moreover, a decline in the population could be predicted to be apparent if human threats are added [2,5]. Currently, marine mammals are designated and protected as marine protection organisms in Korea but research on marine mammals are insufficient, so information on the biological characteristics or distribution status of most species is not well understood [42]. Therefore, in order to effectively conserve marine mammals in Korea, it is necessary to seek realistic management methods and to conduct various research approaches more actively [15]. A failure to do so may result in the local extinction of marine mammal species in Korea, as with the Dokdo sea lion. We should learn an invaluable lesson from the historical human-caused extinction of the Dokdo sea lion.

Author Contributions: Conceptualization, Y.-J.L. and S.-H.L.; methodology, G.C., S.K. and S.-H.L.; validation, Y.-J.L. and S.-H.L.; formal analysis, G.C. and S.K.; investigation, Y.-J.L., N.-Y.K., M.-J.K., G.C., I.H., S.-O.I., H.-M.P., S.-N.K. and D.L.; data curation, Y.-J.L. and S.-H.L.; writing—original draft preparation Y.-J.L. and S.-H.L.; writing—review and editing, S.-H.L.; visualization, Y.-J.L. and S.-H.L.; supervision, S.-H.L.; project administration, I.H., S.-O.I., H.-M.P. and S.-H.L.; funding acquisition, I.H., S.-O.I., H.-M.P., S.-N.K. and S.-H.L. All authors have read and agreed to the published version of the manuscript.

Funding: This research was supported by the "Investigation on the habitat of pinnipeds in the Korean East Sea, 2021" funded by the Ministry of Oceans and Fisheries, Korea.

Institutional Review Board Statement: Not applicable.

Informed Consent Statement: Not applicable.

Data Availability Statement: Not applicable.

Acknowledgments: The authors would like to thank the anonymous reviewers and the handling editors who dedicated their time to providing the authors with constructive and valuable recommendations.

Conflicts of Interest: The authors declare no conflict of interest.

References

1. Rice, D.W. *Marine Mammals of the World: Systematics and Distribution*; Special Publication Number 4, The Society for Marine Mammalogy; Allen Press: Lawrence, KS, USA, 1998; pp. 29–42.
2. Heath, C.B.; Perrin, W.F. California, Galapagos, and Japanese sea lions: *Zalophus californianus, Z. wollebaeki*, and *Z. japonicus*. In *Encyclopedia of Marine Mammals*; Elsevier: Amsterdam, The Netherlands, 2009; pp. 170–176.
3. Sakahira, F.; Niimi, M. Ancient DNA analysis of the Japanese sea lion (*Zalophus californianus* japonicus Peters, 1866): Preliminary results using mitochondrial control-region sequences. *Zoolog. Sci.* **2007**, *24*, 81–85. [CrossRef] [PubMed]
4. Kim, E.-B.; Kim, M.J.; Hwang, I.; Park, H.-M.; Lee, S.H.; Kim, H.-W. The complete mitochondrial genome of Japanese sea lion, *Zalophus japonicus* (Carnivora: Otariidae) analyzed using the excavated skeletal remains from Ulleungdo, South Korea. *Mitochondrial DNA Part B* **2021**, *6*, 3184–3185. [CrossRef] [PubMed]
5. Melin, S.R.; Trillmich, F.; Aurioles-Gamboa, D. California, Galapagos, and Japanese Sea Lions: *Zalophus californianus, Z. wollebaeki*, and *Z. japonicus*. In *Encyclopedia of Marine Mammals*; Elsevier: Amsterdam, The Netherlands, 2018; pp. 153–157.
6. Lee, S.-R.; Kim, Y.-B.; Lee, T. The first molecular evidence of Korean *Zalophus japonicus* (Otariidae: Sea lions) from the archaeological site of Dokdo island, Korea. *Ocean Sci. J.* **2019**, *54*, 497–501. [CrossRef]
7. Choy, K.; Richards, M.P. Isotopic evidence for diet in the Middle Chulmun period: A case study from the Tongsamdong shell midden, Korea. *Archaeol. Anthropol. Sci.* **2010**, *2*, 1–10. [CrossRef]
8. IIzumi, K.; IIzumi, K.; Koda, Y.; Koike, W.; Nishimoto, T.; Ando, H.; Date, M. Latest Pleistocene Japanese Sea lion(Otariidae) fossil from the riverbed of the Hanamurogawa River west of Kasumigaura Lake, Ibaraki Prefecture. *J. Geol. Soc. Japan. Tokyo* **2010**, *116*, 243–251. [CrossRef]
9. Itoo, T. Miscellaneous on the Japanese sea lion Zalophus californianusc japonicus. *Mammal Sci.* **1979**, *19*, 27–39. [CrossRef]

10. Reynolds, J.E., III.; Marsh, H.; Ragen, T.J. Marine mammal conservation. *Endanger. Species Res.* **2009**, *7*, 23–28. [CrossRef]
11. Kovacs, K.M.; Aguilar, A.; Aurioles, D.; Burkanov, V.; Campagna, C.; Gales, N.; Gelatt, T.; Goldsworthy, S.D.; Goodman, S.J.; Hofmeyr, G.J.G. Global threats to pinnipeds. *Mar. Mammal Sci.* **2012**, *28*, 414–436. [CrossRef]
12. Jackson, P.F.R. Hunt for the Japanese Sea-lion. *Environ. Conserv.* **1977**, *4*, 290. [CrossRef]
13. ME (Ministry of Environment). *Study on the Monitoring of Marine Mammals (Pinniped) in the East Sea and Management Plans*; NIER (National Institute of Environmental Research): Incheon, Korea, 2007.
14. NIBR (National Institute of Biological Resources). *Investigation and Network Construction for the Restoration of Endangered Marine Mammals (Pinniped)*; NIBR: Incheon, Korea, 2010.
15. Kim, H.W.; Lee, S.; Sohn, H. A Review on the Status of Pinnipeds in Korea. *Korean J. Fish. Aquat. Sci.* **2021**, *54*, 231–239.
16. Lowry, L. *Zalophus japonicus*. The IUCN Red List of Threatened Species. 2017. Available online: https://www.iucnredlist.org/species/41667/113089431; (accessed on 5 January 2022).
17. Nakamura, K. An essay on the Japanese Sea Lion, *Zalophus californianus* japonicus, living on the seven islands of Izu. *Bull. Kanagawa Prefect. Museum (Nat. Sci.)* **1991**, *20*, 59–66.
18. Joo, K.H. *The History of Extinction of Dokdo Sea Lions*; Seohaemunjip: Seoul, Korea, 2016.
19. Kim, S. Japan [Sea Otter and Fur Seal Fishing Law (臘虎肭獸獵法)] and Dokdo Sea Lion Fishing. *J. Jpn. Cult.* **2016**, *70*, 25–40.
20. Liu, S.; Chen, L.; Agarwal, R. Recent progress on stage-structured population dynamics. *Math. Comput. Model.* **2002**, *36*, 1319–1360. [CrossRef]
21. Jung, S.; Choi, I.; Jin, H.; Lee, D.; Cha, H.; Kim, Y.; Lee, J. Size-dependent mortality formulation for isochronal fish species based on their fecundity: An example of Pacific cod (Gadus macrocephalus) in the eastern coastal areas of Korea. *Fish. Res.* **2009**, *97*, 77–85. [CrossRef]
22. Hilborn, R.; Walters, C.J. *Quantitative Fisheries Stock Assessment: Choice, Dynamics and Uncertainty*; Springer Science Business Media: Berlin/Heidelberg, Germany, 2013; ISBN 1461535980.
23. Brownell, R.L., Jr.; LeBoeuf, B.J. California sea lion mortality: Natural or artifact. In *Biological and Oceanographical Survey of the Santa Barbara Channel Oil spill 1969–1970*; Allan Hancock Foundation: Los Angeles, CA, USA, 1971; pp. 287–305.
24. Aurioles, D.; Sinsel, F. Mortality of California Sea Lion Pups at Los Islotes, Baja California Sur, Mexico. *J. Mammal.* **1988**, *69*, 180–183. [CrossRef]
25. Hernández-Camacho, C.J.; Aurioles-Gamboa, D.; Gerber, L.R. Age-specific birth rates of California sea lions (*Zalophus californianus*) in the Gulf of California, Mexico. *Mar. Mamm. Sci.* **2008**, *24*, 664–676. [CrossRef]
26. Gerber, L.R.; Hilborn, R. Catastrophic events and recovery from low densities in populations of otariids: Implications for risk of extinction. *Mamm. Rev.* **2001**, *31*, 131–150. [CrossRef]
27. Mangel, M.; Tier, C. Four facts every conservation biologists should know about persistence. *Ecology* **1994**, *75*, 607–614. [CrossRef]
28. Diamond, J.M. The present, past and future of human-caused extinctions. *Philos. Trans. R. Soc. London B Biol. Sci.* **1989**, *325*, 469–477.
29. Villegas-Amtmann, S.; Simmons, S.E.; Kuhn, C.E.; Huckstadt, L.A.; Costa, D.P. Latitudinal range influences the seasonal variation in the foraging behavior of marine top predators. *PLoS ONE* **2011**, *6*, e23166. [CrossRef]
30. Laake, J.L.; Lowry, M.S.; DeLong, R.L.; Melin, S.R.; Carretta, J. V Population growth and status of California sea lions. *J. Wildl. Manag.* **2018**, *82*, 583–595. [CrossRef]
31. Joo, H.; Park, J.W.; Son, S.; Noh, J.; Jeong, J.; Kwak, J.H.; Saux-Picart, S.; Choi, J.H.; Kang, C.; Lee, S.H. Long-term annual primary production in the Ulleung Basin as a biological hot spot in the East/Japan Sea. *J. Geophys. Res. Oceans* **2014**, *119*, 3002–3011. [CrossRef]
32. Kahru, M.; Kudela, R.; Manzano-Sarabia, M.; Mitchell, B.G. Trends in primary production in the California Current detected with satellite data. *J. Geophys. Res. Oceans* **2009**, *114*, 1–7. [CrossRef]
33. Roman, J.; McCarthy, J.J. The whale pump: Marine mammals enhance primary productivity in a coastal basin. *PLoS ONE* **2010**, *5*, e13255. [CrossRef]
34. Roman, J.; Estes, J.A.; Morissette, L.; Smith, C.; Costa, D.; McCarthy, J.; Nation, J.B.; Nicol, S.; Pershing, A.; Smetacek, V. Whales as marine ecosystem engineers. *Front. Ecol. Environ.* **2014**, *12*, 377–385. [CrossRef]
35. Kiszka, J.J.; Heithaus, M.R.; Wirsing, A.J. Behavioural drivers of the ecological roles and importance of marine mammals. *Mar. Ecol. Prog. Ser.* **2015**, *523*, 267–281. [CrossRef]
36. Baisre, J.A. Shifting baselines and the extinction of the Caribbean monk seal. *Conserv. Biol.* **2013**, *27*, 927–935. [CrossRef]
37. Park, K.J.; Kim, Z.G.; Zhang, C.I. Abundance estimation of the finless porpoise, Neophocaena asiaeorientalis, using models of the detection function in a line transect. *Korean J. Fish. Aquat. Sci.* **2007**, *40*, 201–209.
38. Park, K.J.; Sohn, H.; An, Y.R.; Kim, H.W.; An, D.H. A new abundance estimate for the finless porpoise Neophocaena asiaeorientalis on the west coast of Korea: An indication of population decline. *Fish. Aquat. Sci.* **2015**, *18*, 411–416. [CrossRef]
39. Jinhai, D.; Feng, S. Estimates of historical population size of harbor seal (*Phoca largha*) in Liaodong Bay. *Mar. Sci.* **1991**, *3*, 26–31.
40. Yan, H.-K.; Wang, N.; Wu, N.; Lin, W. Abundance, habitat conditions, and conservation of the largha seal (*Phoca largha*) during the past half century in the Bohai Sea, China. *Mammal Study* **2018**, *43*, 1–9. [CrossRef]
41. Won, C.; Yoo, B.-H. Abundance, seasonal haul-out patterns and conservation of spotted seals Phoca largha along the coast of Bak-ryoung Island, South Korea. *Oryx* **2004**, *38*, 109–112. [CrossRef]
42. NIBR (National Institute of Biological Resources). *Red Data Book of Endangered Mammals in Korea*; NIBR: Incheon, Korea, 2012.

MDPI
St. Alban-Anlage 66
4052 Basel
Switzerland
Tel. +41 61 683 77 34
Fax +41 61 302 89 18
www.mdpi.com

Journal of Marine Science and Engineering Editorial Office
E-mail: jmse@mdpi.com
www.mdpi.com/journal/jmse